Universitext

Universitext

Universitext is a series of textbooks that presents material from a wide variety of mathematical disciplines at master's level and beyond. The books, often well class-tested by their author, may have an informal, personal even experimental approach to their subject matter. Some of the most successful and established books in the series have evolved through several editions, always following the evolution of teaching curricula, to very polished texts.

Thus as research topics trickle down into graduate-level teaching, first textbooks written for new, cutting-edge courses may make their way into *Universitext*.

More information about this series at http://www.springer.com/series/223

Vydas Čekanavičius

Approximation Methods in Probability Theory

 Springer

Vydas Čekanavičius
Vilnius University
Vilnius, Lithuania

ISSN 0172-5939 ISSN 2191-6675 (electronic)
Universitext
ISBN 978-3-319-34071-5 ISBN 978-3-319-34072-2 (eBook)
DOI 10.1007/978-3-319-34072-2

Library of Congress Control Number: 2016941172

Mathematics Subject Classification (2010): 62E20, 60E10, 60G50, 60F99, 41A25, 41A27

This Springer imprint is published by Springer Nature
The registered company is Springer International Publishing AG Switzerland

Preface

The limit theorems of probability theory are at the core of multiple models used in the broad field of scientific research. Their main function is to replace the initial complicated stochastic model of a phenomenon by its somewhat simpler approximate substitute. As a rule, such substitute is easier to use since its properties are well known. However, it raises a key question: how good is the approximation? For example, even in the famous central limit theorem, the rate of convergence to the normal law can be extremely slow. Therefore, it is important to measure the magnitude of the difference between models or, in other words, to estimate the accuracy of approximation. However, there is a notable lack of books that are specifically focused on teaching how to do it. One could find numerous monographs and textbooks devoted to approximations, especially related to the central limit theorem, but the prime concern of their authors is the impressive results, not the methods that are used in the process. Thus, such books rarely involve more than one method, not to mention an actual comparison of the applicability of several approaches.

This book is loosely based on a course I have been teaching during my visit at Hamburg University in June, 2004, combined with specific methods and proofs accumulated from teaching PhD-level courses since then. It presents a wide range of various well-known and less common methods for estimation of the accuracy of probabilistic approximations. In other words, it is a book on tools for proving approximation theorems. As a rule, we demonstrate the correctness of the 'tool' by providing an appropriate proof. In a few cases, when the proofs are very long or sophisticated or not directly related to the topic of the book, they are omitted, and an appropriate reference to the source of the result is provided.

Our ultimate goal is to teach the reader the correct usage of various methods. Therefore, we intentionally present simple cases, examining in detail all steps required for the proof. We provide one further simplification on placing the emphasis on the order of accuracy of approximation rather than on the magnitude of absolute constants. In order to gain a better command of the presented techniques, various exercises are added at the end of each chapter, with the majority of their solutions at the end of the book. Bibliographical notes provide information on the origins of each method and its more advanced applications.

The specific methods described in this book include a convolution method, which can be easily extended from distributions to the more general commutative objects. In addition, a considerable part of the book is related to the classical characteristic function method. As usual, we present the Esseen-type inversion formulas. On the other hand, we systematically treat the lattice case, since, in our opinion, it is undeservedly rarely considered in the literature. Furthermore, one of the chapters is devoted to the powerful but rarely used triangular function method. Though we usually deal with independent random variables, Heinrich's method for m-dependent variables is also included. Due to the fact that Stein's method has had a uniquely comprehensive coverage in the literature (with a specific emphasis on the method in [7, 50, 127]), only its short version is presented, together with a discussion on some specific methodological aspects, which might be of interest even to experienced users of the method.

Naturally, this book does not contain all known methods for estimating the accuracy of approximation. For example, we use only four metrics and, therefore, the method of metrics is not included. Moreover, it is already comprehensively treated in [116]. Meanwhile, methods for random vectors or elements of infinite-dimensional spaces deserve a separate textbook.

A standard intermediate course in probability should be sufficient for the readers. This book is expected to be especially useful for masters and PhD students. In fact, I wish I had similar book during my PhD studies, because learning various methods from scientific papers is doing it the hard way for no good reason, since a lot of intermediate results are usually omitted and small tricks and twists are not adequately explained. I was lucky to meet many mathematicians who helped me in the process, and I would like to extend my special thanks to A. Bikelis, J. Kruopis, A.Yu. Zaĭtsev, A.D. Barbour and B. Roos.

Vilnius, Lithuania   Vydas Čekanavičius
2016

Contents

1   Definitions and Preliminary Facts 1
    1.1   Distributions and Measures 1
    1.2   Moment Inequalities 5
    1.3   Norms and Their Properties.. 6
    1.4   Fourier Transforms 9
    1.5   Concentration Function 14
    1.6   Algebraic Identities and Inequalities 15
    1.7   The Schemes of Sequences and Triangular Arrays 19
    1.8   Problems 19

2   The Method of Convolutions 21
    2.1   Expansion in Factorial Moments 21
    2.2   Expansion in the Exponent 25
    2.3   Le Cam's Trick 28
    2.4   Smoothing Estimates for the Total Variation Norm 29
    2.5   Estimates in Total Variation via Smoothing........................ 32
    2.6   Smoothing Estimates for the Kolmogorov Norm 37
    2.7   Estimates in the Kolmogorov Norm via Smoothing 38
    2.8   Kerstan's Method... 44
    2.9   Problems 47

3   Local Lattice Estimates 51
    3.1   The Inversion Formula 51
    3.2   The Local Poisson Binomial Theorem 53
    3.3   Applying Moment Expansions 54
    3.4   A Local Franken-Type Estimate 57
    3.5   Involving the Concentration Function 59
    3.6   Switching to Other Metrics 62
    3.7   Local Smoothing Estimates.. 64
    3.8   The Method of Convolutions for a Local Metric 65
    3.9   Problems 67

4   Uniform Lattice Estimates 69
   4.1   The Tsaregradskii Inequality 69
   4.2   The Second Order Poisson Approximation 71
   4.3   Taking into Account Symmetry 74
   4.4   Problems 75

5   Total Variation of Lattice Measures 77
   5.1   Inversion Inequalities 77
   5.2   Examples of Applications.. 79
   5.3   Smoothing Estimates for Symmetric Distributions 85
   5.4   The Barbour-Xia Inequality 86
   5.5   Application to the Wasserstein Norm 89
   5.6   Problems 91

6   Non-uniform Estimates for Lattice Measures 93
   6.1   Non-uniform Local Estimates 93
   6.2   Non-uniform Estimates for Distribution Functions................ 95
   6.3   Applying Taylor Series... 98
   6.4   Problems 100

7   Discrete Non-lattice Approximations 101
   7.1   Arak's Lemma 101
   7.2   Application to Symmetric Distributions 103
   7.3   Problems 105

8   Absolutely Continuous Approximations 107
   8.1   Inversion Formula 107
   8.2   Local Estimates for Bounded Densities 109
   8.3   Approximating Probability by Density 110
   8.4   Estimates in the Kolmogorov Norm 112
   8.5   Estimates in Total Variation.. 113
   8.6   Non-uniform Estimates 117
   8.7   Problems 119

9   The Esseen Type Estimates 121
   9.1   General Inversion Inequalities 121
   9.2   The Berry-Esseen Theorem... 125
   9.3   Distributions with $1 + \delta$ Moment 127
   9.4   Estimating Centered Distributions.................................. 130
   9.5   Discontinuous Distribution Functions.............................. 135
   9.6   Problems 138

10   Lower Estimates 141
   10.1 Estimating Total Variation via the Fourier Transform 141
   10.2 Lower Estimates for the Total Variation........................... 143
   10.3 Lower Estimates for Densities...................................... 146
   10.4 Lower Estimates for Probabilities 147

10.5   Lower Estimates for the Kolmogorov Norm 149
10.6   Problems 151

11 The Stein Method 153
11.1   The Basic Idea for Normal Approximation 153
11.2   The Lattice Case 155
11.3   Establishing Stein's Operator 157
11.4   The Big Three Discrete Approximations 159
11.5   The Poisson Binomial Theorem 161
11.6   The Perturbation Approach 163
11.7   Estimating the First Pseudomoment 167
11.8   Lower Bounds for Poisson Approximation 173
11.9   Problems 175

12 The Triangle Function Method 179
12.1   The Main Lemmas 179
12.2   Auxiliary Tools 184
12.3   First Example 186
12.4   Second Example 195
12.5   Problems 206

13 Heinrich's Method for m-Dependent Variables 207
13.1   Heinrich's Lemma 207
13.2   Poisson Approximation 211
13.3   Two-Way Runs 214
13.4   Problems 220

14 Other Methods 223
14.1   Method of Compositions 223
14.2   Coupling of Variables 229
14.3   The Bentkus Approach 230
14.4   The Lindeberg Method 232
14.5   The Tikhomirov Method 234
14.6   Integrals Over the Concentration Function 236
14.7   Asymptotically Sharp Constants 237

Solutions to Selected Problems 241

Bibliography 266

Index 273

Notation and Abbreviations

$\mathbb{R}$	real numbers
$\mathbb{C}$	complex numbers
$\mathbb{N}$	natural numbers
$\mathbb{Z}$	integers
$\mathbb{Z}_+$	$\mathbb{N} \cup \{0\}$
$\mathcal{B}$	σ-field of one-dimensional Borel subsets
$\mathcal{M}$	set of finite signed measures
$\mathcal{M}_Z$	set of finite signed measures concentrated on $\mathbb{Z}$
$\mathcal{F}$	set of all distributions
$\mathcal{F}_Z$	set of distributions concentrated on $\mathbb{Z}$
$\mathcal{F}_s$	set of symmetric distributions
$\mathcal{F}_+$	set of distributions having nonnegative characteristic functions
$\Phi_{\mu,\sigma}$	normal distribution with mean μ and variance σ^2
Φ_σ, Φ	$\Phi_\sigma \equiv \Phi_{0,\sigma}$, $\Phi \equiv \Phi_1$
I_a	distribution, concentrated at a, $\widehat{I}_a(t) = e^{iat}$
I	$I \equiv I_0$, $\widehat{I}(t) \equiv 1$
F^n	n-fold convolution of F
$\exp\{M\}$	exponential measure
$M^{(-)}$	$M^{(-)}\{X\} = M\{-X\}$, $\widehat{M}^{(-)}(t) = \widehat{M}(-t)$
Θ	any measure satisfying $\parallel \Theta \parallel \leqslant 1$
$\parallel M \parallel$	total variation norm of M
$\mid M \mid_K$	Kolmogorov norm of M
$\parallel M \parallel_\infty$	local norm of M
$\parallel M \parallel_W$	Wasserstein norm of M
$\widehat{M}(t)$	Fourier transform of $M \in \mathcal{M}$
$Q(F, h)$	concentration function
$\nu_k(F)$	factorial moment of F

iid rvs	independent identically distributed random variables
$\mathbb{E}\,\xi$	mean of ξ
CLT	central limit theorem
CP	compound Poisson
θ	any complex number, satisfying $\lvert\theta\rvert \leqslant 1$
i	imaginary unit, $\mathrm{i}^2 = -1$
$\lfloor a \rfloor$	integer part of $a > 0$
$Re\,z$	real part of $z \in \mathbb{C}$
C	absolute constant
Δg	$\Delta g(j) = g(j+1) - g(j)$
$b_n = O(a_n)$	means that $\sup(b_n/a_n) < \infty$
$b_n = o(a_n)$	means that $b_n/a_n \to 0$
$\square$	end of a proof

Chapter 1
Definitions and Preliminary Facts

1.1 Distributions and Measures

Let $\mathbb{R}$ denote the set of real numbers, $\mathbb{C}$ denote the set of complex numbers, $\mathbb{Z}$ denote the set of all integers, $\mathbb{N} = \{1, 2, \ldots\}$ and $\mathbb{Z}_+ = \mathbb{N} \cup \{0\}$. For $a > 0$ we denote by $\lfloor a \rfloor$ its integer part, that is $a = \lfloor a \rfloor + \delta, 0 \leqslant \delta < 1$. Throughout the book we denote by C positive absolute constants. The letter θ stands for any real or complex number satisfying $|\theta| \leqslant 1$. The values of C and θ can vary from line to line, or even within the same line. By $C(\cdot)$ we denote constants depending on the indicated argument only.

Let $\mathcal{M}$ denote the set of all finite signed measures defined on the σ-field $\mathcal{B}$ of one-dimensional Borel subsets; $\mathcal{M}_Z \subset \mathcal{M}$ denotes the set of all finite signed measures concentrated on $\mathbb{Z}$; $\mathcal{F} \subset \mathcal{M}$ denotes the set of all distributions; $\mathcal{F}_s \subset \mathcal{F}$ denotes the set of symmetric distributions, $\mathcal{F}_+ \subset \mathcal{F}_s$ denotes the set of all distributions having nonnegative characteristic functions and $\mathcal{F}_Z \subset \mathcal{F}$ denotes the set of all distributions concentrated on $\mathbb{Z}$. Let $I_a \in \mathcal{F}$ denote the distribution concentrated at a point $a \in \mathbb{R}$, with $I \equiv I_0$. In principle, I_a is the distribution of an indicator variable, since for any Borel set X

$$I_a\{X\} = \begin{cases} 1 & \text{if } a \in X, \\ 0 & \text{if } a \notin X. \end{cases}$$

Observe that if $F \in \mathcal{F}_Z$, then

$$F = \sum_{j=-\infty}^{\infty} F\{j\} I_j.$$

Let the random variable ξ have distribution $F \in \mathcal{F}$. Then, for any Borel set $X \in \mathcal{B}$, $P(\xi \in X) = F\{X\}, P(\xi \leqslant x) = F(x) = F\{(-\infty, x]\}$, etc. In this book, the measure

© Springer International Publishing Switzerland 2016
V. Čekanavičius, *Approximation Methods in Probability Theory*, Universitext,
DOI 10.1007/978-3-319-34072-2_1

notation is usually preferred over the random variable notation. For the distribution of a random variable ξ we also sometimes use the notation $\mathcal{L}(\xi)$. Roughly, the differences among measures, signed measures and distributions are the following: we use the term *measure* for a finite nonnegative measure; *a signed measure* can be expressed as the difference of two measures and a distribution (say F) is a measure satisfying $F\{\mathbb{R}\} = 1$.

A distribution $F \in \mathcal{F}$ is absolutely continuous with the density function $f(x)$ if, for all $A \in \mathcal{B}$,

$$F\{A\} = \int_A f(x)\,dx.$$

Henceforth all integrals are understood as Lebesgue or Lebesgue-Stieljes integrals.

Let $0 \leqslant p \leqslant 1$, $F, G \in \mathcal{F}$. Then $pF + (1 - p)G \in \mathcal{F}$. This property can be extended to the case of more than two distributions. Let $w_j \in [0, 1]$, $F_j \in \mathcal{F}$ ($j \in \mathbb{Z}$) and $\sum_{j\in\mathbb{Z}} w_j = 1$, then

$$\sum_{j\in\mathbb{Z}} w_j F_j \in \mathcal{F}. \tag{1.1}$$

Let $U \in \mathcal{F}$. Then $U^{(-)}$ denotes the distribution, for any Borel set X, satisfying $U^{(-)}\{X\} = U\{-X\}$. Similarly, $U^{(2)}\{X\} = U\{X/2\}$. By $[X]_\tau$ we denote a closed τ-neighborhood of the set X.

We use the notation $\Phi_{\mu,\sigma}$ for the normal distribution with mean $\mu \in \mathbb{R}$ and variance $\sigma^2 > 0$, $\Phi_\sigma \equiv \Phi_{0,\sigma}$, $\Phi \equiv \Phi_{0,1}$. The density function of $\Phi_{\mu,\sigma}$ is equal to

$$\phi_{\mu,\sigma}(x) = \frac{1}{\sigma\sqrt{2\pi}} \exp\left\{-\frac{(x-\mu)^2}{2\sigma^2}\right\}, \quad x \in \mathbb{R}.$$

The Bernoulli distribution $(1 - p)I + pI_1$, $(0 < p < 1)$ is the distribution of a Bernoulli random variable ξ: $P(\xi = 1) = p = 1 - P(\xi = 0)$.

Convolutions All products and powers of (signed) measures are defined *in the convolution sense*, that is, for $F, G \in \mathcal{M}$ and Borel set X

$$FG\{X\} = \int_{\mathbb{R}} F\{X - x\}\, G\{dx\}.$$

We assume that $F^0 \equiv I$. Further on, for any measures M_k, we also assume that $\prod_{j=m}^n M_k \equiv I$, if $m > n$. Observe that $(I_1)^k = I_k$, $\Phi_{\mu,\sigma}^n = \Phi_{n\mu,\sqrt{n}\sigma}$. If $F, G \in \mathcal{M}_{\mathbb{Z}}$, then

$$FG\{k\} = \sum_{j=-\infty}^{\infty} F\{k - j\}G\{j\}, \quad k \in \mathbb{Z}.$$

If $F, G \in \mathcal{F}$ have densities $f(x)$ and $g(x)$, then FG also has the density function $h(x)$:

$$h(x) = \int_{-\infty}^{\infty} f(x - y)g(y)dy.$$

Convolution of distributions is related to the distribution of the sum of independent random variables. For example, F^n is the distribution of $\xi_1 + \xi_2 + \ldots + \xi_n$, where all ξ_j are independent and have the same distribution F.

By the exponential of $M \in \mathcal{M}$ we understand

$$\exp\{M\} = \sum_{m=0}^{\infty} \frac{M^m}{m!}.$$

Exponential measures have some useful properties, for example, $\exp\{M\}\exp\{V\} = \exp\{V + M\}$, $\exp\{aM\}\exp\{bM\} = \exp\{(a + b)M\}$.

Similarly, we define the logarithm of $M \in \mathcal{M}$:

$$\ln(I + M) = \sum_{j=1}^{\infty} \frac{(-1)^{j+1}M^j}{j}.$$

In the above definition we assume that the total variation of M is less than unity.

Convolutions allow us to write some distributions in a convenient way. Let the random variable ξ have a binomial distribution with parameters $n \in \mathbb{N}$ and $p \in (0, 1)$, that is

$$P(\xi = k) = \binom{n}{k}p^k(1 - p)^{n-k}, \quad k = 1, 2, \ldots, n.$$

Then its distribution $Bi(n, p)$ can be written in the following way

$$Bi(n, p) = ((1 - p)I + pI_1)^n = \sum_{k=0}^{n} \binom{n}{k}p^k(1 - p)^k(I_1)^k = \sum_{k=0}^{n} \binom{n}{k}p^k(1 - p)^k I_k.$$

Here, as usual,

$$\binom{n}{k} = \frac{n!}{k!(n - k)!}.$$

Similarly, let η have a Poisson distribution with parameter $\lambda > 0$,

$$P(\eta = k) = \frac{\lambda^k}{k!}e^{-\lambda}, \quad k = 0, 1, 2, \ldots$$

Then its distribution $Po(\lambda)$ is

$$Po(\lambda) = \exp\{\lambda(I_1 - I)\} = \sum_{k=0}^{\infty} \frac{\lambda^k (I_1 - I)^k}{k!} = \sum_{k=0}^{\infty} \frac{\lambda^k}{k!} e^{-\lambda} I_k.$$

Compound measures By *a compound* (signed) measure we understand

$$\varphi(F) = \sum_{m=0}^{\infty} a_m F^m, \quad \text{where} \quad F \in \mathcal{F}, \quad \sum_{m=0}^{\infty} |a_m| < \infty. \tag{1.2}$$

If $a_0 + a_1 + a_2 + a_3 + \ldots = 1, 0 \leq a_j \leq 1$, then $\varphi(F)$ is a compound distribution. Any compound distribution corresponds to the random sum of independent identically distributed random variables (iid rvs) $\xi_1 + \xi_2 + \ldots + \xi_\eta$, where all ξ_j have the same distribution F and η is independent of ξ_j and has distribution $P(\eta = k) = a_k$, $k = 0, 1, \ldots$. As a rule, F in (1.2) is called the compounding distribution.

Example 1.1 Let us assume that a claim occurs with probability p and the amount of a claim is determined by the distribution B. Then the aggregate claims distribution of n individuals is equal to $((1 - p)I + pB)^n$.

Similarly, assuming that the probabilities of a claim's occurrence and the distributions of claims differ from individual to individual, we get the aggregate claims distribution equal to $\prod_{i=1}^{n} H_i$, where $H_i = (1 - p_i)I + p_i B_i$ is the distribution of risk i, p_i is the probability that risk i produces a claim and B_i is the distribution of the claim in risk i, given the claim occurrence in risk i.

Example 1.2 Compound Poisson distribution (CP). Let $F \in \mathcal{F}, \lambda > 0$, then the CP distribution with compounding distribution F is defined as

$$\exp\{\lambda(F - I)\} := \sum_{m=0}^{\infty} \frac{\lambda^m e^{-\lambda}}{m!} F^m = \sum_{m=o}^{\infty} \frac{\lambda^m (F - I)^m}{m!}. \tag{1.3}$$

Note that $\exp\{\lambda(F - I)\}$ is a direct generalization of the Poisson law $\exp\{\lambda(I_1 - I)\}$.

Observe that by (1.1) the convolution of two CP distributions is also a CP distribution. Indeed, if $F, G \in \mathcal{F}, \lambda, \gamma > 0$, then

$$\exp\{\lambda(F - I)\} \exp\{\gamma(G - I)\} = \exp\left\{(\lambda + \gamma)\left(\frac{\lambda}{\lambda + \gamma}F + \frac{\gamma}{\lambda + \gamma}G - I\right)\right\}.$$

Example 1.3 Compound geometric distribution (CG). Let $F \in \mathcal{F}, 0 < q \leq 1$, $p = 1 - q$. By a compound geometric distribution we understand

$$CG(q, F) := \sum_{m=0}^{\infty} pq^m F^m. \tag{1.4}$$

Remarkably, $CG(q, F)$ is also a CP distribution. Indeed, $CG(q, F)$ can be written as

$$CG(q, F) = \exp\left\{\sum_{j=1}^{\infty} \frac{q^j}{j}(F^j - I)\right\}.$$

The last expression allows us to define a compound negative binomial distribution.

Example 1.4 Compound negative binomial (CNB) distribution. Let $F \in \mathcal{F}$, $0 < q < 1$, $p = 1 - q$, $\gamma > 0$. By a compound negative binomial distribution we understand

$$CNB(\gamma, q, F) := \exp\left\{\gamma \sum_{j=1}^{\infty} \frac{q^j}{j}(F^j - I)\right\}. \tag{1.5}$$

1.2   Moment Inequalities

Let ξ be a random variable with distribution F. The moment and absolute moment of order $k \in \mathbb{N}$ and $a > 0$ are respectively defined as

$$\mathbb{E}\,\xi^k = \int_{\mathbb{R}} x^k F\{dx\}, \quad \mathbb{E}\,|x|^a = \int_{\mathbb{R}} |x|^a F\{dx\}.$$

In this book, we use the expressions *moment of* ξ and *moment of F* as synonyms. Frequently, the variance is denoted by $\sigma^2 = \text{Var}\,\xi = \mathbb{E}\,(\xi - \mathbb{E}\,\xi)^2 = \mathbb{E}\,\xi^2 - \mathbb{E}^2\xi$.

Next, we formulate some moment inequalities. It is assumed that ξ and η are random variables and all moments in the formulas below are finite.

- Lyapunov's inequality: for $1 < s \leqslant t$:

$$\mathbb{E}\,|\xi| \leqslant \left(\mathbb{E}\,|\xi|^s\right)^{1/s} \leqslant \left(\mathbb{E}\,|\xi|^t\right)^{1/t}. \tag{1.6}$$

- Hölder's inequality: if $1 < p, q < \infty$, $1/p + 1/q = 1$, then

$$\mathbb{E}\,|\xi\eta| \leqslant \left(\mathbb{E}\,|\xi|^p\right)^{1/p}\left(\mathbb{E}\,|\eta|^q\right)^{1/q}.$$

- Minkowski's inequality: if $s \geqslant 1$, then

$$\left(\mathbb{E}\,|\xi + \eta|^s\right)^{1/s} \leqslant \left(\mathbb{E}\,|\xi|^s\right)^{1/s} + \left(\mathbb{E}\,|\eta|^s\right)^{1/s}.$$

- Jensen's inequality: let $g(x)$ be a Borel-measurable convex function, then

$$g(\mathbb{E}\,\xi) \leqslant \mathbb{E}\,g(\xi), \quad \text{e.g.,} \quad e^{\mathbb{E}\,\xi} \leqslant \mathbb{E}\,e^{\xi}.$$

- Markov's inequality: if $a > 0$, $P(|\xi| \geqslant a) \leqslant \mathbb{E}|\xi|/a$.
- Chebyshev's inequality: if $a > 0$ and σ^2 is the variance of ξ, then $P(|\xi - \mathbb{E}\xi| \geqslant a) \leqslant a^{-2}\sigma^2$.
- Generalized Chebyshev inequality: let $a, b > 0$ and $h(x) \geqslant 0$ be a non-decreasing function. Then

$$h(a)P(\xi \geqslant a) \leqslant \mathbb{E}h(\xi), \quad \text{e.g.,} \quad P(\xi \geqslant a) \leqslant e^{-ba}\mathbb{E}e^{b\xi}.$$

- Rosenthal's inequality: let $S = \xi_1 + \xi_2 + \cdots + \xi_n$, where ξ_i are independent random variables, $\mathbb{E}\xi_i = 0$, $(i = 1, \ldots, n)$ and $t \geqslant 2$. Then

$$\mathbb{E}|S|^t \leqslant C(t) \max \left(\sum_{i=1}^{n} \mathbb{E}|\xi_i|^t, \left(\sum_{i=1}^{n} \mathbb{E}\xi_i^2 \right)^{t/2} \right).$$

We recall that a function $g(x)$ is convex if $g(\alpha x + (1-\alpha)y) \leqslant \alpha g(x) + (1-\alpha)g(y)$ for all $x, y \in \mathbb{R}$, $\alpha \in [0, 1]$. A twice differentiable function on some interval function $g(x)$ is convex on that interval if and only if $g''(x) \geqslant 0$.

1.3 Norms and Their Properties

We need some tools to measure the closeness of distributions. Instead of the frequently used terms *metric, semi-metric* (for the precise definitions see [116], Chapter 2) we, in all cases, use the same term *norm*. For example, we write Kolmogorov norm instead of Kolmogorov metric.

Total variation norm Let $M \in \mathcal{M}$. A measurable set $A \in \mathcal{B}$ is called the support of M if it is the complement of the largest open set with the measure zero. The support of M is denoted by $supp\, M$. Thus, $M\{B\} = 0$, for any $B \subset \mathcal{B} \setminus supp\, M$. The Jordan-Hahn decomposition states that M can be expressed as $M = M^+ - M^-$, where nonnegative measures M^+ and M^- have different supports, that is $supp\, M^+ \cap supp\, M^- = \emptyset$. The total variation norm (total variation) of M is defined as

$$\|M\| := M^+\{\mathbb{R}\} + M^-\{\mathbb{R}\},$$

or equivalently

$$\|M\| = \int_{\mathbb{R}} 1 |M\{dx\}|.$$

If $M \in \mathcal{M}_Z$ and $F, G \in \mathcal{F}$ have densities $f(x), g(x)$, then respectively

$$\|M\| = \sum_{k=-\infty}^{\infty} |M\{k\}|, \qquad \|F - G\| = \int_{-\infty}^{\infty} |f(x) - g(x)| \, dx.$$

The total variation norm is invariant with respect to scale transformation: $\| I_a M^{(b)} \| = \| M \|$. In other words, the total variation between two distributions remains the same if we multiply both respective random variables by b and add a. Obviously, $\| - M \| = \| M \|$. The following estimates hold for any $M, V \in \mathcal{M}$, $a \in \mathbb{R}$:

$$\| MV \| \leqslant \| M \| \| V \|, \quad \| \exp\{M\} \| \leqslant \exp\{\| M \|\}, \quad \| I_a M \| = \| M \|. \quad (1.7)$$

Next we introduce the measure analogue of θ. Further on Θ stands for any signed measure satisfying $\| \Theta \| \leqslant 1$. The expression of Θ can vary from line to line, or even within the same line.

Observe that, for any $F \in \mathcal{F}$, $\| F \| = 1$. This property is especially useful when we deal with CP distributions.

Example 1.5 Let $\lambda > 0$, $F \in \mathcal{F}$. Then a direct application of (1.7) gives us

$$\| \exp\{\lambda(F - I)\} \| \leqslant \exp\{\lambda \| F - I \|\} \leqslant \exp\{\lambda(\| F \| + \| I \|)\} = \exp\{2\lambda\}.$$

If λ is large, this estimate is very rough. It can be improved if we notice that $\exp\{\lambda(F - I)\}$ is a distribution. Since the total variation of any distribution equals 1, we conclude that

$$\| \exp\{\lambda(F - I)\} \| = 1. \quad (1.8)$$

Note that (1.8) is not true for $\lambda < 0$, since we are dealing with signed measures.

In probability theory, the accuracy of approximation is also estimated by the total variation distance. The total variation distance between $F, G \in \mathcal{F}$ is defined as

$$d_{TV}(F, G) := \sup_{X \in B} | F\{X\} - G\{X\} |.$$

The total variation norm is equivalent to $d_{TV}(\cdot, \cdot)$ in the sense that, for $F, G \in \mathcal{F}$, we have

$$\| F - G \| = 2 d_{TV}(F, G).$$

Moreover, for any $M \in \mathcal{M}$,

$$\frac{1}{2} \| M \| \leqslant \sup_{X \in B} | M\{X\} | \leqslant \| M \|.$$

In this book, we prefer to use the total variation norm, since it is more convenient to write $\| F(G - I)^2 \|$ instead of $d_{TV}(FG^2 + F, 2FG)$.

For discrete and absolutely continuous distributions other expressions of the total variation norm exist. Let $a^+ = \max(a, 0)$, $a^- = -\min(a, 0)$, so that $a = a^+ - a^-$.

Let $F, G \in \mathcal{F}_{\mathbb{Z}}$. Then

$$\| F - G \| = 2 \sum_{k \in \mathbb{Z}} (F\{k\} - G\{k\})^+, \quad \| F - G \| = 2 - 2 \sum_{k \in \mathbb{Z}} \min(F\{k\}, G\{k\}).$$
(1.9)

We will prove (1.9). On one hand, we deal with distributions and, consequently, $\sum_{k \in \mathbb{Z}} (F\{k\} - G\{k\}) = 1 - 1 = 0$. On the other hand,

$$\sum_{k \in \mathbb{Z}} (F\{k\} - G\{k\}) = \sum_{k \in \mathbb{Z}} (F\{k\} - G\{k\})^+ - \sum_{k \in \mathbb{Z}} (F\{k\} - G\{k\})^-.$$

Therefore

$$\sum_{k \in \mathbb{Z}} (F\{k\} - G\{k\})^- = \sum_{k \in \mathbb{Z}} (F\{k\} - G\{k\})^+.$$
(1.10)

Observe that

$$| F\{k\} - G\{k\} | = (F\{k\} - G\{k\})^+ + (F\{k\} - G\{k\})^-.$$

Therefore, taking into account (1.10), we prove

$$\| F - G \| = \sum_{k \in \mathbb{Z}} | F\{k\} - G\{k\} | = \sum_{k \in \mathbb{Z}} (F\{k\} - G\{k\})^+$$

$$+ \sum_{k \in \mathbb{Z}} (F\{k\} - G\{k\})^- = 2 \sum_{k \in \mathbb{Z}} (F\{k\} - G\{k\})^+.$$

Next, observe that $(F\{k\} - G\{k\})^+ = F\{k\} - \min(F\{k\}, G\{k\})$ and therefore

$$\sum_{k \in \mathbb{Z}} (F\{k\} - G\{k\})^+ = \sum_{k \in \mathbb{Z}} F\{k\} - \sum_{k \in \mathbb{Z}} \min(F\{k\}, G\{k\}) = 1 - \sum_{k \in \mathbb{Z}} \min(F\{k\}, G\{k\}).$$

Obviously, (1.9) follows from the last expression.

Similar expressions hold for absolutely continuous distributions. Let $F, G \in \mathcal{F}$ respectively have densities $f(x)$ and $g(x)$. Then

$$\| F - G \| = \int_{-\infty}^{\infty} | f(x) - g(x) | \, dx = 2 \int_{-\infty}^{\infty} (f(x) - g(x))^+ dx$$

$$= 2 - 2 \int_{-\infty}^{\infty} \min(f(x), g(x)) dx.$$

Note that for the proof of (1.10) the assumption $F, G \in \mathcal{F}$ is crucial. There is no direct analogue of (1.10) for an arbitrary measure $M \in \mathcal{M}_{\mathbb{Z}}$.

Kolmogorov norm The (uniform) Kolmogorov norm of $M \in \mathcal{M}$ is defined as the supremum over all intervals

$$|M|_K := \sup_{x \in \mathbb{R}} |M\{(-\infty, x]\}| = \sup_{x \in \mathbb{R}} |M(x)|.$$

Let $V, M \in \mathcal{M}, F \in \mathcal{F}, a \in \mathbb{R}$, then

$$|M|_K \leqslant \|M\|, \quad |VM|_K \leqslant |V|_K \|W\|, \quad |I_a M|_K = |M|_K, \quad |F|_K = 1.$$

The Kolmogorov norm is invariant with respect to scale transformation.

Local norm Let $M \in \mathcal{M}_Z$, i.e. let M be concentrated on $\mathbb{Z}$. Then the local norm of M is defined in the following way

$$\|M\|_\infty := \sup_{j \in \mathbb{Z}} |M\{j\}|.$$

If $M, V \in \mathcal{M}_Z$ then

$$\|MV\|_\infty \leqslant \|M\| \|V\|_\infty, \quad \|M\|_\infty \leqslant \|M\|. \tag{1.11}$$

Note that the analogue of $\|M\|_\infty$ can be defined for absolutely continuous measures as a supremum of their densities.

Wasserstein norm Apart from the main three norms defined above, the so-called Wasserstein (also known as Kantorovich, Dudley or Fortet-Mourier) norm is sometimes used. The Wasserstein norm for $M \in \mathcal{M}_Z$ is defined in the following way

$$\|M\|_W := \sum_{k=-\infty}^{\infty} |M\{(-\infty, k]\}|.$$

Note that, in the literature, other more general definitions of the Wasserstein norm for $M \in \mathcal{M}$ are also available.

1.4 Fourier Transforms

General properties Let $M \in \mathcal{M}$. Then its Fourier transform is defined as

$$\widehat{M}(t) = \int_{-\infty}^{\infty} e^{itx} M\{dx\}.$$

Here $\exp\{itx\} = \cos(tx) + i\sin(tx)$ and i is the imaginary unit, i.e. $i^2 = -1$. If $F \in \mathcal{F}$ has a density function $f(x)$ and $M \in \mathcal{M}_Z$ then respectively

$$\widehat{F}(t) = \int_{-\infty}^{\infty} f(x)e^{itx}dx, \quad \widehat{M}(t) = \sum_{k=-\infty}^{\infty} e^{itk}M\{k\}.$$

The main properties:

$$\widehat{I}_a(t) = e^{ita}, \quad \widehat{I}(t) = 1, \quad |\widehat{M}(t)| \leq \|M\|,$$

$$\widehat{MV}(t) = \widehat{M}(t)\widehat{V}(t), \quad \widehat{M}(t)\widehat{M}(-t) = |\widehat{M}(t)|^2,$$

$$\widehat{\varphi}(F)(t) = \varphi(\widehat{F}(t)), \quad \widehat{\exp\{M\}}(t) = \exp\{\widehat{M}(t)\}.$$

Here $\varphi(F)$ is the compound measure defined by (1.2).

If $F \in \mathcal{F}$, i.e. F is the distribution of some random variable ξ, then $\widehat{F}$ is called *the characteristic function* of ξ. In this case,

$$\widehat{F}(t) = \mathbb{E}\, e^{i\xi t}.$$

Example 1.6 Let $F \in \mathcal{F}$, $0 < p < 1$, $q = 1 - p$, $\lambda, \gamma > 0$. Then the characteristic functions of CP, CG and CNB distributions (see (1.3), (1.4) and (1.5)) are respectively

$$\exp\{\lambda(\widehat{F}(t) - 1)\}, \qquad \frac{p}{1 - q\widehat{F}(t)}, \qquad \left(\frac{p}{1 - q\widehat{F}(t)}\right)^{\gamma}.$$

Note that if $\widehat{F}(t) = e^{it}$ ($F \equiv I_1$), then the same formulas give us the characteristic functions of Poisson, geometric and negative binomial distributions.

If $\widehat{F}(t)$ is a characteristic function then

(1) $|\widehat{F}(t)| \leq 1$, $\widehat{F}(0) = 1$.
(2) $\widehat{F}^k(t)$, $\widehat{F}(-t)$, $Re\,\widehat{F}(t)$ and $|\widehat{F}(t)|^2$ are also characteristic functions. Here $Re\widehat{F}(t)$ denotes the real part of $\widehat{F}(t)$ and $k \in \mathbb{Z}_+$.
(3) Let the random variable ξ have the characteristic function $\widehat{F}_\xi(t)$. Then, for any $a, b \in \mathbb{R}$, $\widehat{F}_{a\xi+b}(t) = \exp\{itb\}\widehat{F}_\xi(at)$.
(4) If the random variables ξ and η are independent, then $\widehat{F}_{\xi+\eta}(t) = \widehat{F}_\xi(t)\widehat{F}_\eta(t)$.
(5) $\widehat{F}(t) \in \mathbb{R} \Leftrightarrow \xi$ is symmetric.
(6) The random variable ξ is integer-valued $\Leftrightarrow |\widehat{F}(2\pi)| = 1$.
(7) If, for some $k \geq 0$, $\mathbb{E}|\xi|^k < \infty$. Then, for $j \leq k$, $\widehat{F}^{(j)}(0) = i^j\mathbb{E}\,\xi^j$.

(8) Let $\mathbb{E}\,|\,\xi\,|^s < \infty$ for some $s \in \mathbb{N}$. Then the following moment expansions hold

$$\widehat{F}(t) = 1 + (it)\,\mathbb{E}\,\xi + \frac{(it)^2}{2}\,\mathbb{E}\,\xi^2 + \frac{(it)^3}{3!}\,\mathbb{E}\,\xi^3 + \ldots + \frac{(it)^{s-1}}{(s-1)!}\,\mathbb{E}\,\xi^{s-1}$$

$$+\theta\frac{|\,t\,|^s}{s!}\,\mathbb{E}\,|\,\xi\,|^s, \tag{1.12}$$

$$\widehat{F}'(t) = i\left\{\mathbb{E}\,\xi + (it)\,\mathbb{E}\,\xi^2 + \frac{(it)^2}{2!}\,\mathbb{E}\,\xi^3 + \ldots + \frac{(it)^{s-2}}{(s-2)!}\,\mathbb{E}\,\xi^{s-1}\right\}$$

$$+\theta\frac{|\,t\,|^{s-1}}{(s-1)!}\,\mathbb{E}\,|\,\xi\,|^s. \tag{1.13}$$

(9) If F has a finite s-th moment ($s \in \mathbb{N}$), then

$$\ln\widehat{F}(t) = \sum_{j=1}^{s} \frac{\gamma_j(F)}{j!}(it)^j + o(|\,t\,|^j), \quad (t \to 0).$$

The coefficients $\gamma_j(F)$ are called the cumulants (semi-invariants) of F.
(10) For all $t \in \mathbb{R}$ the following estimate holds:

$$|\,1 - \widehat{F}(t)\,|^2 \leqslant 2(1 - Re\,\widehat{F}(t)). \tag{1.14}$$

(11) Let $c < 1$ and b be some positive constants. Let, for $|\,t\,| \geqslant b$, $|\,\widehat{F}(t)\,| \leqslant c$.
    Then, for $|\,t\,| < b$, $|\,\widehat{F}(t)\,| \leqslant 1 - (1 - c^2)t^2/(8b^2)$.
(12) If $\limsup_{|t|\to\infty}|\,\widehat{F}(t)\,| < 1$ (Cramer's (C) condition), then, for any $\varepsilon > 0$, there
    exists a $c < 1$ such that $|\,\widehat{F}(t)\,| \leqslant c$ for $|\,t\,| \geqslant \varepsilon$.
(13) Let $F \in \mathcal{F}$ be non-degenerate ($F \neq I_a$). Then there exist $\varepsilon = \varepsilon(F) > 0$ and
    $C(F) > 0$ such that, for $|\,t\,| \leqslant \varepsilon$, $|\,\widehat{F}(t)\,| \leqslant 1 - C(F)t^2$.

Integer-valued random variables   Let $F \in \mathcal{F}_Z$. Then

$$\widehat{F}(t) = \sum_{k\in\mathbb{Z}} e^{itk}F\{k\}.$$

If F is concentrated on $0, 1, 2, \ldots$, then it is more natural to use the expansion of
$\widehat{F}(t)$ in powers of $(e^{it} - 1)$. In such an expansion, the role of moments is played by
the so-called factorial moments. The k-th *factorial moment* of F is defined in the

following way:

$$v_k(F) = \sum_{j=1}^{\infty} j(j-1)(j-2)\ldots(j-k+1)F\{j\}, \quad k = 1, 2, \ldots \tag{1.15}$$

Note that $v_1(F)$ is the mean of F. In general, there is no analogue of (1.6) for factorial moments.

Let $F \in \mathcal{F}$ be concentrated on $\mathbb{Z}_+$ and let, for some integer $s \geqslant 1$, $v_s(F) < \infty$. Then

$$\widehat{F}(t) = \sum_{j=0}^{\infty} F\{j\}e^{itj} = 1 + (e^{it} - 1)\, v_1(F) + \frac{(e^{it} - 1)^2}{2!}\, v_2(F) + \frac{(e^{it} - 1)^3}{3!}\, v_3(F) +$$

$$+ \ldots + \frac{(e^{it} - 1)^{s-1}}{(s-1)!}\, v_{s-1}(F) + \theta \frac{|e^{it} - 1|^s}{s!}\, v_s(F), \tag{1.16}$$

$$\widehat{F}'(t) = i\sum_{j=1}^{\infty} F\{j\}je^{itj} = ie^{it}\left(v_1(F) + (e^{it} - 1)\, v_2(F) + \frac{(e^{it} - 1)^2}{2!}\, v_3(F) \right.$$

$$\left. + \ldots + \frac{(e^{it} - 1)^{s-2}}{(s-2)!}\, v_{s-1}(F) \right) + \theta \frac{|e^{it} - 1|^{s-1}}{(s-1)!}\, v_s(F). \tag{1.17}$$

We can express $v_k(F)$ through moments and vice versa. However, the magnitudes of $v_k(F)$ and moments can be very different.

Example 1.7 Let ξ be a Bernoulli variable, that is $P(\xi = 1) = p = 1 - P(\xi = 0)$. It is obvious that the characteristic function of ξ can be expressed in two ways:

$$(1-p) + pe^{it} = 1 + p(e^{it} - 1) = 1 + (it)p + \frac{(it)^2}{2}p + \frac{(it)^3}{3!}p + \ldots$$

All factorial moments, except the first one, are equal to zero. All moments are of the same order $O(p)$.

Sometimes an expansion of $\ln F(t)$ in powers of $(e^{it} - 1)$ is needed. Then, instead of moments, we deal with *factorial cumulants*. Let F be concentrated on $\mathbb{Z}_+$. Then $\kappa_j(F)$ is its k-th factorial cumulant:

$$\ln \widehat{F}(t) = \sum_{j=1}^{\infty} \frac{\kappa_j(F)}{j!} (e^{it} - 1)^j.$$

In general, there are no analogues of (1.12) and (1.6) for factorial cumulants. As a rule, additional assumptions are required for the estimation of the remainder term in the series.

Example 1.8 Let us consider the negative binomial distribution with $q < 1/4$, $s \in \mathbb{N}$. Then

$$\left(\frac{p}{1-qe^{it}}\right)^{\gamma} = \left(1 - \frac{q}{p}(e^{it}-1)\right)^{-\gamma} = \exp\left\{-\gamma \ln\left(1 - \frac{q}{p}(e^{it}-1)\right)\right\}$$

$$= \exp\left\{\gamma \sum_{j=1}^{\infty} \frac{1}{j}\left(\frac{q}{p}\right)^{j}(e^{it}-1)^{j}\right\}$$

$$= \exp\left\{\gamma \sum_{j=1}^{s} \frac{1}{j}\left(\frac{q}{p}\right)^{j}(e^{it}-1)^{j} + \theta\gamma \frac{3|e^{it}-1|^{s+1}}{s+1}\left(\frac{q}{p}\right)^{s+1}\right\}.$$

Therefore the k-th factorial cumulant is equal to

$$\kappa_k(NB) = (k-1)!\gamma\left(\frac{q}{p}\right)^{k}.$$

Observe that the above series converges absolutely. Indeed, since $q < 1/4$, then $q/p < 1/3$ and

$$\left|\sum_{j=s+1}^{\infty} \frac{1}{j}\left(\frac{q}{p}\right)^{j}(e^{it}-1)^{j}\right| \leq \frac{|e^{it}-1|^{s+1}}{s+1}\left(\frac{q}{p}\right)^{s+1} \sum_{j=s+1}^{\infty}\left(\frac{1}{3}\right)^{j-s-1}(|e^{it}|+1)^{j-s-1}$$

$$= \frac{|e^{it}-1|^{s+1}}{s+1}\left(\frac{q}{p}\right)^{s+1} \sum_{j=s+1}^{\infty}\left(\frac{2}{3}\right)^{j-s-1} = \frac{3|e^{it}-1|^{s+1}}{s+1}\left(\frac{q}{p}\right)^{s+1}.$$

Fourier transforms of integrable functions We say that $f : \mathbb{B} \to \mathbb{R}$ belongs to the space $L_r(\mathbb{R})$, $r \geq 1$, if

$$\int_{-\infty}^{\infty} |f(x)|^r \, dx < \infty.$$

The Fourier transform for $f \in L_1(\mathbb{R})$ is defined by

$$\hat{f}(t) = \int_{-\infty}^{\infty} e^{itx}f(x) \, dx.$$

If f is continuous on $\mathbb{R}$ and $f, \hat{f} \in L_1(\mathbb{R})$, then

$$f(x) = \frac{1}{2\pi}\int_{-\infty}^{\infty} \hat{f}(t)e^{-itx} \, dt.$$

and for $F \in \mathcal{F}$,

$$\int_{-\infty}^{\infty} f(x) F\{dx\} = \frac{1}{2\pi} \int_{-\infty}^{\infty} \hat{f}(t)\widehat{F}(-t)\,dt. \tag{1.18}$$

The last relation is known as Parseval's identity. Another version of Parseval's identity states that if $f \in L_1(\mathbb{R}) \cap L_2(\mathbb{R})$, then

$$\int_{-\infty}^{\infty} |f(x)|^2 dx = \frac{1}{2\pi} \int_{-\infty}^{\infty} |\hat{f}(t)|^2 dt. \tag{1.19}$$

1.5   Concentration Function

Let $F \in \mathcal{F}, h \geqslant 0$. The Lévy concentration function is defined in the following way

$$Q(F, h) = \sup_{x \in \mathbb{R}} F\{[x, x + h]\}, \quad Q(F, 0) = \max_{x \in \mathbb{R}} F\{x\}.$$

It is obvious that $Q(F, h) \leqslant 1$. Moreover, let $F, G \in \mathcal{F}, h > 0, a > 0$. Then

$$Q(FG, h) \leqslant \min\{Q(F, h), Q(G, h)\}, \quad Q(F, h) \leqslant (h/a + 1)Q(F, a). \tag{1.20}$$

The Kolmogorov-Rogozin inequality defines a relation between a distribution and its n-fold convolution:

$$Q(F^n, h) \leqslant \frac{C}{\sqrt{n(1 - Q(F, h))}}. \tag{1.21}$$

The Le Cam inequality estimates the concentration function of a CP distribution through its compounding distribution:

$$Q(\exp\{a(F - I)\}, h) \leqslant \frac{1}{\sqrt{2ea \max[F\{(-\infty, h)\}, F\{(h, \infty)\}]}}$$

$$\leqslant \frac{C}{\sqrt{aF\{x : |x| > h\}}}. \tag{1.22}$$

Two inequalities establish relations between concentration and characteristic functions:

$$Q(F, h) \leqslant Ch \int_{|t|<1/h} |\widehat{F}(t)| \, dt; \quad F \in \mathcal{F}, \tag{1.23}$$

$$h \int_{|t|<1/h} \widehat{F}(t) \, dt \leqslant CQ(F, h), \quad F \in \mathcal{F}_+. \tag{1.24}$$

The following lemma formulates another relation between concentration and characteristic functions.

Lemma 1.1 *Let $F \in \mathcal{F}_+$ and let g be a measurable bounded function. Then, for any $h > 0$, the following inequality holds*

$$\left| \int_{-\infty}^{\infty} g(t)\widehat{F}(t) \, dt \right| \leqslant 13Q(F, h) \left(\frac{\sup_t |g(t)|}{h} + \int_0^{\infty} \sup_{s:|s|\geqslant|t|} |g(s)| \, dt \right).$$

1.6 Algebraic Identities and Inequalities

In this section, various identities and inequalities for real or complex numbers and general commutative objects are collected. We assume that $\sum_{j=k}^{m} \equiv 0$ and $\prod_{j=k}^{m} \equiv 1$ and $\binom{m}{k} \equiv 0$ if $k > m$.

General facts about complex numbers Let i denote the imaginary unit, i.e. $i^2 = -1$. Any $z \in \mathbb{C}$ can be expressed as $z = a + ib$, $a, b \in \mathbb{R}$. Here a and b are called the real and imaginary parts of z, respectively, ($a = Rez$, $b = Imz$). The conjugate of z is defined as $\bar{z} = a - ib$. The absolute value of z can be calculated from its square:

$$|z|^2 = z\bar{z} = a^2 + b^2, \qquad |z| = \sqrt{a^2 + b^2}. \tag{1.25}$$

Obviously, $|z| \leqslant |a| + |b|$. Division by $c + id \neq 0$ is understood in the following way:

$$\frac{a + ib}{c + id} = \frac{(ac + bd) + i(bc - ad)}{c^2 + d^2}.$$

The partial sum of a geometric series is similar to the case of real numbers:

$$1 + z + z^2 + \cdots + z^{k-1} = \frac{1 - z^k}{1 - z}, \quad k \in \mathbb{N}. \tag{1.26}$$

Estimates related to Euler's formula We denote by e^{it} a complex exponential function. By Euler's formula, for all $t \in \mathbb{R}$,

$$e^{it} = \cos t + i \sin t. \tag{1.27}$$

Note that $|e^{it}| = 1$ and $e^{itk} = (e^{it})^k$, $k \in \mathbb{Z}$.

Let $k = 0, 1, 2, \ldots, \lambda > 0$. Then

$$|e^{it} - 1| = 2|\sin(t/2)|, \qquad |\sin(kt/2)| \leq k|\sin(t/2)|, \tag{1.28}$$

$$(e^{it} - 1) + (e^{-it} - 1) = -(e^{it} - 1)(e^{-it} - 1), \tag{1.29}$$

$$|(e^{-it} - 1)^k - (-1)^k (e^{it} - 1)^k| \leq k|e^{it} - 1|^{k+1}, \tag{1.30}$$

$$\int_{-\pi}^{\pi} \left| \sin\frac{t}{2} \right|^k \exp\left\{ -2\lambda \sin^2\frac{t}{2} \right\} \, dt \leq C(k)\lambda^{-(k+1)/2}. \tag{1.31}$$

Taylor series Let $t \in \mathbb{R}$, $k \in \mathbb{Z}_+$. Then

$$e^{it} = 1 + (it) + \frac{(it)^2}{2!} + \cdots + \frac{(it)^k}{k!} + \theta \frac{|t|^{k+1}}{(k+1)!}. \tag{1.32}$$

More generally, let M be a real or complex number, $k \in \mathbb{Z}_+$. Then

$$e^M = 1 + M + \frac{M^2}{2!} + \cdots + \frac{M^k}{k!} + \frac{M^{k+1}}{k!} \int_0^1 e^{\tau M}(1 - \tau)^k \, d\tau. \tag{1.33}$$

We give one example of how (1.33) can be applied.

Example 1.9 Let $Re\, a < 0$. Then

$$|e^a - 1| \leq |a|, \quad |e^a - 1 - a| \leq \frac{|a|^2}{2}. \tag{1.34}$$

Indeed, we have

$$|e^a - 1| = \left| a \int_0^1 e^{\tau a} \, dt \right| \leq |a| \int_0^1 |e^{\tau a}| \, dt = |a| \int_0^1 e^{\tau Re\, a} \, dt \leq |a|,$$

since $e^{\tau Re\, a} \leq 1$. Similarly,

$$|e^a - 1 - a| = \left| a^2 \int_0^2 e^{\tau a}(1 - \tau) d\tau \right| \leq |a|^2 \int_0^1 e^{\tau Re\, a}(1 - \tau) d\tau \leq \frac{|a|^2}{2}.$$

Note that (1.33) can also be applied to exponential measures. Let $M \in \mathcal{M}$, $s \in \mathbb{Z}_+$. Then

$$\exp\{M\} = I + M + \frac{M^2}{2!} + \ldots + \frac{M^s}{s!} + \frac{M^{s+1}}{s!} \int_0^1 \exp\{\tau M\}(1 - \tau)^s \, d\tau$$

$$= I + \sum_{j=1}^{s} \frac{M^j}{j!} + \frac{1}{(s+1)!} \| M \|^{s+1} \exp\{\| M \|\}\Theta. \tag{1.35}$$

In this book, we constantly apply the following Taylor series: let z be a complex number such that $|z| < 1$. Then

$$\frac{1}{1-z} = \sum_{j=0}^{\infty} z^j, \qquad \ln(1 + z) = \sum_{j=1}^{\infty} \frac{(-1)^{j+1}}{j} z^j. \tag{1.36}$$

Bergström's identity Let $a, b > 0$, $n \in \mathbb{N}$. Then

$$a^n = b^n + \sum_{m=1}^{s} \binom{n}{m} b^{n-m}(a - b)^m + r_n(s + 1). \tag{1.37}$$

Here

$$r_n(s + 1) = \sum_{m=s+1}^{n} \binom{m-1}{s} a^{n-m}(a - b)^{s+1} b^{m-s-1}. \tag{1.38}$$

Note that

$$\sum_{m=s+1}^{n} \binom{m-1}{s} = \binom{n}{s+1}. \tag{1.39}$$

Bergström's identity (1.37) can be extended to products. Particularly,

$$\prod_{k=1}^{n} a_k - \prod_{k=1}^{n} b_k = \sum_{k=1}^{n}(a_k - b_k) \prod_{j=1}^{k-1} a_j \prod_{j=k+1}^{n} b_j, \tag{1.40}$$

$$\prod_{k=1}^{n} a_k - \prod_{k=1}^{n} b_k - \sum_{k=1}^{n}(a_k - b_k) \prod_{j \neq k} b_j$$

$$= \sum_{l=2}^{n}(a_l - b_l) \prod_{j=l+1}^{n} a_l \sum_{k=1}^{l-1}(a_k - b_k) \prod_{j=1, j \neq k}^{l-1} b_j. \tag{1.41}$$

Bergström's identity remains valid for characteristic functions. It can also be applied to all commutative objects. For example, one can apply (1.37) and (1.41) to convolutions of finite measures. The following Bergström expansion was used for the proof of (1.16). Let $n \in \mathbb{N}$, $s \in \mathbb{Z}_+$. Then

$$e^{itn} = 1 + \sum_{m=1}^{s} \binom{n}{m}(e^{it} - 1)^m + (e^{it} - 1)^{s+1} \sum_{m=s+1}^{n} \binom{m-1}{s} e^{it(n-m)}$$

$$= 1 + \sum_{m=1}^{s} \binom{n}{m}(e^{it} - 1)^m + \theta \, | \, e^{it} - 1 \, |^{s+1} \binom{n}{s+1}. \tag{1.42}$$

Stirling's formula Let $n \in \mathbb{N}$. Then

$$n! = n^n e^{-n} \sqrt{2\pi n} \, e^{\delta(n)}, \qquad \frac{1}{12n + 1} < \delta(n) < \frac{1}{12n}. \tag{1.43}$$

Abel's partial summation formula

$$\sum_{k=M}^{N} a_k b_k = A_N b_N - \sum_{k=M}^{N-1} A_k (b_{k+1} - b_k). \tag{1.44}$$

Here

$$A_k = \sum_{m=M}^{k} a_m.$$

Other inequalities and identities To estimate of the difference of Fourier transforms we repeatedly use the following simple inequality

$$|a^n - b^n| \leqslant n| a - b| \max\{| a |^{n-1}, | b |^{n-1}\}. \tag{1.45}$$

If $a \in \mathbb{R}$ and $b > 0$, then

$$\sum_{k \in \mathbb{Z}} \left(1 + \left(\frac{k - a}{b}\right)^2\right)^{-1} \leqslant 1 + b\pi. \tag{1.46}$$

If $0 < p < 1$, $n \geqslant 1$, $r \in [1, \infty]$, then

$$\sum_{k=0}^{n} \binom{n}{k} p^k (1 - p)^{n-k} (k + 1)^{-1/r} \leqslant p^{-1/r}(n + 1)^{-1/r}. \tag{1.47}$$

If $x > 0, \alpha > 0$, then

$$x^\alpha e^{-x} \le \alpha^\alpha e^{-\alpha}. \tag{1.48}$$

If $x \in \mathbb{R}$, then

$$1 + x \le e^x. \tag{1.49}$$

If $|t| \le \pi$, then

$$|\sin(t/2)| \ge |t|/\pi. \tag{1.50}$$

If $0 < y < 1, k \in \mathbb{N}$, then

$$\sum_{j=k}^{\infty} \binom{j}{k} y^j = \frac{y^k}{(1-y)^{k+1}}. \tag{1.51}$$

1.7 The Schemes of Sequences and Triangular Arrays

Let us consider a sequence of iid rvs $\xi_1, \xi_2, \ldots$, having distribution F. Let $S_n = \xi_1 + \xi_2 + \cdots + \xi_n$. The distribution of S_n is equal to F^n. Note that the characteristics of F (probabilities, means etc.) do not depend on n. In the literature, this is known as *a scheme of sequences*.

A generalization of the scheme of sequences is the so-called scheme of series or triangular array. It means that, for each n, we consider different sets of random variables $\xi_{1n}, \xi_{2n}, \ldots, \xi_{nn}, S_{n-1} = \xi_{1,n-1} + \ldots + \xi_{n-1,n-1}, S_n = \xi_{1n} + \ldots + \xi_{nn}$.

In a triangular array the distribution of the k-th summand in the n-th series may depend on n, that is, we deal with F_n^n. For example, the binomial distribution with parameters n and $p = 1/2$ corresponds to the scheme of sequences. Meanwhile, the binomial distribution with parameters n and $p = n^{-1/3}$ corresponds to the scheme of triangular arrays.

The notation F_n^n is rather inconvenient. Therefore, in all cases, we use the same notation F^n. Typically the case of sequences is distinguished by the remark that *the distribution of F does not depend on n*.

1.8 Problems

1.1 Prove that $|\sin(kt/2)| \le k|\sin(t/2)|, t \in \mathbb{R}, k = 1, 2, \ldots$.

1.2 Let $F_i, G_i \in \mathcal{F}, i = 1, 2$, and let $|F_1 - G_1|_K \le a_1, |F_2 - G_2|_K \le a_2$. Prove that $|F_1 F_2 - G_1 G_2|_K \le a_1 + a_2$.

1.3 Prove (1.11).

1.4 Prove (1.14).

1.5 Prove (1.16).

1.6 Prove that $\widehat{M}(t)\widehat{M}(-t) = |\widehat{M}(t)|^2$.

1.7 Let $M \in \mathcal{M}$. Prove that $\| e^M - I \| \leqslant \exp\{\| M \|\} - 1 \leqslant \| M \| e^{\| M \|}$.

1.8 Let $M \in \mathcal{M}$ and $M\{\mathbb{R}\} = 0$. Prove that $\| M \| = 2 \sup_{A \in \mathcal{B}} | M\{A\} |$.

1.9 Let $F \in \mathcal{F}_Z$ and let, for k=1,2,…,

$$v_k^+(F) = \sum_{j=k}^{\infty} j(j-1)\cdots(j-k+1)F\{j\}, \quad v_k^-(F) = \sum_{j=k}^{\infty} j(j-1)\cdots(j-k+1)F\{-j\}.$$

Prove that, if $v_4^+(F) + v_4^-(F) < \infty$, then

$$\widehat{F}(t) = 1 + v_1^+(F)(e^{it} - 1) + v_1^-(F)(e^{-it} - 1) + \frac{v_2^+(F)(e^{it} - 1)^2}{2}$$

$$+ \frac{v_2^-(F)(e^{-it} - 1)^2}{2} + \frac{(v_3^+(F) - v_3^-(F))(e^{it} - 1)^3}{3!}$$

$$+ \theta[12v_3^-(F) + v_4^+(F) + v_4^-(F)]\frac{|e^{it} - 1|^4}{4!}.$$

Bibliographical Notes

Most of the material of this chapter is quite standard and can be found in [5, 106, 141]. For Lebesgue-Stieltjes integrals and their properties, see [141], Chapter II. A comprehensive discussion on metrics is presented in [114–116], see also [16, 54], Appendix A1. Distributions and characteristic functions are discussed in the first chapters of [5, 106]. Properties of the concentration function are given in [5], Chapter II, and in [106], Section 1.5. For further research related to supports with known algebraic structure or vector spaces see [56–58].

The proof of Rosenthal's inequality can be found in [106], Theorem 2.9. Further improvements are obtained in [81]. For more advanced Hoeffding's type moment inequalities, see [18, 60, 107]. For Jensen's type inequalities, see [76] and the references therein.

Estimate (1.22) was proved by Roos [125], p. 541. Factorial moments and their generalizations are discussed in [63, 89, 137]. In [161] similar expansions were proposed for probabilities. Bergström's identity is proved in [25] and generalized in [35].

In the literature, the Kolmogorov norm is also denoted by $| \cdot |$, see [5]. The notation $\| \cdot \|_\infty$ corresponds to that used for the ℓ_2 norm in calculus. The positive and negative factorial moments of the last problem were introduced in [89].

Research on the accuracy of approximations is in no way restricted to the four metrics considered in this book, see, for example, [28, 167].

Chapter 2
The Method of Convolutions

In this chapter, we show how to apply properties of the norms for compound approximation. The method of convolutions is also called Le Cam's operator method, since, in principle, it can be applied to all commutative operators. For convenience, we repeat the definition of a compound measure:

$$\varphi(F) = \sum_{m=0}^{\infty} a_m F^m, \qquad F \in \mathcal{F}, \qquad \sum_{m=0}^{\infty} |a_m| < \infty. \tag{2.1}$$

If $a_m \in [0, 1]$ and $a_0 + a_1 + \cdots = 1$, then $\varphi(F)$ is a compound distribution. Observe that the difference of approximated and approximating compound distributions can be expressed as a signed measure of the form (2.1). Therefore, in general, our aim is to estimate $\varphi(F)$. In this chapter, we consider the cases $F \in \mathcal{F}$ and $F \in \mathcal{F}_s$ only. The method of convolutions for $F \in \mathcal{F}_Z$ is discussed in the chapters devoted to lattice approximations.

2.1 Expansion in Factorial Moments

In this section, an expansion of $\varphi(F)$ in powers of $(F - I)$ is given. We generalize (1.16) for compound measures, which may be preferable if one does not apply the characteristic function method. To make our notation shorter we write

$$\alpha_k(\varphi) = \sum_{j=0}^{\infty} j(j-1) \cdots (j-k+1) a_j,$$

$$\beta_k(\varphi) = \sum_{j=0}^{\infty} j(j-1) \cdots (j-k+1) |a_j|.$$

© Springer International Publishing Switzerland 2016
V. Čekanavičius, *Approximation Methods in Probability Theory*, Universitext,
DOI 10.1007/978-3-319-34072-2_2

It is clear that $\alpha_k(\varphi)$ and $\beta_k(\varphi)$ are very similar to the factorial moments ν_k defined in the previous chapter. Indeed, let $a_j \in (0, 1)$, $a_0 + a_1 + \ldots = 1$. Then $\varphi(F)$ is a compound *distribution* and $\alpha_k(\varphi)$ is the k-th factorial moment for $\varphi(I_1)$. In this case, we can give a probabilistic interpretation. Let us recall the fact that a compound distribution can be viewed as a distribution of random sum of random variables. Then $\alpha_k(\varphi)$ is the k-th factorial moment of the number of summands. The following lemma generalizes (1.16).

Lemma 2.1 *Let $\varphi(F)$ be defined by (2.1), $a_1 + a_2 + \ldots = 1$, and let $\beta_{s+1}(\varphi) < \infty$ for some $s \geqslant 1$. Then*

$$\varphi(F) = I + \sum_{m=1}^{s} \alpha_m(\varphi)\frac{(F-I)^m}{m!} + \beta_{s+1}(\varphi)\frac{(F-I)^{s+1}}{(s+1)!}\Theta. \tag{2.2}$$

Proof Applying (1.37) and (1.38) we get

$$\varphi(F) = \sum_{m=0}^{\infty} a_m F^m = \sum_{m=0}^{\infty} a_m \left\{ \sum_{j=0}^{s} \binom{m}{j}(F-I)^j + F^m - \sum_{j=0}^{s} \binom{m}{j}(F-I)^j \right\}$$

$$= \sum_{m=0}^{\infty} a_m \sum_{j=0}^{s} \binom{m}{j}(F-I)^j + (F-I)^{s+1} \sum_{m=s+1}^{\infty} a_m \sum_{j=s+1}^{m} \binom{j-1}{s} F^{m-j}$$

$$=: J_1 + (F-I)^{s+1} J_2.$$

Changing the order of summation we obtain

$$J_1 = I + \sum_{m=0}^{\infty} a_m \sum_{j=1}^{s} \binom{m}{j}(F-I)^j = I + \sum_{j=1}^{s}(F-I)^j \frac{\alpha_j(\varphi)}{j!}.$$

Next, observe that by (1.39)

$$\| J_2 \| \leqslant \left\| \sum_{m=s+1}^{\infty} a_m \sum_{j=s+1}^{m} \binom{j-1}{s} F^{m-j} \right\| \leqslant \sum_{m=s+1}^{\infty} |a_m| \sum_{j=s+1}^{m} \binom{j-1}{s}$$

$$= \sum_{m=s+1}^{\infty} |a_m| \binom{m}{s+1} = \frac{\beta_{s+1}(\varphi)}{(s+1)!},$$

that is,

$$J_2 = \frac{\beta_{s+1}(\varphi)}{(s+1)!}\Theta.$$

Combining all the expressions we complete the proof of Lemma 2.1. $\square$

Example 2.1 We apply Lemma 2.1 to a CP distribution. Let $\lambda > 0, 0 < p \leqslant 1$, $F \in \mathcal{F}, s \in \{0, 1, \ldots\}$. Then

$$\exp\{\lambda(F - I)\} = I + \sum_{m=1}^{s} \frac{\lambda^m (F - I)^m}{m!} + (F - I)^{s+1} \frac{\lambda^{s+1}}{(s + 1)!} \Theta. \qquad (2.3)$$

Example 2.2 Let us consider a compound geometric distribution. We recall that $CG(q, F) = \sum_{m=0}^{\infty} pq^m F^m, 0 < q < 1, p = 1 - q, F \in \mathcal{F}$, see (1.4). Therefore

$$a_m = pq^m, \qquad \alpha_k(CG) = \sum_{j=0}^{\infty} j(j - 1) \cdots (j - k + 1)pq^j.$$

It is possible to calculate α_k in the following way. Let $0 < x < 1$, then it is not difficult to prove by induction that

$$\left(\frac{1}{1 - x}\right)^{(k)} = \frac{k!}{(1 - x)^{k+1}}.$$

Therefore, taking into account the geometric series in (1.36), we prove

$$\sum_{j=0}^{\infty} j(j-1) \cdots (j-k+1)x^{j-k} = \sum_{j=0}^{\infty} (x^j)^{(k)} = \left(\sum_{j=0}^{\infty} x^j\right)^{(k)} = \left(\frac{1}{1 - x}\right)^{(k)} = \frac{k!}{(1 - x)^{k+1}}$$

and

$$\alpha_k(CG) = pq^k \sum_{j=0}^{\infty} j(j - 1) \cdots (j - k + 1)q^{j-k}$$

$$= pq^k \frac{k!}{(1 - q)^{k+1}} = k! \frac{q^k}{p^k}.$$

Applying Lemma 2.1, for $s \in \{0, 1, \ldots\}$, we obtain

$$CG(q, F) = I + \left(\frac{q}{p}\right)(F - I) + \left(\frac{q}{p}\right)^2 (F - I)^2 + \cdots + \left(\frac{q}{p}\right)^s (F - I)^s$$

$$+ \left(\frac{q}{p}\right)^{s+1} (F - I)^{s+1} \Theta.$$

Remarks on application The main benefit is that expansions are applied to the difference of compound distributions, not to their convolutions. In short, we expand in powers of $(F - I)$ the measure $\varphi(F)$ instead of $\varphi^n(F)$. Frequently, the following identity is instrumental for this task:

$$\prod_{j=1}^{n} \varphi_j(F) - \prod_{j=1}^{n} \psi_j(F) = \sum_{j=1}^{n} (\varphi_j(F) - \psi_j(F)) \prod_{k=1}^{j-1} \varphi_k(F) \prod_{k=j+1}^{n} \psi_k(F). \qquad (2.4)$$

Typical application. Let us assume that we want to estimate

$$\Big\| \prod_{j=1}^{n} \varphi_j(F) - \prod_{j=1}^{n} \varphi_j(F) \Big\|, \quad \varphi_j(F) = \sum_{m=0}^{\infty} a_{mj} F^m, \ \psi_j(F) = \sum_{m=0}^{\infty} b_{mj} F^m.$$

- Apply (2.4).
- If $\varphi_k(F), \psi_k(F) \in \mathcal{F}$, then

$$\Big\| \prod_{k=1}^{j-1} \varphi_k(F) \prod_{k=j}^{n} \psi_k(F) \Big\| = 1,$$

since convolution of distributions is also a distribution and the total variation norm of any distribution equals unity. Otherwise, estimates

$$\| \varphi_k(F) \| \leq \sum_{m=0}^{\infty} |a_{mk}|, \quad \| \psi_k(F) \| \leq \sum_{m=0}^{\infty} |b_{mk}|$$

can be applied.
- Expand $\varphi_k(F) - \psi_k(F)$ in powers of $(F - I)$ and note that $\| F - I \| \leq 2$.

Advantages. The method can be applied in very general cases. No assumptions on compounding distribution. Simple to use.

Drawbacks. Possible effect of $n - 1$ convolution is neglected. The method cannot be applied to distributions of sums of dependent random variables.

The Le Cam inequality We demonstrate how (2.2) can be used to estimate the closeness of two compound distributions. The following classical result is usually associated with the names of Khintchin, Döblin or Le Cam.

Theorem 2.1 *Let* $0 \leqslant p_k \leqslant 1$, $F_k \in \mathcal{F}$ $(k = 1, 2, \ldots, n)$. *Then*

$$\left\| \prod_{k=1}^{n} \left((1 - p_k)I + p_k F_k \right) - \exp\left\{ \sum_{k=1}^{n} p_k (F_k - I) \right\} \right\| \leqslant 2 \sum_{k=1}^{n} p_k^2. \qquad (2.5)$$

Proof The proof of (2.5) is based on (2.3). For the sake of brevity let

$$H_k = (1 - p_k)I + p_k F_k = I + p_k(F_k - I), \quad D_k = \exp\{ p_k(F_k - I) \}.$$

Then, applying (2.3) with $s = 1$ and recalling that $\| F_k \| = \| I \| = 1$ since $F_k, I \in \mathcal{F}$, we obtain

$$\| D_k - H_k \| = \| \exp\{ p_k(F_k - I) \} - I - p_k(F_k - I) \|$$
$$\leqslant \frac{p_k^2}{2} \| F_k - I \|^2 \leqslant \frac{p_k^2}{2} (\| F_k \| + \| I \|)^2 = 2p_k^2.$$

Next, recall that a convolution of distributions is also a distribution. Therefore, for any $0 \leqslant k \leqslant n$,

$$\left\| \prod_{j=1}^{k-1} H_j \prod_{j=k+1}^{n} D_j \right\| = 1.$$

Taking into account both estimates in the above and applying (2.4) we prove

$$\left\| \prod_{k=1}^{n} H_k - \prod_{k=1}^{n} D_k \right\| \leqslant \sum_{k=1}^{n} \| H_k - D_k \| \leqslant 2 \sum_{k=1}^{n} p_k^2.$$

$\square$

In general, (2.5) cannot be improved. On the other hand, if $F_k \equiv F$, then better estimates than (2.5) exist. We will prove them in other sections.

2.2 Expansion in the Exponent

In some cases, it is more convenient to use an expansion of $\ln \varphi(F)$ in powers of $(F - I)$. Such an expansion corresponds to an expansion in factorial cumulants, see Sect. 1.3. Formally

$$\ln \varphi(F) = \kappa_1(F - I) + \frac{\kappa_2}{2!}(F - I)^2 + \frac{\kappa_3}{3!}(F - I)^3 + \cdots$$

However, in general, it is difficult to write such an expansion with a useful remainder term. Usually, additional assumptions are needed for absolute convergence of the series.

Example 2.3 Let us consider a CNB distribution with $0 < q < 1, p = 1 - q, \gamma > 0$, $F \in \mathcal{F}$. Note that $\gamma = 1$ corresponds to the compound geometric distribution. Taking into account the definition of (1.5) and Bergström's identity (1.37), for $s \in \mathbb{N}$ we obtain

$$
CNB(\gamma, q, F) = \exp\left\{ \gamma \sum_{j=1}^{\infty} \frac{q^j}{j}(F^j - I) \right\} = \exp\left\{ \gamma \sum_{j=1}^{\infty} \frac{q^j}{j}\left((I + (F - I))^j - I \right) \right\}
$$

$$
= \exp\left\{ \gamma \sum_{j=1}^{\infty} \frac{q^j}{j}\Big[\sum_{m=1}^{s} \binom{j}{m}(F - I)^m + (F - I)^{s+1} \sum_{m=s=1}^{j} \binom{m-1}{s} F^{m-s-1} \Big] \right\}
$$

$$
= \exp\left\{ \gamma \sum_{m=1}^{s} (F - I)^m \sum_{j=m}^{\infty} \frac{q^j}{j}\binom{j}{m} \right\}
$$

$$
\times \exp\left\{ \gamma (F - I)^{s+1} \sum_{j=1}^{\infty} \frac{q^j}{j} \sum_{m=s+1}^{j} \binom{m-1}{s} F^{m-s-1} \right\}. \tag{2.6}
$$

Taking into account (1.51), we prove

$$
\sum_{j=m}^{\infty} \frac{q^j}{j}\binom{j}{m} = \frac{q}{m} \sum_{j=m}^{\infty} q^{j-1}\binom{j-1}{m-1} = \frac{q}{m} \sum_{j=m-1}^{\infty} q^j \binom{j}{m-1} = \frac{q}{m} \frac{q^{m-1}}{p^m} = \frac{q^m}{m p^m}
$$

and, since $\| F \| = 1$,

$$
\Big\| \sum_{j=1}^{\infty} \frac{q^j}{j} \sum_{m=s+1}^{j} \binom{m-1}{s} F^{m-s-1} \Big\| \leqslant \sum_{j=1}^{\infty} \frac{q^j}{j} \sum_{m=s+1}^{j} \binom{m-1}{s}
$$

$$
= \sum_{j=s+1}^{\infty} \frac{q^j}{j}\binom{j}{s+1} = \sum_{j=s+1}^{\infty} \frac{q^j}{s+1}\binom{j-1}{s} = \frac{q^{s+1}}{(s+1)p^{s+1}}.
$$

Substituting the last two expressions into (2.6) we obtain

$$
CNB(\gamma, q, F) = \exp\left\{ \gamma \sum_{m=1}^{s} \frac{1}{m}\left(\frac{q}{p} \right)^m (F - I)^m + \gamma \frac{(F - I)^{s+1}}{s+1}\left(\frac{q}{p} \right)^{s+1} \Theta \right\}. \tag{2.7}
$$

It is obvious that (2.7) also holds for the characteristic function. Observe that an application of Bergström's identity allows us to use weaker assumptions on q than in Example 1.8.

Example 2.4 We show that expansion in the exponent also holds for the compound binomial distribution. Let $0 < p < 1/2, F \in \mathcal{F}, n, s \in \mathbb{Z}_+$. Then

$$((1-p)I + pF)^n = \exp\{n \ln(I + p(F-I))\} = \exp\left\{n \sum_{j=1}^{\infty} \frac{(-1)^{j+1}p^j}{j}(F-I)^j\right\}$$

$$= \exp\left\{n \sum_{j=1}^{s} \frac{(-1)^{j+1}p^j}{j}(F-I)^j + n\frac{p^{s+1}}{s+1}(F-I)^{s+1}\frac{1}{1-2p}\Theta\right\}.$$

For the last equality we used routine estimates $\| F - I \| \leqslant \| F \| + \| I \| = 2$ and

$$\sum_{k=s+1}^{\infty} p^{k-s-1} \frac{\| F - I \|^{k-s-1}}{k} \leqslant \frac{1}{s+1} \sum_{k=s+1}^{\infty} (2p)^{k-s-1} = \frac{1}{(s+1)(1-2p)}.$$

The following lemma can be usefully combined with the expansions in the exponent.

Lemma 2.2 *Let $M_1, M_2 \in \mathcal{M}$. Then*

$$\| \exp\{M_1\} - \exp\{M_2\} \| \leqslant \| M_1 - M_2 \| \| \exp\{M_1\} \| \exp\{\| M_1 - M_2 \|\}. \qquad (2.8)$$

Proof By the properties of total variation and (1.35)

$$\| \exp\{\mathcal{M}_1\} - \exp\{M_2\} \| = \| \exp\{M_1\}(I - \exp\{M_2 - M_1\}) \|$$

$$\leqslant \| \exp\{M_1\} \| \| M_2 - M_1 \| \exp\{\| M_2 - M_1 \|\}.$$

$\square$

Typical application. Let us assume that we want to estimate

$$\left\| \prod_{j=1}^{n} \varphi_j(F) - \prod_{j=1}^{n} \varphi_j(F) \right\|, \quad \varphi_j(F) = \sum_{m=0}^{\infty} a_{mj}F^m, \psi_j(F) = \sum_{m=0}^{\infty} b_{mj}F^m.$$

- Let $M_1 = \sum_{k=1}^{n} \ln \varphi_k(F), M_2 = \sum_{k=1}^{n} \ln \psi_k(F)$.
- Apply (2.8).
- Expand $M_1 - M_2$ in powers of $(F - I)$ and note that $\| F - I \| \leqslant 2$.

Advantages. No assumptions on compounding distribution. Simple to use.

Drawbacks. From convolution of n distributions, possible effect of $n - 1$ convolution is neglected. Additional assumptions for expansions of logarithms of distributions in converging series are needed.

The Hipp inequality If we want to approximate the binomial distribution by some CP measure, then it is natural to match as many factorial cumulants as possible. The same logic applies to approximation of the compound binomial distribution.

Theorem 2.2 *Let* $0 < p < 1/2$, $F \in \mathcal{F}$, $n, s \in \mathbb{Z}_+$. *Then*

$$\left\| ((1-p)I + pF)^n - \exp\left\{ n \sum_{k=1}^{s} \frac{(-1)^{k+1}p^k}{k}(F-I)^k \right\} \right\|$$

$$\leq n\frac{(2p)^{s+1}}{(s+1)(1-2p)} \exp\left\{ n\frac{(2p)^{s+1}}{(s+1)(1-2p)} \right\}.$$

Proof Let

$$M_1 = \ln((1-p)I + pF)^n = M_2 + W,$$

$$M_2 = n\sum_{k=1}^{s}(-1)^k\frac{p^k}{k}(F-I)^k, \quad W = n\sum_{k=s+1}^{\infty}(-1)^k\frac{p^k}{k}(F-I)^k.$$

Observe that $\exp\{M_1\} = ((1-p)I + pF)^n \in \mathcal{F}$. Therefore $\| \exp\{M_1\} \| = 1$. From Example 2.4 it follows that

$$\| M_1 - M_2 \| = \| W \| \leq n\frac{p^{s+1}}{s+1}\| F - I \|^{s+1}\frac{1}{1-2p} \leq n\frac{(2p)^{s+1}}{(s+1)(1-2p)}.$$

It remains to apply (2.8). $\square$

Observe that D in the above is not a distribution, but some signed measure.

2.3 Le Cam's Trick

One of the best known properties of the total variation norm allows us to switch from the general compound distributions to the integer-valued ones. Though the corresponding inequality is usually associated with Le Cam, one can find it (without explicit comments) in earlier papers of Prokhorov and Kolmogorov.

The following estimate holds for all compound measures.

$$\| \varphi(F) \| \leq \| \varphi(I_1) \|. \tag{2.9}$$

Indeed, taking into account the fact that the m-th convolution $F^m \in \mathcal{F}$ and, therefore, $\| F^m \| = 1$, we prove

$$\| \varphi(F) \| \leq \sum_{m=0}^{\infty} | a_m | \| F^m \| = \sum_{m=0}^{\infty} | a_m | = \left\| \sum_{m=0}^{\infty} a_m I_m \right\| = \left\| \sum_{m=0}^{\infty} a_m I_1^m \right\| = \| \varphi(I_1) \|.$$

Example 2.5 We demonstrate how the difference between compound binomial and CP distributions can be reduced to the difference between binomial and Poisson distributions. For $F \in \mathcal{F}, 0 < p < 1$, we obtain

$$\| ((1-p)I + pF)^n - \exp\{np(F-I)\} \| \leq \| ((1-p)I + pI_1)^n - \exp\{np(I_1-I)\} \|$$

$$= \sum_{k=0}^{\infty} \left| \binom{n}{k} p^k (1-p)^{n-k} - \frac{(np)^k}{k!} e^{-np} \right|.$$

It is interesting to note that from (2.9) it follows that

$$\| \varphi(I_1) \| \leq \sup_{F \in \mathcal{F}} \| \varphi(F) \| \leq \| \varphi(I_1) \|,$$

that is,

$$\sup_{F \in \mathcal{F}} \| \varphi(F) \| = \| \varphi(I_1) \|.$$

In other words, $F \equiv I_1$ corresponds to the worst possible case. This means that by applying (2.9) one can lose additional information, which might significantly improve the accuracy of approximation. For example, if F is a symmetric distribution then by applying (2.9) we lose possible benefits due to the symmetry of distributions.

Le Cam's trick does not guarantee the estimation of the accuracy of approximation. It only simplifies the initial problem and, as a rule, must be combined with other methods.

2.4 Smoothing Estimates for the Total Variation Norm

Let us compare two convolutions $F^n M$ and $F^n V$, where $F, M, V \in \mathcal{F}$. For large n, it is natural to expect that distributions of both convolutions should be similar. If this is the case, we say that F^n has a smoothing effect on M and V. There are not so many results for smoothing in total variation.

Lemma 2.3 *Let* $F \in \mathcal{F}, a \in (0, \infty), p = 1 - q \in (0, 1),$ *and* $k, n \in \mathbb{N}$. *Then*

$$\| (F-I) \exp\{a(F-I)\} \| \leq \sqrt{\frac{2}{ae}}, \tag{2.10}$$

$$\| (F-I)^2 \exp\{a(F-I)\} \| \leq \frac{3}{ae}, \tag{2.11}$$

$$\| (F - I)^k \exp\{a(F - I)\} \| \leq \frac{\sqrt{k!}}{a^{k/2}}, \tag{2.12}$$

$$\| (F - I)^k (qI + pF)^n \| \leq \binom{n+k}{k}^{-1/2} (pq)^{-k/2} \leq \frac{C(k)}{(npq)^{k/2}}. \tag{2.13}$$

The proofs of (2.10), (2.11), and (2.12) are quite technical and therefore omitted. The estimates in Lemma 2.3 are very sharp. On the other hand, it is not difficult to get similar rough estimates. Indeed, applying (2.9) we obtain

$$\| (F - I) \exp\{a(F - I)\} \| \leq \| (I_1 - I) \exp\{a(I_1 - I)\} \| = \left\| \sum_{k=0}^{\infty} \frac{a^k e^{-a}}{k!} I_k (I_1 - I) \right\|$$

$$= \left\| \sum_{k=0}^{\infty} \frac{a^k e^{-a}}{k!} I_{k+1} - \sum_{k=0}^{\infty} \frac{a^k e^{-a}}{k!} I_k \right\| = \left\| \sum_{k=0}^{\infty} \frac{a^{k-1} e^{-a}}{k!} (k - a) I_k \right\|$$

$$= \sum_{k=0}^{\infty} \frac{a^{k-1} e^{-a}}{k!} | k - a | = \frac{1}{a} \sum_{k=0}^{\infty} \frac{a^k e^{-a}}{k!} | k - a | = \frac{1}{a} \mathbb{E} | \xi - a |.$$

Here ξ is a Poisson random variable with parameter a. By (1.6) we prove

$$\mathbb{E} | \xi - a | = \mathbb{E} | \xi - \mathbb{E} \xi | \leq \sqrt{\mathbb{E} (\xi - \mathbb{E} \xi)^2} = \sqrt{\operatorname{Var} \xi} = \sqrt{a}.$$

Therefore

$$\| (F - I) \exp\{a(F - I)\} \| \leq \frac{1}{\sqrt{a}}. \tag{2.14}$$

Observe that

$$\| (F - I)^k \exp\{a(F - I)\} \| \leq \left\| (F - I) \exp\left\{ \frac{a}{k}(F - I) \right\} \right\|^k \leq \frac{C(k)}{a^{k/2}}. \tag{2.15}$$

Similarly,

$$\| (I_1 - I)(qI + pI_1)^n \| = \left\| \sum_{k=1}^{n} \left[\binom{n}{k-1} p^{k-1} q^{n+1-k} - \binom{n}{k} p^k q^{n-k} \right] I_k + p^n I_{n+1} - q^n I \right\|$$

$$= \sum_{k=1}^{n} \binom{n+1}{k} p^k q^{n+1-k} \frac{| k - (n+1)p |}{(n+1)pq} + p^n + q^n$$

$$= \sum_{k=0}^{n+1} \binom{n+1}{k} p^k q^{n+1-k} \frac{|k-(n+1)p|}{(n+1)pq} = \frac{\mathbb{E}\,|\eta - \mathbb{E}\,\eta|}{(n+1)pq}$$

$$\leq \frac{\sqrt{\mathrm{Var}\,\eta}}{(n+1)pq} = \frac{\sqrt{(n+1)pq}}{(n+1)pq} = \frac{1}{\sqrt{(n+1)pq}}.$$

Here η is the binomial random variable with parameters $(n+1)$ and p. Let m be the integer part of n/k. Then $(m+1) > n/k$ and

$$\| (F-I)^k (qI+pF)^n \| \leq \| (I_1 - I)(qI + pI_1)^m \|^k \leq \frac{1}{((m+1)pq)^{k/2}} \leq \frac{k^{k/2}}{(npq)^{k/2}}. \tag{2.16}$$

If we are not trying to obtain minimal constants, the estimates (2.14), (2.15) and (2.16) are sufficient.

For smoothing estimates to be effective, the difference of n-fold convolutions must be written in a form suitable for application of Lemma 2.3. The following two lemmas can be useful in achieving this goal.

Lemma 2.4 *Let* $M_1, M_2 \in \mathcal{M}$. *Then*

$$\| \exp\{M_1\} - \exp\{M_2\} \| \leq \sup_{0 \leq \tau \leq 1} \| (M_1 - M_2) \exp\{\tau M_1 + (1-\tau)M_2\} \|. \tag{2.17}$$

Proof We have

$$\| \exp\{M_1\} - \exp\{M_2\} \| = \| \exp\{M_2\}(\exp\{M_1 - M_2\} - I) \|$$

$$= \| \exp\{M_2\} \int_0^1 (\exp\{\tau(M_1 - M_2)\})' d\tau \|$$

$$= \| \exp\{M_2\} \int_0^1 (M_1 - M_2) \exp\{\tau(M_1 - M_2)\} d\tau \|$$

$$\leq \int_0^1 \| (M_1 - M_2) \exp\{\tau M_1 + (1-\tau)M_2\} \| d\tau.$$

$\square$

Lemma 2.5 *Let* $a > 0$, $F \in \mathcal{F}$. *Then*

$$\left\| \exp\left\{ a(F-I) + \frac{2a}{7}(F-I)^2 \Theta \right\} \right\| \leq C. \tag{2.18}$$

Proof Taking into account that $\| \exp\{a(F - I)\} \| = 1$, Stirling's formula (1.43) and (2.11) we obtain

$$\left\| \exp\left\{ a(F - I) + \frac{2a}{7}(F - I)^2\Theta \right\} \right\| = \left\| \sum_{j=0}^{\infty} e^{a(F-I)} \frac{(2a(F - I)^2\Theta)^j}{7^j j!} \right\|$$

$$\leqslant 1 + \sum_{j=1}^{\infty} \frac{2^j a^j}{7^j j!} \left\| \exp\left\{ \frac{a}{j}(F - I) \right\} (F - I)^2 \right\|^j \leqslant 1 + \sum_{j=1}^{\infty} \frac{2^j a^j e^j}{7^j j^j \sqrt{2\pi j}} \frac{3^j j^j}{a^j e^j}$$

$$= 1 + \sum_{j=1}^{\infty} \frac{1}{\sqrt{2\pi j}} \left(\frac{6}{7} \right)^j \leqslant 1 + \frac{6}{\sqrt{2\pi}}.$$

□

Observe that $2a/7$ in (2.18) can be replaced by any other number less than $a/3$.

2.5 Estimates in Total Variation via Smoothing

Remarks on application Differences of n-fold convolutions should be rewritten in a form allowing application of Lemma 2.3. In this sense, all proceedings are almost the same as described in Sect. 2.2.

Typical application. Let us assume that we want to estimate

$$\left\| \prod_{j=1}^{n} \varphi_j(F) - \prod_{j=1}^{n} \varphi_j(F) \right\|, \quad \varphi_j(F) = \sum_{m=0}^{\infty} a_{mj}F^m, \ \psi_j(F) = \sum_{m=0}^{\infty} b_{mj}F^m.$$

- Apply (2.4) or (2.17) with $M_1 = \sum_{j=1}^{n} \ln \varphi_j(F)$, $M_2 = \sum_{j=1}^{n} \ln \psi_j(F)$.
- Then apply Lemma 2.3 and, if needed, (2.18).
- In case of asymptotic expansions, one should apply Bergström's identity or (1.33).

Advantages. No assumptions on compounding distribution.

Drawbacks. Additional assumptions for expansions of logarithms of distributions in converging series might be needed.

Note that, so far, we have described just two possible approaches. There exist others, see, for example, Sect. 2.8 below.

CP approximation to a compound binomial distribution From (2.5) it follows that the binomial and Poisson distributions are close, when np^2 is small. We show that for the closeness of distributions it suffices that $p = o(1)$.

Theorem 2.3 *Let* $p = 1 - q \in (0, 1)$, $F \in \mathcal{F}$, $n \in \mathbb{N}$. *Then*

$$\| ((1 - p)I + pF)^n - \exp\{np(F - I)\} \| \leqslant Cp. \qquad (2.19)$$

Proof Without loss of generality we can assume that $n \geqslant 6$ and $p < 1/2$. Indeed, if $n < 6$, then (2.19) follows from (2.5). If $p \geqslant 1/2$, then

$$\|((1-p)I+pF)^n-\exp\{np(F-I)\}\| \leqslant \|((1-p)I+pF)^n\|+\| \exp\{np(F-I)\}\| = 2 \leqslant 4p.$$

We used the fact that norm of any distribution equals 1. Let us denote the integer part of $2n/3$ by m. Then

$$m \geqslant \frac{2n}{3} - 1 \geqslant \frac{n}{2}, \quad \frac{n}{3} - 1 \geqslant \frac{n}{6}, \quad n - j \geqslant m, \text{ if } j < n/3.$$

$$H = qI + pF, \qquad D = \exp\{ p(F - I)\}.$$

Then by (2.3)

$$H - D = p^2(F - I)^2\Theta/2.$$

Moreover, $\| H \| = \| D \| = 1$, since $H, D \in \mathcal{F}$. Therefore

$$\| H^n - D^n \| \leqslant \sum_{j=1}^n \| (H - D)H^{n-j}D^{j-1} \| = \sum_{j<n/3} \| (H - D)H^m \|$$

$$+ \sum_{j \geqslant n/3} \| (H - D)D^{n/6} \| \leqslant np^2\| (F - I)^2H^m \| + np^2\| (F - I)^2D^{n/6} \|.$$

By (2.13) and (2.10) we obtain

$$\| (F - I)^2H^m \| \leqslant \frac{C}{mpq} \leqslant \frac{C}{npq}, \qquad \| (F - I)^2D^{n/6} \| \leqslant \frac{C}{np}.$$

Combining the last two estimates, we complete the proof of theorem. $\qquad \square$

Though we used general compound distributions, the comments in Sect. 2.3 demonstrate that (2.19) is equivalent to Poisson approximation to the binomial law.

CP approximation to a compound Poisson binomial distribution We will generalize (2.19) assuming small p.

Theorem 2.4 *Let* $\max_{1\leqslant j\leqslant n} p_j \leqslant 1/4$. *Then*

$$\left\| \prod_{j=1}^n ((1-p_j)I + p_j F) - \exp\left\{\sum_{j=1}^n p_j(F-I)\right\} \right\| \leqslant C \frac{\sum_{j=1}^n p_j^2}{\sum_{j=1}^n p_j}. \qquad (2.20)$$

Proof Let

$$M_1 = \sum_{j=1}^n \ln((1-p_j)I + p_j F), \qquad M_2 = \sum_{j=1}^n p_j(F-I), \qquad \lambda = \sum_{j=1}^n p_j.$$

Example 2.4 for $s = n = 1$ gives

$$M_1 = \lambda(F-I) + \sum_{j=1}^n \frac{p_j^2(F-I)^2}{2(1-2p_j)}\Theta = \lambda(F-I) + \frac{1}{4}\lambda(F-I)^2\Theta,$$

so that

$$M_1 - M_2 = C\sum_{j=1}^n p_j^2(F-I)^2\Theta$$

and

$$\exp\{\tau M_1 + (1-\tau)M_2\} = \exp\left\{\frac{7\tau\lambda}{8}(F-I) + \frac{2}{7}\frac{7\tau\lambda}{8}(F-I)^2\Theta\right\} \exp\left\{\frac{\lambda}{8}(F-I)\right\}$$

$$\times \exp\left\{\frac{7(1-\tau)\lambda}{8}(F-I)\right\}.$$

Applying (2.18) with $a = 7\tau\lambda/8$ we prove

$$\left\| \exp\left\{\frac{7\tau\lambda}{8}(F-I) + \frac{2}{7}\frac{7\tau\lambda}{8}(F-I)^2\Theta\right\} \right\| \leqslant C.$$

If $0 \leqslant \tau \leqslant 1$, then $\exp\{7(1-\tau)(\lambda/8)(F-I)\}$ is a CP distribution and, therefore, $\| \exp\{7(1-\tau)(\lambda/8)(F-I)\} \| = 1$.

Taking into account all these estimates and applying (2.17) we obtain

$$\| \exp\{M_1\} - \exp\{M_2\} \| \leqslant \sup_{0\leqslant\tau\leqslant 1} \| (M_1 - M_2)\exp\{\tau M_1 + (1-\tau)M_2\} \|$$

$$\leqslant C\left\| (M_1 - M_2)\exp\left\{\frac{\lambda}{8}(F-I)\right\} \right\|$$

$$\leqslant C\left\| \sum_{j=1}^n p_j^2(F-I)^2 \exp\left\{\frac{\lambda}{8}(F-I)\right\} \right\|.$$

To complete the proof of the theorem it remains to apply (2.11).   □

The main difference between (2.19) and (2.20) is the assumptions on the smallness of p_j. The method of proof we used in Theorem 2.4 cannot be applied to the proof of (2.19) if $p/(2 - 4p) > 1/3$ (that is, $p > 2/7$), see the remark after Lemma 2.5.

Compound negative binomial approximation We shall demonstrate that (2.18) is not always necessary, even if (2.17) is applied. We consider an approximation of an n-fold convolution of compound geometric distributions by a CNB distribution. Let $0 < q_k < 1$, $p_k = 1 - q_k$, $k = 1, 2, \ldots, n$, $F \in \mathcal{F}$. We recall the definitions of compound geometric and compound negative binomial distributions

$$CG(q_k, F) = \exp\left\{\sum_{j=1}^{\infty} \frac{q_k^j}{j}(F^j - I)\right\}, \qquad CNB(\gamma, q, F) = \exp\left\{\gamma \sum_{j=1}^{\infty} \frac{q^j}{j}(F^j - I)\right\}.$$

$$(2.21)$$

How to choose the parameters of approximation? For smoothing to be effective, we match both expansions in powers of $(F - I)$ as much as possible. Let

$$M_1 = \sum_{k=1}^{n} \ln CG(q_k, F), \qquad M_2 = \ln CNB(\gamma, q, F).$$

Setting for the sake of brevity

$$a_1 = \sum_{k=1}^{n} \frac{q_k}{p_k}, \qquad a_2 = \sum_{k=1}^{n}\left(\frac{q_k}{p_k}\right)^2, \qquad a_3 = \sum_{k=1}^{n}\left(\frac{q_k}{p_k}\right)^3$$

and applying (2.7) we obtain

$$M_1 = a_1(F - I) + \frac{a_2}{2}(F - I)^2 + \frac{a_3}{3}(F - I)^3\Theta, \tag{2.22}$$

$$M_2 = \gamma\frac{q}{p}(F - I) + \frac{\gamma}{2}\left(\frac{q}{p}\right)^2(F - I)^2 + \frac{\gamma}{3}\left(\frac{q}{p}\right)^3(F - I)^3\Theta. \tag{2.23}$$

Next, we choose γ and q to match the coefficients of $(F - I)$ and $(F - I)^2$ in both expansions:

$$\begin{cases} \gamma q/p &= a_1, \\ \gamma q^2/p^2 &= a_2. \end{cases}$$

Solving the system we get

$$\gamma = \frac{a_1^2}{a_2}, \qquad \frac{q}{p} = \frac{a_2}{a_1}, \qquad q = \frac{a_2}{a_1 + a_2}. \tag{2.24}$$

Now we can estimate the accuracy of the approximation.

Theorem 2.5 *Let $0 < q_k \leqslant C_0 < 1$, $(k = 1, \ldots, n)$, $F \in \mathcal{F}$ and let γ, q be defined as in (2.24). Then*

$$\left\| \prod_{k=1}^{n} CG(q_k, F) - CNB(\gamma, q, F) \right\| \leqslant C \frac{\sum_{k=1}^{n} q_k^3}{\left(\sum_{k=1}^{n} q_k \right)^{3/2}}. \qquad (2.25)$$

Proof Taking into account (2.22), (2.23) and (2.24) we obtain

$$M_1 - M_2 = (F - I)^3 \left(\frac{a_3}{3} + \gamma \left(\frac{q}{p} \right)^3 \right) \Theta.$$

Then by (2.17)

$$\| \exp\{M_1\} - \exp\{M_2\} \|$$

$$\leqslant C \frac{1}{3} \left(a_3 + \gamma \left(\frac{q}{p} \right)^3 \right) \sup_{0 \leqslant \tau \leqslant 1} \| (F - I)^3 \exp\{\tau M_1 + (1 - \tau) M_2\} \|. \qquad (2.26)$$

In view of (2.21) we can write

$$\exp\{\tau M_1 + (1 - \tau) M_2\} = \exp\left\{ \tau \sum_{k=1}^{n} \sum_{j=1}^{\infty} \frac{q_k{}^j}{j} (F^j - I) + (1 - \tau) \gamma \sum_{j=1}^{\infty} \frac{q^j}{j} (F^j - I) \right\}$$

$$= \exp\left\{ \left(\tau \sum_{k=1}^{n} q_k + (1 - \tau) \gamma q \right) (F - I) \right\}$$

$$\times \prod_{k=1}^{n} \exp\left\{ \tau \sum_{j=2}^{\infty} \frac{q_k{}^j}{j} (F^j - I) \right\} \exp\left\{ (1 - \tau) \gamma \sum_{j=2}^{\infty} \frac{q^j}{j} (F^j - I) \right\}$$

$$= \exp\left\{ \left(\tau \sum_{k=1}^{n} q_k + (1 - \tau) \gamma q \right) (F - I) \right\} D.$$

Here all remaining convolutions are denoted by D. Observe that $\exp\{\tau(q_k{}^j/j)(F^j - I)\}$ is a CP *distribution*, since $F^j \in \mathcal{F}$ and $q_k{}^j/j > 0$. The convolution of CP distributions is also a CP distribution. Therefore $D \in \mathcal{F}$ and, consequently, $\| D \| = 1$.

Applying (2.12) we prove that

$$\| (F - I)^3 \exp\{\tau M_1 + (1 - \tau) M_2\} \| \leqslant \left\| (F - I)^3 \exp\left\{ \left(\tau \sum_{k=1}^{n} q_k + (1 - \tau) \gamma q \right) (F - I) \right\} \right\|$$

$$\leqslant C \left(\min \left(\sum_{k=1}^{n} q_k, \gamma q \right) \right)^{-3/2}. \qquad (2.27)$$

Observe that

$$\sum_{k=1}^{n} q_k^m \leqslant a_m \leqslant \frac{1}{(1-C_0)^m}\sum_{k=1}^{n} q_k^m, \qquad m=1,2,3,$$

and by Hölder's inequality $a_2^2 \leqslant a_1 a_3$. Therefore

$$\gamma q = \frac{a_1^2}{a_1+a_2} \geqslant C\sum_{k=1}^{n} q_k$$

and

$$a_3 + \gamma\left(\frac{q}{p}\right)^3 = a_3 + \frac{a_2^2}{a_1} \leqslant 2a_3 \leqslant C\sum_{k=1}^{n} q_k^3.$$

Substituting the last two estimates and (2.27) into (2.26) we complete the proof of the theorem.   □

Note that assumption $q \leqslant C_0$ is not essential to the above proof. Such an assumption is only needed to write the estimate in a simpler form. Observe that if all $q, q_k \equiv C$, then the estimate in (2.25) is of the order $n^{-1/2}$. Meanwhile, (2.19) is not trivial only if p is very small. This result is typical of all two-parametric approximations. By matching two factorial moments (cumulants, or simply multipliers to $(F-I)$ and $(F-I)^2$) in the factorial moments expression (2.2) we can expect to achieve an accuracy of approximation of the order $O(n^{-1/2})$. Note that such an order is more common for approximations by the normal law.

2.6   Smoothing Estimates for the Kolmogorov Norm

Due to the fact that $|M|_K \leqslant \|M\|$, all smoothing estimates from Lemma 2.3 hold. On the other hand, if $F \in \mathcal{F}_s$ or $F \in \mathcal{F}_+$ then significant improvements of those estimates are possible. For the sake of convenience we formulate smoothing estimates for symmetric distributions in a separate lemma. Observe that no assumptions about the finite moment existence are needed.

Lemma 2.6 Let $F \in \mathcal{F}_+$, $G \in \mathcal{F}_s$, $a \in (0,\infty)$, and let $n,k,m \in \mathbb{N}$, $p = 1-q \in (0,1)$. Then

$$|F^n(F-I)^k(G-I)^m \exp\{a(G-I)\}|_K \leqslant \frac{C(k,m)}{n^k a^m}, \qquad (2.28)$$

$$|F^n(F-I)^k|_K \leqslant \frac{C(k)}{n^k}, \qquad (2.29)$$

$$|(G-I)^m \exp\{a(G-I)\}|_K \leqslant \frac{C(m)}{a^m}, \tag{2.30}$$

$$|(G-I)G^n|_K \leqslant \frac{C}{\sqrt{n}}, \tag{2.31}$$

$$|(G-I)^m(qI+pG)^n|_K \leqslant \frac{C(m)}{q(npq)^m}. \tag{2.32}$$

We omit the proof of the lemma, noting only that the triangular function method is used. Estimates (2.29) and (2.30) are special cases of (2.28). They are formulated separately just for convenience of application. The proof of (2.29) via the triangular function method is given in Chap. 12. The smoothing estimate (2.32) is proved in Sect. 2.8.

It is obvious that (2.30) holds for $F \in \mathcal{F}_+$, since $\mathcal{F}_+ \subset \mathcal{F}_s$. On the other hand, there is no need to prove the general estimate (2.30) via the sophisticated triangular function method if only $F \in \mathcal{F}_+$ are of interest. An analogue of (2.30) then follows from (2.29) via a combinatorial argument. We illustrate this approach.

Example 2.6 Let $\lambda > 0$, and let $s, n \in \mathbb{N}$. If $F \in \mathcal{F}_+$ then

$$|(F-I)^s \exp\{\lambda(F-I)\}|_K \leqslant C(s)\lambda^{-s}. \tag{2.33}$$

Indeed, from the definition of the exponential measure and (2.29) it follows that

$$|(F-I)^s \exp\{\lambda(F-I)\}|_K \leqslant \sum_{k=0}^{\infty} e^{-\lambda} \frac{\lambda^k}{k!} |F^k(F-I)^s|_K \leqslant 2^s e^{-\lambda}$$

$$+C(s) \sum_{k=1}^{\infty} e^{-\lambda} \frac{\lambda^k}{k!\, k^s} \leqslant C(s) \sum_{k=0}^{\infty} e^{-\lambda} \frac{\lambda^k}{(k+s)!}$$

$$= C(s)\lambda^{-s} \sum_{k=0}^{\infty} e^{-\lambda} \frac{\lambda^{k+s}}{(k+s)!} \leqslant C(s)\lambda^{-s}.$$

Here we used the elementary inequalities $(k+1)(k+2)\cdots(k+s) \leqslant k^s(s+1)!$ and $|F^k(F-I)^s|_K \leqslant \|F^k(F-I)^s\| \leqslant \|F^k\|(\|F\|+\|I\|)^s = 2^s$.

2.7   Estimates in the Kolmogorov Norm via Smoothing

In this section, we present examples of the method of convolutions for symmetric compounding distributions. It is easy to check that (2.4) and (2.17) also hold for the Kolmogorov norm. Thus, if $M_1, M_2 \in \mathcal{M}$ then

$$|\exp\{M_1\} - \exp\{M_2\}|_K \leqslant \sup_{0 \leqslant \tau \leqslant 1} |(M_1-M_2)\exp\{\tau M_1 + (1-\tau)M_2\}|_K. \tag{2.34}$$

Typical application. We use the same methods as applied for the total variation norm, combining them with Lemma 2.6 for smoothing estimates.

- Apply (2.4) or (2.34) or any other expression allowing for smoothing estimates.
- Then apply Lemma 2.6 and, if needed, (2.18).
- In the case of asymptotic expansions, apply Bergström's identity or (1.33).

Advantages. Very general estimates. Good accuracy. No moment assumptions. No assumptions on the structure of distributions except for symmetry.

Drawbacks. No explicit absolute constant.

We begin with one extension of smoothing estimates to the difference of powers of a symmetric distribution. The tricky part is to get larger parameter in the denominator of the estimate.

Theorem 2.6 *Let* $G \in \mathcal{F}_s$, $F \in \mathcal{F}_+$, $a > 0$, $b > 0$, $n, m \in \mathbb{N}$. *Then*

$$|F^n - F^m|_K \leq C\frac{|m-n|}{n}, \tag{2.35}$$

$$|G^n - G^m|_K \leq \frac{C}{\sqrt{n}} + C\frac{|m-n|}{n}, \tag{2.36}$$

$$|\exp\{a(G-I)\} - \exp\{b(G-I)\}|_K \leq C\frac{|a-b|}{b}. \tag{2.37}$$

Proof It is not difficult to prove (2.35) for $n \leq m$. Indeed, by (2.29)

$$|F^n - F^m|_K = |F^n(I - F^{m-n})|_K = |F^n(I-F)(I + F + \cdots + F^{m-n-1})|_K$$

$$\leq (m-n)|F^n(I-F)|_K \leq C\frac{|m-n|}{n}.$$

We want to prove the same estimate for $n > m$. This can be done by employing the fact that if n is very large, then $|m - n|/n$ is close to unity.

Let $m < n \leq 2m$. Then, arguing as above, we prove

$$|F^n - F^m|_K \leq C\frac{|m-n|}{m} \leq 2C\frac{|m-n|}{n}.$$

Now, let $2m < n$. Then

$$\frac{|m-n|}{n} = \frac{n-m}{n} = 1 - \frac{m}{n} > 1 - \frac{1}{2} = \frac{1}{2}$$

and, consequently,

$$1 < \frac{2|m-n|}{n}.$$

It remains to observe that

$$|F^n - F^m|_K \leq \|F^n\| + \|F^m\| = 2 \leq \frac{4|m-n|}{n}.$$

Estimate (2.35) is completely proved. Estimate (2.37) is proved similarly. For $b \leq a$, by (2.34) and (2.30)

$$|\exp\{a(G-I)\} - \exp\{b(G-I)\}|_K$$
$$\leq \sup_{0 \leq \tau \leq 1} |(a-b)(G-I)\exp\{(\tau a + (1-\tau)b)(G-I)\}|_K$$
$$\leq \sup_{0 \leq \tau \leq 1} \frac{C|a-b|}{\tau a + (1-\tau)b} \leq \frac{C|a-b|}{b}.$$

For the cases $a < b \leq 2a$ and $2a \leq b$, we repeat the above argument.

For the proof of (2.36) let us consider four possible situations: (a) when n and m are even; (b) when n and m are odd; (c) when n is odd and m is even; (d) when n is even and m is odd.

Let $n = 2j$ and $m = 2i$. Taking into account that $G^2 \in \mathcal{F}_+$ and applying (2.35) we prove

$$|(G^2)^j - (G^2)^i|_K \leq C\frac{|i-j|}{j} = C\frac{|2i-2j|}{2j} = C\frac{|n-m|}{n}.$$

Next, consider the situation when $n = 2j+1$, $m = 2i+1$. Then

$$|G(G^2)^j - G(G^2)^i|_K \leq \|G\||(G^2)^j - (G^2)^i|_K \leq C\frac{|i-j|}{j}$$
$$= C\frac{|2j+1-2i-1|}{2j+1-1} \leq C\frac{|n-m|}{n-1} \leq 2C\frac{|n-m|}{n}.$$

Let $n = 2j+1$ and $m = 2i$. Then, taking into account (2.31) and (2.35), we obtain

$$|G^{2j+1} - G^{2i}|_K \leq |G^{2j+1} - G^{2j}|_K + |G^{2j} - G^{2i}|_K$$
$$= |(G-I)G^{2j}|_K + |(G^2)^j - (G^2)^i|_K$$

$$\leq \frac{C}{\sqrt{2j}} + \frac{C|j - i|}{j} = \frac{C}{\sqrt{n-1}} + \frac{C|n - m - 1|}{n - 1}$$

$$\leq \frac{2C}{\sqrt{n}} + \frac{2C(|n - m| + 1)}{n} \leq \frac{C}{\sqrt{n}} + \frac{C|n - m|}{n}.$$

Let $n = 2j \geq 3$ and $m = 2i + 1$. Then

$$|G^{2j} - G^{2i+1}|_K \leq \|G\| |G^{2j-1} - G^{2i}|_K$$

and we can apply the previous estimate. If $n < 3$, then we simply note that $|G^n - G^m|_K \leq 2 \leq 2\sqrt{3}/\sqrt{n}$.   □

Smoothing estimates are instrumental in approximation by CP distribution $\exp\{n(F - I)\}$. Note that, in the literature, $\exp\{n(F - I)\}$ is also called the accompanying law.

Theorem 2.7 *Let* $F \in \mathcal{F}_+$, $n \in \mathbb{N}$. *Then*

$$|F^n - \exp\{n(F - I)\}|_K \leq Cn^{-1}. \tag{2.38}$$

Proof Without loss of generality, we can assume that $n > 10$. Indeed, if $n \leq 10$, then

$$|F^n - \exp\{n(F - I)\}|_K \leq \|F^n\| + \|\exp\{n(F - I)\}\| = 2 \leq 20n^{-1}.$$

Let $m = \lfloor n/3 \rfloor$. Observe that $m \geq n/3 - 1 > n/5$. Then consequently applying (2.4), (2.3), properties of the norms, (2.29), and (2.30), we obtain

$$|F^n - \exp\{n(F - I)\}|_K \leq \sum_{k=1}^n |(F - \exp\{F - I\})F^{k-1} \exp\{(n - k)(F - I)\}|_K$$

$$\leq C\|\Theta\| \sum_{k=1}^n |(F - I)^2 F^{k-1} \exp\{(n - k)(F - I)\}|_K$$

$$\leq C \sum_{k \leq n/3} \|F^{k-1}\| |(F - I)^2 \exp\{(n - k)(F - I)\}|_K$$

$$+ C \sum_{k > n/3} \|\exp\{(n - k)(F - I)\}\| |(F - I)^2 F^{k-1}|_K$$

$$\leq Cn |(F - I)^2 \exp\{m(F - I)\}|_K + Cn |(F - I)^2 F^m|_K$$

$$\leq Cnm^{-2} \leq Cn^{-1}.$$

□

For symmetric distributions the accuracy of the CP approximation is much weaker than for distributions with nonnegative characteristic functions. On the other hand, it is comparable with the accuracy of the normal approximation, see the Berry-Esseen theorem 9.1 below.

Theorem 2.8 *Let $G \in \mathcal{F}_s$, $n \in \mathbb{N}$. Then*

$$| G^n - \exp\{n(G - I)\} |_K \leq \frac{C}{\sqrt{n}}.$$

Proof Let $n \leq j$. Let η be a Poisson random variable with parameter n ($\mathbb{E}\,\eta = \text{Var}\,\eta = n$). Then, taking into account (2.36) and the Cauchy inequality, we obtain

$$| G^n - \exp\{n(G - I)\} |_K = \left| G^n - \sum_{j=0}^{\infty} e^{-n} \frac{n^j}{j!} G^j \right|_K = \left| \sum_{j=0}^{\infty} e^{-n} \frac{n^j}{j!} (G^n - G^j) \right|_K$$

$$\leq \sum_{j=0}^{\infty} e^{-n} \frac{n^j}{j!} | G^n - G^j |_K \leq \sum_{j=0}^{\infty} e^{-n} \frac{n^j}{j!} \left(\frac{C}{\sqrt{n}} + \frac{C|j - n|}{n} \right)$$

$$= \frac{C}{\sqrt{n}} + \frac{C}{n} \mathbb{E}\,| \eta - n | \leq \frac{C}{\sqrt{n}} + \frac{C}{n} \sqrt{\text{Var}\,\eta} \leq \frac{C}{\sqrt{n}}.$$

□

We once more demonstrate how the fact $F^k \in \mathcal{F}_+$ can be employed in the proof. The following result shows that an accuracy of approximation similar to (2.38) can be achieved by CP distributions different from the accompanying ones.

Theorem 2.9 *Let $F \in \mathcal{F}_+$, $m, m \in \mathbb{N}$. Then*

$$\left| F^n - \exp\left\{ \frac{n}{m} (F^m - I) \right\} \right|_K \leq C \frac{m}{n}. \tag{2.39}$$

Proof Let $k = \lceil n/m \rceil$, that is the smallest natural number greater than or equal to n/m. Then

$$\left| F^n - \exp\left\{ \frac{n}{m} (F^m - I) \right\} \right|_K \leq \left| \exp\left\{ k(F^m - I) \right\} - \exp\left\{ \frac{n}{m} (F^m - I) \right\} \right|_K$$

$$+ | F^n - F^{km} |_K + | F^{km} - \exp\{k(F^m - I)\} |_K. \tag{2.40}$$

Observe that $k = n/m + \delta$, for some $0 \leq \delta \leq 1$. Therefore $|km - n| \leq m$. This estimate and (2.29) give

$$|F^n - F^{km}|_K = |F^n(I - F^{km-n})|_K = |F^n(I - F)(I + F + \cdots + F^{km-n-1})|_K$$
$$\leq (km - n)|F^n(I - F)|_K \leq \frac{Cm}{n}.$$

Observe that $F^k \in \mathcal{F}_+$. Therefore from (2.38) it follows that

$$|F^{km} - \exp\{k(F^m - I)\}|_K \leq \frac{C}{k} \leq \frac{Cm}{n}.$$

By (2.37)

$$\left| \exp\left\{k(F^m - I)\right\} - \exp\left\{\frac{n}{m}(F^m - I)\right\} \right|_K \leq \frac{C|k - n/m|}{n/m} \leq \frac{Cm}{n}.$$

Submitting the last three estimates into (2.40) we complete the proof of (2.39). □

Finally, we demonstrate that step by step repetition of the proofs for total variation via smoothing can easily result in estimates for the Kolmogorov norm. All the above estimates were proved for n-th convolutional powers of distributions. Next, we consider the more general case of n-fold convolutions.

Theorem 2.10 *Let* $\max_{1 \leq j \leq n} p_j \leq 1/4$, $G \in \mathcal{F}_s$. *Then*

$$\left| \prod_{j=1}^n ((1 - p_j)I + p_jG) - \exp\left\{\sum_{j=1}^n p_j(G - I)\right\} \right|_K \leq C\frac{\sum_{j=1}^n p_j^2}{\left(\sum_{j=1}^n p_j\right)^2}. \tag{2.41}$$

Proof Observe that one cannot apply (2.38). Repeating the proof of (2.4) for the Kolmogorov norm and applying (2.34), we obtain

$$\left| \prod_{j=1}^n ((1-p_j)I+p_jG)-\exp\left\{\sum_{j=1}^n p_j(G-I)\right\} \right|_K \leq C\sum_{j=1}^n p_j^2 \left|(G-I)^2 \exp\left\{\frac{\lambda}{8}(G-I)\right\}\right|_K.$$

Here $\lambda = \sum_{j=1}^n p_j$. It remains to apply (2.30). □

Note that (2.41) can be much more accurate than (2.20). Indeed, if $p_j \equiv C$ the accuracy of approximation in (2.41) is of the order $O(n^{-1})$ as compared to $O(1)$ of (2.20).

2.8   Kerstan's Method

Kerstan's method is a much more sophisticated version of (2.17). Its main idea is to replace all convolutions by smoothing estimates of the form $(F - I)^2 \exp\{a(F - I)\}$. The method can be applied for compound measures with sufficiently large probabilistic mass at zero only. Let

$$g(x) = 2\sum_{s=2}^{\infty} \frac{x^{s-2}(s-1)}{s!} = \frac{2e^x(e^{-x} - 1 + x)}{x^2}, \quad x \neq 0,$$

$$\lambda = \sum_{j=1}^{n} p_j, \quad p_j \in (0,1), \quad F \in \mathcal{F},$$

$$L_i = (I + p_i(F-I))\exp\{-p_i(F-I)\} - I = -p_i^2(F-I)^2 \sum_{s=2}^{\infty} (-p_i(F-I))^{s-2}\frac{s-1}{s!},$$

$$\alpha = \sum_{i=1}^{n} g(2p_i)p_i^2 \min\left(1, \frac{3}{2e\lambda}\right).$$

It is known that

$$g(2p_i) \leq g(2) \leq 4.1946. \tag{2.42}$$

Kerstan's method is based on the following estimate

$$\left\| \prod_{j=1}^{n}((1 - p_j)I + p_jF) - \exp\{\lambda(F - I)\} \right\|$$

$$= \left\| \left(\prod_{j=1}^{n}[(I + p_j(F - I))\exp\{-p_j(F - I)\}] - I \right)\exp\{\lambda(F - I)\} \right\|$$

$$= \left\| \sum_{j=1}^{n} \sum_{1 \leq i_1 < i_2 \cdots < i_n \leq n} \prod_{s=1}^{j} [L_{i_s}\exp\{(\lambda/j)(F - I)\}] \right\|$$

$$\leq \sum_{j=1}^{n} \sum_{1 \leq i_1 < i_2 \cdots < i_n \leq n} \prod_{s=1}^{j} \|L_{i_s}\exp\{(\lambda/j)(F - I)\}\|$$

$$\leq \sum_{j=1}^{n} \frac{1}{j!}\left[\sum_{i=1}^{n} \left\| L_i \exp\left\{\frac{\lambda}{j}(F - I)\right\} \right\| \right]^j. \tag{2.43}$$

Taking into account that $\| F - I \| \leq 2$, the definition of $L_i(F)$ and the smoothing estimate (2.11) we obtain

$$\sum_{i=1}^{n} \left\| L_i \exp\left\{ \frac{\lambda}{j}(F - I) \right\} \right\| \leq \sum_{i=1}^{n} \frac{g(2p_i)}{2} p_i^2 \left\| (F - I)^2 \exp\left\{ \frac{\lambda}{j}(F - I) \right\} \right\|$$

$$\leq \sum_{i=1}^{n} \frac{g(2p_i)}{2} p_i^2 \min\left(2, \frac{3j}{e\lambda} \right) \leq j\alpha.$$

The last estimate and Stirling's formula (1.43) allows to write

$$\sum_{j=1}^{n} \frac{1}{j!} \left[\sum_{i=1}^{n} \left\| L_i \exp\left\{ \frac{\lambda}{j}(F - I) \right\} \right\| \right]^j$$

$$\leq \sum_{j=1}^{n} \frac{(j\alpha)^{j-1}}{j!} \sum_{i=1}^{n} \frac{g(2p_i)}{2} p_i^2 \left\| (F - I)^2 \exp\left\{ \frac{\lambda}{j}(F - I) \right\} \right\|$$

$$\leq \frac{1}{2\sqrt{2\pi}} \sum_{j=1}^{n} \frac{\alpha^{j-1}}{j^{3/2}} \sum_{i=1}^{n} g(2p_i) p_i^2 \left\| (F - I)^2 \exp\left\{ \frac{\lambda}{j}(F - I) \right\} \right\|.$$

Taking into account the last two estimates and (2.43) we get

$$\left\| \prod_{j=1}^{n} ((1 - p_j)I + p_j F) - \exp\{\lambda(F - I)\} \right\|$$

$$\leq \frac{1}{2\sqrt{2\pi}} \sum_{j=1}^{n} \frac{\alpha^{j-1}}{j^{3/2}} \sum_{i=1}^{n} g(2p_i) p_i^2 \left\| (F - I)^2 \exp\left\{ \frac{\lambda}{j}(F - I) \right\} \right\| \qquad (2.44)$$

$$\leq \frac{1}{\sqrt{2\pi}} \sum_{j=1}^{n} \alpha^j. \qquad (2.45)$$

It is not difficult to check that, in (2.44), the total variation norm can be replaced by the Kolmogorov norm:

$$\left| \prod_{j=1}^{n} ((1 - p_j)I + p_j F) - \exp\{\lambda(F - I)\} \right|_K$$

$$\leq \frac{1}{2\sqrt{2\pi}} \sum_{j=1}^{n} \frac{\alpha^{j-1}}{j^{3/2}} \sum_{i=1}^{n} g(2p_i) p_i^2 \left| (F - I)^2 \exp\left\{ \frac{\lambda}{j}(F - I) \right\} \right|_K. \qquad (2.46)$$

Estimate (2.44) is still quite general and allows improvements if additional information about the structure of F is known. Moreover, observe that $\alpha < 1$ if the p_i are small. Indeed, taking into account (2.42), we obtain

$$\alpha \leq 4.1946 \sum_{i=1}^{n} p_i^2 \frac{3}{\lambda 2e} \leq \frac{12.584}{2e} \max_{1 \leq i \leq n} p_i \leq 2.32 \max_{1 \leq i \leq n} p_i. \qquad (2.47)$$

Thus, if $p_i < 1/2.32$, then $\sum_{j=1}^{\infty} \alpha^j \leq \alpha/(1-\alpha)$.

Typical application. We seek to estimate

$$\prod_{j=1}^{n}((1-p_j)I + p_j F) - \exp\left\{\sum_{j=1}^{n} p_j(F-I)\right\}.$$

- Apply (2.44), (2.45) or (2.46).
- Take into account (2.42) and (2.47) and assume $\alpha < 1$.
- If the approximation has a more complicated structure, then (2.18) might help.

Advantages. The method can be extended without many changes to the multivariate case.

Drawbacks. Very much depends on the specific CP structure of approximation. Usage for other approximations is problematic.

Theorem 2.11 *Let* $\max_{1 \leq j \leq n} p_j \leq 0.4$, $F \in \mathcal{F}$. *Then*

$$\left\| \prod_{j=1}^{n}((1-p_j)I + p_j F) - \exp\left\{\sum_{j=1}^{n} p_j(F-I)\right\} \right\| \leq 12.83 \frac{\sum_{j=1}^{n} p_j^2}{\sum_{j=1}^{n} p_j}.$$

Proof We apply (2.47) and (2.45)

$$\frac{1}{\sqrt{2\pi}} \sum_{j=1}^{n} \alpha^j \leq \frac{\alpha}{(1-\alpha)\sqrt{2\pi}} \leq \frac{1}{\sqrt{2\pi}} \frac{3 \cdot 4.1946}{2e} \frac{1}{1-0.4 \cdot 2.32} \frac{\sum_{j=1}^{n} p_j^3}{\sum_{j=1}^{n} p_j} \leq 12.83 \frac{\sum_{j=1}^{n} p_j^3}{\sum_{j=1}^{n} p_j}.$$

$$\square$$

Comparing this result with (2.20) we see that our assumptions for Kerstan's method are much weaker ($p_i \leq 0.4$ vs $p_i \leq 0.25$). We cannot assume $p_i = 0.4$ in Lemma 2.5.

Kerstan's method was specifically designed for CP approximation. On the other hand the same idea can be applied to other kinds of estimates too. An additional lemma is needed.

Lemma 2.7 *Let $F \in \mathcal{F}$ and*

$$\tilde{g}(x) = \frac{1}{(x-1)^2}\Big[\prod_{j=1}^{n}(1 + p_j(x-1)) - e^{\lambda(x-1)}\Big]e^{-r\lambda(x-1)}.$$

If $b = \sum_{j=1}^{n} p_j^2/\lambda \leq 1 - r$, then

$$\|\tilde{g}(F)\| \leq \min\Big\{\frac{\lambda_2}{2\sqrt{1 - b/(1-r)}}, 1.62\sum_{j=1}^{n} p_j^2\Big\}.$$

Note that $\tilde{g}(F)$ does not mean that we put $(F - I)$ in the denominator of the expression (there is no division operation for convolutions of measures). The function $\tilde{g}(x)$ can be expressed as a converging series of $(x - 1)$ powers and $\tilde{g}(F)$ is understood as the same series with $(x - 1)$ replaced by $(F - I)$.

We demonstrate how to apply Lemma 2.7 by proving (2.32). Let $r = q = 1 - p$. Observe that

$$\tilde{g}(x)(x-1)^{m+2}\exp\{npq(x-1)\} = (x-1)^m[(1 + p(x-1))^n - \exp\{np(x-1)\}].$$

Therefore

$$(G-I)^m(qI+pG)^n = (G-I)^m\exp\{np(G-I)\} + \tilde{g}(G)(G-I)^{m+2}\exp\{npq(G-I)\}.$$

Next, we assume that $G \in \mathcal{F}_s$ and apply (2.30) and Lemma 2.7:

$$|(G-I)^m(qI+pG)^n|_K \leq |(G-I)^m\exp\{np(G-I)\}|_K$$

$$+\|\tilde{g}(G)\|\|(G-I)\|\,|(G-I)^{m+1}\exp\{npq(G-I)\}|_K$$

$$\leq \frac{C(m)}{(np)^m} + \frac{C(m)np^2}{(npq)^{m+1}} \leq \frac{C(m)}{q(npq)^m},$$

which is the required smoothing estimate (2.32).

2.9 Problems

2.1 Let $a > 0, b > 0, F \in \mathcal{F}$. Prove that

$$\|\exp\{a(F - I)\} - \exp\{b(F - I)\}\| \leq 2|a - b|.$$

2.2 Let $n \in \mathbb{N}$,

$$H = \sum_{m=0}^{\infty} \left(\frac{1}{2}\right)^{m+1} F^m.$$

Prove that

$$\| H^n - F^n \| \leqslant n \| F - I \|^2.$$

2.3 Let $M \in \mathcal{M}$. Prove that

$$\left\| I + M - e^M \left(I - \frac{M^2}{2}\right) \right\| \leqslant \frac{2}{3} \| M \|^3 e^{\| M \|}.$$

2.4 Let $0 < p \leqslant 1/5, F \in F, n \in \mathbb{N}$. Prove that

$$\left\| ((1 - p)I + pF)^n - \exp\left\{ np(F - I) - \frac{np^2}{2}(F - I)^2 - \frac{np^3}{3}(F - I)^3 \right\} \right\|$$

$$\leqslant C \min \left(np^4, \frac{p^2}{n} \right).$$

2.5 Let $0 < p \leqslant 1/5, n \in \mathbb{N}, np \geqslant 1, F \in \mathcal{F}$. Prove that

$$\left\| ((1 - p)I + pF)^n - \exp\left\{ np(F - I) - \frac{np^2(F - I)^2}{2} \right\} \right.$$

$$\left. \times \left(I + \frac{np^3(F - I)^3}{3} \right) \right\| \leqslant \frac{Cp^2}{n}.$$

2.6 Let $F \in \mathcal{F}_+, n \in \mathbb{N}$. Prove that

$$| F^n - \exp\{n(F - I)\}(I - n(F - I)^2/2) |_K \leqslant Cn^{-2}.$$

2.7 Let $F \in \mathcal{F}_+, n \in \mathbb{N}$. Prove that

$$\left| \left(\frac{1}{2}I + \frac{1}{2}F^2 \right)^n - \exp\{n(F - I)\} \right|_K \leqslant Cn^{-2}. \qquad (2.48)$$

2.8 Let ξ and η be two independent integer-valued random variables $P(\xi = k) = p_k, P(\eta = k) = q_k, (k=0,1,\dots)$ and let

$$\varphi(F) = \sum_{k=0}^{\infty} p_k F^k, \quad \psi(F) = \sum_{k=0}^{\infty} q_k F^k.$$

Let $F \in \mathcal{F}_+$. Prove that

$$| \varphi(F) - \psi(F) |_K \leqslant \mathbb{E} \, \frac{|\xi - \eta|}{\max\{\xi, \eta\} + 1}.$$

2.9 Let $F \in \mathcal{F}, p \leqslant 1/2, n, m \in \mathbb{N}, n \leqslant m$. Prove that

$$\| (qI + pF)^n - (qI + pF)^m \| \leqslant \frac{C(m - n)\sqrt{p}}{\sqrt{n}}.$$

2.10 Let $F \in \mathcal{F}_s, p \leqslant 1/2, n, m \in \mathbb{N}$. Prove that

$$| (qI + pF)^n - (qI + pF)^m |_K \leqslant \frac{C|m - n|}{n}.$$

Bibliographical Notes

The origin of the method of convolutions is usually associated with Le Cam's papers [91, 92]. The Hipp inequality was proved in [75]. Though Le Cam's trick was used in the fifties (see, for example, [85]), the first explicit comment on (the rediscovered) Le Cam's trick can be found in [99].

Expansions in powers of $(F - I)$, such as those in Lemma 2.1 and the examples, can be found in [35]. Approximation by CNB distribution is examined in detail in [155] and [157]. The estimate (2.17), though not formulated explicitly, was used in [41]. Various versions of Lemma 2.5 appeared in [41, 42, 123]. Estimate (2.25) is a simpler version of Theorem 2.2 from [157]. Proofs of (2.10), (2.11), and (2.12) can be found respectively in [54, 121], Lemma 3, [124], Lemma 4 and [120], Lemma 4. Estimates (2.33) and (2.48) and (2.28) are special cases of more general results from [33]. All other smoothing estimates for the Kolmogorov norm and Theorems 2.6, 2.7 and 2.8 can be found in Chapter 5, Sections 4 and 5 of [5]. Theorem 2.9 is a special case of Lemma 2 from [31].

Kerstan's method was introduced in [84] and generalized in [123, 159]. It was extended to the multivariate case by Roos [118, 122], see also [90]. Kerstan's method for the negative binomial approximation was introduced in [157]. Inequality (2.42) was proved in [118]. Lemma 2.7 was proved in [42]. For application to measures defined on Abelian groups, see [126]. In [129], for comparison of two Poisson laws instead of (2.17) the Stirling formula was applied directly. Some aspects of the convolution method combined with the properties of concentration functions were used in [74, 125].

Chapter 3
Local Lattice Estimates

In this chapter, we consider measures $M \in \mathcal{M}_{\mathbb{Z}}$. Note that if a measure is concentrated on a lattice different from $\mathbb{Z}$, then it can be reduced to the integer-lattice case by a simple linear transform.

3.1 The Inversion Formula

In general, the characteristic function method means that we estimate differences between distributions (measures) through differences of their characteristic functions (Fourier transforms). The method is usually associated with Esseen type inequalities (see Chap. 9). Meanwhile, it is more natural and more explicit when applied to probabilities of integer-valued random variables.

Let $M \in \mathcal{M}_{\mathbb{Z}}$. By definition of the Fourier transform

$$\widehat{M}(t) = \sum_{k=-\infty}^{\infty} e^{itk} M\{k\}.$$

Taking into account Euler's formula (1.27) we observe that

$$\int_{-\pi}^{\pi} e^{it(k-m)} \, dt = \int_{-\pi}^{\pi} (\cos(k-m)t + i\sin(k-m)t) \, dt = \begin{cases} 0, & \text{if } k \neq m, \\ 2\pi, & \text{if } k = m. \end{cases}$$

Therefore, for any $k \in \mathbb{Z}$,

$$M\{k\} = \frac{1}{2\pi} \int_{-\pi}^{\pi} e^{-itk} \widehat{M}(t) \, dt. \tag{3.1}$$

© Springer International Publishing Switzerland 2016
V. Čekanavičius, *Approximation Methods in Probability Theory*, Universitext,
DOI 10.1007/978-3-319-34072-2_3

Recalling that $|e^{itk}| = 1$, we easily obtain the local inversion formula

$$\| M \|_\infty \leq \frac{1}{2\pi} \int\limits_{-\pi}^{\pi} |\widehat{M}(t)| \, dt. \tag{3.2}$$

Typical application. Let us assume that we want to estimate

$$\left\| \prod_{j=1}^{n} F_j - \prod_{j=1}^{n} G_j \right\|_\infty, \qquad F_j, G_j \in \mathcal{F}_Z.$$

- Apply

$$\left| \prod_{j=1}^{n} \widehat{F}_j(t) - \prod_{j=1}^{n} \widehat{G}_j(t) \right| \leq \sum_{j=1}^{n} \left| \widehat{F}_j(t) - \widehat{G}_j(t) \right| \prod_{k=1}^{j-1} |\widehat{F}_k(t)| \prod_{k=j+1}^{n} |\widehat{G}_k(t)|.$$

- Expand $\widehat{F}_j(t) - \widehat{G}_j(t)$ in moments or factorial moments.
- Estimate from above, for all $|t| \leq \pi$,

$$\prod_{k=1}^{j-1} |\widehat{F}_k(t)| \prod_{k=j+1}^{n} |\widehat{G}_k(t)|$$

by some function having a smoothing effect on $(e^{it} - 1)$.
- Substitute all estimates into (3.2) and apply (1.48), (1.50) and (1.28), (1.29), (1.30), and (1.31) or (1.22), (1.23), and (1.24).

Advantages. The method can be applied in very general situations and can be generalized for multidimensional distributions.
Drawbacks. Upper bound estimates for n-fold products of Fourier transforms are needed.

In the above, we outlined a typical approach. Exponential estimates can also be used. Let $M_1, M_2 \in \mathcal{M}_Z$. Then for all $|t| \leq \pi$

$$| \exp\{\widehat{M}_1(t)\} - \exp\{\widehat{M}_2(t)\} |$$
$$\leq |\widehat{M}_1(t) - \widehat{M}_2(t)| \sup_{0 \leq \tau \leq 1} \exp\{\tau Re\widehat{M}_1(t) + (1 - \tau)Re\widehat{M}_2(t)\}. \tag{3.3}$$

Here *Re* means the real part of a complex number. The proof of (3.3) is practically identical to the proof of (2.17). Note that, for any complex number $a + ib$

$$| \exp\{a + ib\} | = \exp\{a\}, \tag{3.4}$$

since $| \exp\{ib\} | = 1$.

Observe that $\widehat{F}_j(t), \widehat{G}_j(t)$ are 2π-periodic functions. Therefore an estimate for $|t| \leqslant \pi$ is equivalent to an estimate for $t \in \mathbb{R}$.

3.2 The Local Poisson Binomial Theorem

We illustrate the approach outlined above by considering a local version of (2.20). Observe that there is no assumption on the smallness of p_j.

Theorem 3.1 *Let* $p_j \in [0, 1]$, $q_j = 1 - p_j$, $j = 1, 2, \ldots, n$. *Then*

$$\left\| \prod_{j=1}^{n}(q_j I + p_j I_1) - \exp\left\{ \sum_{j=1}^{n} p_j(I_1 - I) \right\} \right\|_{\infty} \leqslant C \sum_{j=1}^{n} p_j^2 \min\left\{ 1, \left(\sum_{j=1}^{n} p_j q_j \right)^{-3/2} \right\}. \tag{3.5}$$

Proof Let

$$F_j = q_j I + p_j I_1, \qquad G_j = \exp\{ p_j(I_1 - I) \},$$

that is,

$$\widehat{F}_j(t) = q_j + p_j e^{it}, \qquad \widehat{G}_j(t) = \exp\{ p_j(e^{it} - 1) \}.$$

It is easy to check that the proof of Lemma 2.1 also holds for Fourier transforms. Consequently, from analogues of (2.3) and (1.28) it follows that

$$| \widehat{F}_j(t) - \widehat{G}_j(t) | = | [1 + p_j(e^{it} - 1)] - [1 + p_j(e^{it} - 1) + \theta p_j^2 | e^{it} - 1 |^2 / 2] |$$
$$\leqslant 2p_j^2 \sin^2(t/2). \tag{3.6}$$

The next step of the proof is to estimate the characteristic functions from above. Applying (3.4) we obtain

$$| \widehat{G}_j(t) | = | \exp\{ p_j(\cos t - 1) + ip_j \sin t \} | = | \exp\{ -2p_j \sin^2(t/2) \} |.$$

Let us recall that for any complex number $|a + ib|^2 = a^2 + b^2$. Therefore, taking into account Euler's formula (1.27), we obtain

$$|q_j + p_j e^{it}|^2 = (q_j + p_j \cos t)^2 + p_j^2 \sin^2 t = q_j^2 + 2p_j q_j \cos t + p_j^2$$
$$= (q_j^2 + 2p_j q_j + p_j^2) + 2p_j q_j (\cos t - 1) = 1 - 4p_j^2 \sin^2(t/2)$$
$$\leqslant \exp\{-4p_j q_j \sin^2(t/2)\}. \tag{3.7}$$

Combining the last two estimates, for all $t \in \mathbb{R}$, we prove

$$\max(|\widehat{F}_j(t)|, |\widehat{G}_j(t)|) \leqslant \exp\{-2p_j q_j \sin^2(t/2)\}. \tag{3.8}$$

Observing that $2p_j q_j \leqslant 0.5$ we get

$$\prod_{k=1}^{j-1} |\widehat{F}_k(t)| \prod_{k=j+1}^{n} |\widehat{G}_k(t)| \leqslant \exp\left\{-2 \sum_{k \neq j} p_k q_k \sin^2(t/2)\right\}$$
$$= \exp\left\{-2 \sum_{k=1}^{n} p_k q_k \sin^2(t/2) + p_k q_k \sin^2(t/2)\right\}$$
$$\leqslant \sqrt{e} \exp\left\{-2 \sum_{k=1}^{n} p_k q_k \sin^2(t/2)\right\}. \tag{3.9}$$

Substituting (3.9) and (3.6) into (3.2) and applying (1.31) we prove

$$\left\| \prod_{j=1}^{n} F_j - \prod_{j=1}^{n} G_j \right\|_{\infty} \leqslant C \sum_{j=1}^{n} p_j^2 \int_{-\pi}^{\pi} \sin^2(t/2) \exp\left\{-2 \sum_{k=1}^{n} p_j q_j \sin^2(t/2)\right\} dt$$
$$\leqslant C \sum_{j=1}^{n} p_j^2 \left(\sum_{k=1}^{n} p_k q_k\right)^{-3/2}.$$

To complete the proof observe that the above integral is less than 2π.   □

3.3 Applying Moment Expansions

We can use an expansion in powers of (it), combining it with (1.28) and (1.50). We illustrate this approach by considering an approximation of the binomial distribution by a shifted Poisson distribution. The inversion formula (3.2) is applicable to measures from $\mathcal{M}_Z$ only. Therefore the shift must be by an integer. Let $q = 1 - p$

and let $\lfloor np^2 \rfloor$ as usual denote the integer part of np^2, that is,

$$np^2 = \lfloor np^2 \rfloor + \delta, \qquad 0 \leqslant \delta < 1,$$

$$H = ((1-p)I + pI_1)^n, \quad D = I_{\lfloor np^2 \rfloor} \exp\{npq(I_1 - I)\}.$$

Now we can formulate our result.

Theorem 3.2 *Let $p \leqslant 1/4$, $n \in \mathbb{N}$. Then*

$$\|H - D\|_\infty \leqslant C\sqrt{\frac{p}{n}} + \frac{C}{np}. \tag{3.10}$$

Proof Let $0 \leqslant \tau \leqslant 1$. Observing that $|e^{it} - 1| \leqslant |e^{it}| + 1 = 2$ and taking into account (1.28) we obtain

$$\tau \ln H(t) = \tau n \ln(1 + p(e^{it} - 1)) = \tau n \sum_{j=1}^{\infty} \frac{(-1)^{j+1}}{j} p^j (e^{it} - 1)^j$$

$$= \tau np(e^{it} - 1) + \tau n\theta \frac{p^2 |e^{it} - 1|^2}{2} \sum_{j=2}^{\infty} p^{j-2} |e^{it} - 1|^{j-2}$$

$$= \tau np(e^{it} - 1) + \tau\theta 2np^2 \sin^2(t/2) \sum_{j=1}^{\infty} (0.5)^j = \tau np(e^{it} - 1) + \tau\theta 4np^2 \sin^2(t/2)$$

$$= \tau np(\cos t - 1) + i\tau np \sin t + \tau\theta np \sin^2(t/2)$$

$$= -\tau 2np \sin^2(t/2) + \tau\theta np \sin^2(t/2) + i\tau np \sin t.$$

Similarly,

$$(1 - \tau) \ln \widehat{D}(t) = (1 - \tau)\lfloor np^2 \rfloor it + (npq + \delta)(\cos t - 1 + i \sin t)$$

$$= -(1 - \tau)2npq \sin^2(t/2) + \theta(1 - \tau)\delta 2 + i[\lfloor np^2 \rfloor + (npq + \delta) \sin t].$$

Observing that $2npq > np$, for $|t| \leqslant \pi$, we prove

$$\exp\{\tau Ren \ln \widehat{H}(t) + (1 - \tau)Re \ln \widehat{D}(t)\}$$

$$\leqslant \exp\{-\tau 2np \sin^2(t/2) + \tau np \sin^2(t/2) - (1 - \tau)2npq \sin^2(t/2) + 2\}$$

$$\leqslant e^2 \exp\{-np \sin^2(t/2)\} \leqslant e^2 \exp\{-npt^2/\pi^2\}. \tag{3.11}$$

The last estimate follows from (1.50). Next, we expand $\ln \widehat{H}(t)$ in powers of it by (1.32). There is one small trick involved, we do not expand $(e^{it} - 1)$

$$\ln \widehat{H}(t) = np(e^{it} - 1) - \frac{np^2}{2}(e^{it} - 1)^2 + n\theta \frac{p^3 |e^{it} - 1|^3}{3} \sum_{j=3}^{\infty} (2p)^{j-3}$$

$$= np(e^{it} - 1) - np^2 \frac{(it)^2}{2} + \theta C(np^2 + np^3)|t|^3$$

$$= np(e^{it} - 1) - np^2 \frac{(it)^2}{2} + \theta Cnp^2 |t|^3.$$

Similarly, for $|t| \leqslant \pi$,

$$\ln \widehat{D}(t) = \lfloor np^2 \rfloor it + np(e^{it} - 1) + (\delta - np^2)(e^{it} - 1)$$

$$= np(e^{it} - 1) + \lfloor np^2 \rfloor it + (\delta - np^2)\left((it) + \frac{(it)^2}{2} + \theta C|t|^3\right)$$

$$= np(e^{it} - 1) + (it)(\lfloor np^2 \rfloor it + \delta - np^2) - np^2 \frac{(it)^2}{2} + \theta C\delta t^2 + \theta C(\delta + np^2)|t|^3$$

$$= np(e^{it} - 1) - np^2 \frac{(it)^2}{2} + C\theta\delta t^2 + C\theta\delta|t|^3 + C\theta np^2 |t|^3.$$

Combining the last two estimates we obtain

$$| \ln \widehat{H}(t) - \ln \widehat{D}(t) | \leqslant C\delta t^2 + C\delta t^2 \pi + Cnp^2 |t|^3 \leqslant Ct^2 + Cnp^2 |t|^3. \qquad (3.12)$$

Observe that, had we expanded $(e^{it} - 1)$ in powers of (it), the remainder term would have been $Ct^2 + Cnp|t|^3$. Thus, the benefit from our approach is the higher power of p.

Substituting (3.11) and (3.12) into (3.3), for all $|t| \leqslant \pi$, we get

$$| \widehat{H}(t) - \widehat{D}(t) | \leqslant C(t^2 + np^2 |t|^3) \exp\{-npt^2/\pi^2\}.$$

Substituting the last estimate into (3.2) we prove

$$\| H - D \|_\infty \leqslant C \int_{-pi}^{\pi} (t^2 + np^2 |t|^3) \exp\{-npt^2/\pi^2\} dt$$

$$\leqslant C \int_0^\infty (t^2 + np^2 t^3) \exp\{-npt^2/\pi^2\} dt \leqslant \frac{C}{np} + C\frac{np^2}{np\sqrt{np}},$$

which is equivalent to (3.10). $\qquad\qquad\qquad\qquad\qquad\qquad\qquad\qquad\qquad\qquad\quad \square$

3.4 A Local Franken-Type Estimate

As was noted in Sect. 3.1, we need to estimate products of characteristic functions for all $|t| \leqslant \pi$. For a Bernoulli random variable we have estimate (3.7). It is natural to expect similar estimates to hold for measures differing slightly from $q_j I + p_j I_1$. In this section, we prove one such estimate.

Let $F \in \mathcal{F}_{\mathbb{Z}}$ be concentrated on nonnegative integers. Its factorial moments are denoted by $\nu_k(F)$, see (1.15). Let

$$\lambda(F) = \nu_1(F) - \nu_1^2(F) - \nu_2(F) > 0. \tag{3.13}$$

Condition (3.13) is known as Franken's condition. In general, it is quite restrictive. It means that $F\{0\}, F\{1\}$ are relatively large and $F\{2\}, F\{3\}, \ldots$ are relatively small. Moreover, the mean of F and the mean of G are both less than unity. Indeed,

$$\nu_1(F) > \nu_2(F) \Rightarrow F\{1\} - \sum_{k=3}^{\infty} k(k-2)F\{k\} > 0$$

and

$$\lambda(F) > 0 \Rightarrow \nu_1(F) - \nu_1^2(F) > 0 \Rightarrow \nu_1(F) < 1 \Rightarrow \lambda(F) < 1.$$

Franken's condition is sufficient for an analogue of (3.7) to hold.

Lemma 3.1 *Let $F \in \mathcal{F}$ be concentrated on nonnegative integers, $\lambda(F) > 0$. Then*

$$|\widehat{F}(t)| \leqslant \exp\left\{-2\lambda(F) \sin^2 \frac{t}{2}\right\}. \tag{3.14}$$

Proof Let us write for brevity ν_1, ν_2 instead of $\nu_1(F), \nu_2(F)$. From (1.16) and (1.28) it follows that

$$|\widehat{F}(t)| \leqslant |1 + \nu_1(e^{it}-1)| + \frac{\nu_2}{2}|e^{it}-1|^2 = |1 + \nu_1(e^{it}-1)| + 2\nu_2 \sin^2 \frac{t}{2}. \tag{3.15}$$

Next, observe that by (1.25) and (1.27)

$$
\begin{aligned}
|1 + \nu_1(e^{it}-1)|^2 &= |(1 + \nu_1(\cos t - 1)) + i\nu_1 \sin t|^2 \\
&= [1 + \nu_1(\cos t - 1)]^2 + \nu_1^2 \sin^2 t \\
&= 1 + \nu_1^2(\cos^2 t - 2\cos t + 1 + \sin^2 t) + 2\nu_1(\cos t - 1) \\
&= 1 + 2[\nu_1 - \nu_1^2](\cos t - 1) \\
&= 1 - 4[\nu_1 - \nu_1^2]\sin^2 \frac{t}{2}.
\end{aligned}
$$

Taking into account that if $0 \leqslant u \leqslant 1$ then $\sqrt{1-u} \leqslant 1 - u/2$, we prove

$$| 1 + v_1(e^{it} - 1) | \leqslant \sqrt{1 - 4[v_1 - v_1^2] \sin^2 \frac{t}{2}} \leqslant 1 - 2[v_1 - v_1^2] \sin^2 \frac{t}{2}.$$

Substituting the last estimate into (3.15) we get

$$| \widehat{F}(t) | \leqslant 1 - 2[v_1 - v_1^2] \sin^2 \frac{t}{2} + 2v_2 \sin^2 \frac{t}{2} = 1 - 2\lambda(F) \sin^2 \frac{t}{2}.$$

To complete the lemma's proof it suffices to apply the trivial inequality $1 - x \leqslant \exp\{-x\}$ to the last estimate. $\square$

Theorem 3.3 *Let $F, G \in \mathcal{F}_Z$ satisfy (3.13) and, for fixed integer $s > 1$,*

$$v_k(F) = v_k(G), \quad (k = 1, 2, \ldots, s - 1), \qquad v_{s+1}(F), v_{s+1}(G) < \infty.$$

Then

$$\| F^n - G^n \|_\infty \leqslant C(s)n[v_s(F) + v_s(G)] \min \left(1, \frac{1}{(n\lambda)^{(s+1)/2}} \right). \tag{3.16}$$

Here $\lambda = \min(\lambda(F), \lambda(G))$.

Proof From the expansion in factorial moments (1.16) we have

$$| \widehat{F}(t) - \widehat{G}(t) | \leqslant [v_s(F) + v_s(G)] | e^{it} - 1 |^s / s! = [v_s(F) + v_s(G)] \frac{2^s}{s!} \left| \sin \frac{t}{2} \right|^s.$$

Moreover, by (3.14),

$$\max\{| \widehat{F}(t) |, | \widehat{G}(t) |\} \leqslant \exp\left\{-2\lambda \sin^2 \frac{t}{2}\right\}.$$

Consequently, substituting the last two estimates into (1.45) and (3.2) and taking into account that $\lambda < 1$, we prove

$$\| F^n - G^n \|_\infty \leqslant C \int_{-\pi}^{\pi} n| \widehat{F}(t) - \widehat{G}(t) | \max(| \widehat{F}(t) |^{n-1}, | \widehat{G}(t) |^{n-1})dt$$

$$\leqslant C(s)n[v_s(F) + v_s(G)] \int_{-\pi}^{\pi} \left| \sin \frac{t}{2} \right|^s \exp\left\{-2(n - 1)\lambda \sin^2 \frac{t}{2}\right\}$$

$$\leqslant C(s)n[v_s(F) + v_s(G)] \int_{-\pi}^{\pi} \left| \sin \frac{t}{2} \right|^s \exp\left\{-2n\lambda \sin^2 \frac{t}{2}\right\} dt.$$

The desired estimate follows from (1.31). $\square$

Example 3.1 Let F^n be the binomial distribution with parameters $p \leqslant 1/4$ and $n \in \mathbb{N}$. Then

$$F = (1-p)I + pI_1 = I + p(I_1 - I), \quad \nu_1(F) = p, \quad \nu_2(F) = 0.$$

Similarly, if $G^n = \exp\{np(I_1 - I)\}$ is the corresponding Poisson law, then

$$G = \exp\{p(I_1 - I)\}, \quad \nu_1(G) = p, \quad \nu_2(G) = p^2.$$

It is easy to check that $\nu_2(F) + \nu_2(G) = p^2$ and $\lambda = \nu_1(G) - \nu_1^2(G) - \nu_2(G) = p - 2p^2 > p/2$. Substituting these estimates into (3.16) we obtain

$$\| F^n - G^n \|_\infty \leqslant C \min\left(np^2, \sqrt{\frac{p}{n}} \right).$$

3.5 Involving the Concentration Function

Another method to guarantee estimates for all $|t| \leqslant \pi$ is to switch to compound distributions. Any $F \in \mathcal{F}_Z$ can be expressed as $F\{0\}I + (1 - F\{0\})V$, where $V \in \mathcal{F}$ and $V\{k\} = F\{k\}/(1 - F\{0\}), k \neq 0$. Now the structure of the distribution is similar to $qI + pI_1$ and one can expect to get an estimate similar to (3.9). For the sake of convenience, we avoid the notation $F\{k\}$, preferring q_k instead.

Let $F \in \mathcal{F}_Z$ and $F\{k\} = q_k$, that is

$$\widehat{F}(t) = \sum_{k=-\infty}^{\infty} q_k e^{itk}. \tag{3.17}$$

Note that all $q_k \in [0, 1]$ and their sum equals 1. Later in this section we assume that $q_0 > 0$.

Lemma 3.2 *Let $\widehat{F}(t)$ be defined by (3.17). Then, for all t,*

$$|\widehat{F}(t)| \leqslant \exp\{q_0(\mathrm{Re}\,\widehat{F}(t) - 1)\}. \tag{3.18}$$

Proof We write

$$\widehat{F}(t) = q_0 + (1 - q_0)\widehat{V}(t), \quad \widehat{V}(t) = \sum_{\substack{k \in \mathbb{Z} \\ k \neq 0}} e^{itk} \frac{q_k}{1 - q_0}. \tag{3.19}$$

It is obvious that $\widehat{V}(t)$ is a characteristic function. But all characteristic functions satisfy the inequality

$$|\widehat{V}(t)| \leqslant 1.$$

Therefore, recalling (1.25), we obtain

$$(Re \, \widehat{V}(t))^2 + (Im \, \widehat{V}(t))^2 \leqslant 1.$$

For the proof of the lemma we use the same idea as in (3.14). We have

$$|\widehat{F}(t)|^2 = |q_0 + (1 - q_0)Re \, \widehat{V}(t) + i(1 - q_0)Im \, \widehat{V}(t)|^2 =$$

$$q_0^2 + 2q_0(1 - q_0)Re \, \widehat{V}(t) + (1 - q_0)^2\left((Re \, \widehat{V}(t))^2 + (Im \, \widehat{V}(t))^2\right) \leqslant$$

$$q_0^2 + 2q_0(1 - q_0)Re \, \widehat{V}(t) + (1 - q_0)^2 = 1 + 2q_0(1 - q_0)(Re \, \widehat{V}(t) - 1) \leqslant$$

$$\exp\{2q_0(1 - q_0)(Re \, \widehat{V}(t) - 1)\} = \exp\{2q_0(Re \, \widehat{F}(t) - 1)\}.$$

For the last step, one should note that from (3.19) it follows that

$$(1 - q_0)(\widehat{V}(t) - 1) = \widehat{F}(t) - 1, \quad (1 - q_0)(Re \, \widehat{V}(t) - 1) = Re \, \widehat{F}(t) - 1.$$

$$\square$$

Comparing (3.18) with the estimates in the previous section, we see that $Re\widehat{F}(t)$ plays a role similar to that of $\sin^2(t/2)$. Therefore some estimate similar to (1.30) is necessary.

Lemma 3.3 *Let F be defined by (3.17). Then*

$$\int_{-\pi}^{\pi} \exp\left\{\lambda(Re \, \widehat{F}(t) - 1)\right\} dt = \int_{-\pi}^{\pi} \exp\left\{-2\lambda \sum_{j=-\infty}^{\infty} q_j \sin^2 \frac{tk}{2}\right\} dt \leqslant \frac{C}{\sqrt{\lambda(1 - q_0)}}.$$

$$(3.20)$$

Proof The proof of (3.20) is based on the properties of the concentration function. First, note that

$$\lambda(Re \, \widehat{F}(t) - 1) = \lambda(1 - q_0)(Re \, \widehat{V}(t) - 1).$$

Here $\widehat{V}(t)$ is defined by (3.19). Next, observe that $Re\widehat{V}(t) = 0.5\widehat{V}(t) + 0.5\widehat{V}(-t)$ is a real characteristic function and, therefore, $\exp\{\lambda(1 - q_0)(Re \, \widehat{V}(t) - 1)\}$ is a nonnegative characteristic function. Applying (1.24) and (1.22) we obtain

$$\int_{-\pi}^{\pi} \exp\left\{\lambda(Re\,\widehat{F}(t) - 1)\right\} dt = \int_{-\pi}^{\pi} \exp\left\{\lambda(1 - q_0)(Re\,\widehat{V}(t) - 1)\right\} dt$$

$$\leqslant CQ(\tilde{V}, \pi^{-1}) \leqslant \frac{C}{\sqrt{\lambda(1 - q_0)\tilde{V}\{x : |x| > \pi^{-1}\}}}$$

$$= \frac{C}{\sqrt{\lambda(1 - q_0)}}.$$

Here $\tilde{V} = 0.5V + 0.5V^{(-)}$, that is, its characteristic function equals $Re\,\widehat{V}(t)$. Observe also that $\tilde{V}$ is concentrated at $\mathbb{Z}\setminus\{0\}$ and, therefore, $\tilde{V}\{x : |x| > \pi^{-1}\} = \tilde{V}\{\mathbb{R}\} = 1$.

□

As an application of Lemmas 3.2 and 3.3 we consider the approximation of F^n by its accompanying law $\exp\{n(F - I)\}$. Note that

$$\exp\{F - I\} = \exp\left\{\sum_{k=-\infty}^{\infty} q_k(I_k - I)\right\},$$

i.e. we have a convolution of various Poisson distributions concentrated on different sub-lattices of $\mathbb{Z}$.

Theorem 3.4 Let F be defined by (3.17), $n \in \mathbb{N}$. Then

$$\| F^n - \exp\{n(F - I)\} \|_\infty \leqslant C \min\left\{n(1 - q_0)^2, q_0^{-3/2}\sqrt{\frac{1 - q_0}{n}}\right\}.$$

Proof If $q_0 = 1$ then $F = \exp\{F - I\} \equiv I$ and the difference of distributions is equal to 0. Therefore we further assume that $q_0 < 1$. Applying (1.45) we get

$$|\widehat{F^n}(t) - \exp\{n(\widehat{F}(t) - 1)\}| \leqslant n|\widehat{F}(t) - \exp\{\widehat{F}(t) - 1\}|$$
$$\times \max\{|\widehat{F}(t)|^{n-1}, |\exp\{\widehat{F}(t) - 1\}|^{n-1}\}.$$

$$(3.21)$$

Taking into account (3.4) we get

$$|\exp\{\widehat{F}(t) - 1\}| = \exp\{Re\,\widehat{F}(t) - 1\} \leqslant \exp\{q_0(Re\,\widehat{F}(t) - 1)\}.$$

Combining this estimate with (3.18) we prove

$$\max\{|\widehat{F}(t)|^{n-1}, |\exp\{\widehat{F}(t)-1\}|^{n-1}\} \leq \exp\{(n-1)q_0(Re\,\widehat{F}(t)-1)\}$$
$$\leq \exp\{q_0(|Re\,\widehat{F}(t)|+1)\}\exp\{nq_0(Re\,\widehat{F}(t)-1)\}$$
$$\leq e^2\exp\{nq_0(Re\,\widehat{F}(t)-1)\}.$$

Next, we apply (1.33)

$$|\exp\{\widehat{F}(t)-1\}-1-(\widehat{F}(t)-1)| \leq |\widehat{F}(t)-1|^2\int_0^1|\exp\{\tau(\widehat{F}(t)-1)\}|(1-\tau)d\tau$$
$$\leq |\widehat{F}(t)-1|^2\int_0^1(1-\tau)d\tau = \frac{|\widehat{F}(t)-1|^2}{2}.$$

Here we used the fact that $\exp\{\tau(\widehat{F}(t)-1)\}$ is a characteristic function and its absolute value is less than or equal to 1. Substituting all estimates into (3.21) and, consequently applying (1.14) and (1.48), we prove

$$|\widehat{F^n}(t)-\exp\{n(\widehat{F}(t)-1)\}| \leq Cn|\widehat{F}(t)-1|^2\exp\{nq_0(Re\,\widehat{F}(t)-1)\}$$
$$= Cn(1-q_0)^2|\widehat{V}(t)-1|^2\exp\left\{n\frac{q_0}{2}(1-q_0)(Re\,\widehat{V}(t)-1)\right\}\exp\left\{n\frac{q_0}{2}(Re\,\widehat{F}(t)-1)\right\}$$
$$\leq Cn(1-q_0)^2|Re\,\widehat{V}(t)-1|\exp\left\{n\frac{q_0}{2}(1-q_0)(Re\,\widehat{V}(t)-1)\right\}\exp\left\{n\frac{q_0}{2}(Re\,\widehat{F}(t)-1)\right\}$$
$$\leq C\min\left\{n(1-q_0)^2, \frac{(1-q_0)}{q_0}\right\}\exp\left\{n\frac{q_0}{2}(Re\,\widehat{F}(t)-1)\right\}.$$

Next, from (3.2) we get

$$\|F^n-\exp\{n(F-I)\}\|_\infty \leq C\min\left\{n(1-q_0)^2, \frac{(1-q_0)}{q_0}\right\}\int_{-\pi}^\pi\exp\left\{n\frac{q_0}{2}(Re\,\widehat{F}(t)-1)\right\}dt.$$

It remains to apply (3.20). $\square$

3.6   Switching to Other Metrics

Let $M \in \mathcal{M}_Z$. Then

$$\|M\|_\infty \leq \|M\|. \tag{3.22}$$

Inequality (3.22) allows us to switch from the local norm to the total variation. Sometimes such an approach is sufficiently effective. For example, when applying (2.5) to $p_i = O(n^{-1})$. However, in many situations, estimate (3.22) is too rough. Indeed, let us compare (2.20) and (3.5) for $p_i \equiv p = O(1)$. Then (2.20) is of the order $O(1)$. Meanwhile, (3.5) is of the order $O(n^{-1/2})$. Therefore the following relations are more useful:

$$\| M \|_\infty = | (I_1 - I)M |_K \leqslant \| (I_1 - I)M \|. \tag{3.23}$$

Relations (3.23) are most effective when combined with the method of convolutions for the total variation norm and smoothing estimates from Sect. 5.4. We prove the first equality of (3.23). From the definition of a discrete measure we have

$$M(I_1 - I) = \sum_{k=-\infty}^{\infty} M\{k\}I_k(I_1 - I) = \sum_{k=-\infty}^{\infty} M\{k\}I_{k+1} - \sum_{k=-\infty}^{\infty} M\{k\}I_k$$

$$= \sum_{k=-\infty}^{\infty} (M\{k-1\} - M\{k\})I_k.$$

Hence

$$| (I_1 - I)M |_K = \sup_{x \in \mathbb{Z}} \left| \sum_{k=-\infty}^{x} (M\{k-1\} - M\{k\}) \right| = \sup_{x \in \mathbb{Z}} | M\{x\} | = \| M \|_\infty.$$

Example 3.2 Let $a > 0$. Then by (3.23) and (2.10)

$$\| \exp\{a(I_1 - I)\} \|_\infty \leqslant \| (I_1 - I)\exp\{a(I_1 - I)\} \| \leqslant \sqrt{\frac{2}{ae}}.$$

Example 3.3 Let $0 < q_k \leqslant C_0 < 1$, $(k = 1,\ldots,n)$ and let γ, q be defined as in (2.24). Then

$$\left\| \prod_{k=1}^{n} CG(q_k, I_1) - CNB(\gamma, q, I_1) \right\|_\infty \leqslant C \frac{\sum_{k=1}^{n} q_k^3}{\left(\sum_{k=1}^{n} q_k \right)^2}.$$

Indeed, proceeding similarly to the proof of (2.25) we obtain

$$\left\| \prod_{k=1}^{n} CG(q_k, I_1) - CNB(\gamma, q, I_1) \right\|_\infty$$

$$\leqslant C \sum_{k=1}^{\infty} q_k^3 \left\| (I_1 - I)^3 \exp\left\{ (\tau \sum_{k=1}^{n} q_k + (1 - \tau)\gamma q)(I_1 - I) \right\} \right\|_\infty$$

$$= C \sum_{k=1}^{\infty} q_k^3 \left\| (I_1 - I)^4 \exp\left\{ \left(\tau \sum_{k=1}^{n} q_k + (1-\tau)\gamma q \right)(I_1 - I) \right\} \right\|$$

$$\leqslant C \sum_{k=1}^{\infty} q_k^3 \left(\min\left(\sum_{k=1}^{n} q_k, \gamma q \right) \right)^{-2} \leqslant C \sum_{k=1}^{\infty} q_k^3 \left(\sum_{k=1}^{n} q_k \right)^{-2}.$$

3.7 Local Smoothing Estimates

One can apply the method of convolutions combining local smoothing estimates with the properties of norms. Applying (3.23) to the estimates in Lemma 2.3, for $a \in (0, \infty), p = 1 - q \in (0, 1)$, and $k, n \in \mathbb{N}$, we get

$$\| (I_1 - I)^k \exp\{a(I_1 - I)\} \|_{\infty} \leqslant \frac{\sqrt{(k+1)!}}{a^{(k+1)/2}},$$

$$\| (I_1 - I)^k (qI + pI_1)^n \|_{\infty} \leqslant \frac{C(k)}{(npq)^{(k+1)/2}}.$$

More accurate estimates can be obtained for symmetric distributions.

Lemma 3.4 Let $V \in \mathcal{F}_s$ be concentrated on $\pm 1, \pm 2, \ldots, a > 0, j \in \mathbb{N}$. Then

$$\| (V - I)^j \exp\{a(V - I)\} \|_{\infty} \leqslant 2 \left(\frac{j + 1/2}{ae} \right)^{j+1/2}. \tag{3.24}$$

Proof We begin with the proof of the discrete version of the Parseval identity. Let $M \in \mathcal{M}_z$. We recall that $M^{(-)}\{k\} = M\{-k\}$ and $\widehat{M}^{(-)}(t) = \widehat{M}(-t)$. By the definition of convolution

$$MM^{(-)}\{k\} = \sum_{j=-\infty}^{\infty} M^{(-)}\{k - j\}M\{j\} = \sum_{j=-\infty}^{\infty} M\{j - k\}M\{j\}.$$

On the other hand, by the inversion formula (3.1) and (1.12)

$$MM^{(-)}\{k\} = \frac{1}{2\pi} \int_{-\pi}^{\pi} \widehat{M}(-t)\widehat{M}(t)e^{-itk}dt = \frac{1}{2\pi} \int_{-\pi}^{\pi} |\widehat{M}(t)|^2 e^{-itk}dt.$$

For $k = 0$ we get

$$\sum_{j=-\infty}^{\infty} M^2\{j\} = \frac{1}{2\pi} \int_{-\pi}^{\pi} |\widehat{M}(t)|^2 dt. \tag{3.25}$$

If $G \in \mathcal{F}_Z$, then

$$\frac{1}{2\pi} \int_{-\pi}^{\pi} |\widehat{G}(t)|^2 dt = \sum_{j=-\infty}^{\infty} G^2\{j\} \leqslant \|G\|_\infty \sum_{j=-\infty}^{\infty} G\{j\} = \|G\|_\infty.$$

Therefore, for arbitrary λ, applying (1.22), we obtain

$$\int_{-\pi}^{\pi} \exp\{\lambda(\widehat{F}(t) - 1)\} dt = \int_{-\pi}^{\pi} \exp\left\{\frac{\lambda}{2}(\widehat{F}(t) - 1)\right\} \exp\left\{\frac{\lambda}{2}(\widehat{F}(t) - 1)\right\} dt$$

$$\leqslant 2\pi \left\| \exp\left\{\frac{\lambda}{2}(F - I)\right\} \right\|_\infty \leqslant 2\pi \sqrt{\frac{2}{\lambda e}}. \qquad (3.26)$$

Observe that $\widehat{V}(t) \in \mathbb{R}$. Indeed, due to the symmetry of V

$$\widehat{V}(t) = \sum_{j=-\infty}^{\infty} V\{j\} e^{itj} = V\{0\} + \sum_{j=1}^{\infty} V\{j\}(e^{itj} + e^{-itj}) = 2\sum_{j=1}^{\infty} V\{j\} \cos tj.$$

Applying (3.26) and (3.1) we prove

$$\| (V - I)^j \exp\{a(V - I)\} \|_\infty \leqslant \frac{1}{2\pi} \int_{-\pi}^{\pi} (1 - \widehat{V}(t))^j \exp\{a(\widehat{V}(t) - 1)\} dt$$

$$= \frac{1}{2\pi} \int_{-\pi}^{\pi} (1 - \widehat{V}(t))^j \exp\left\{\frac{aj}{j + 1/2}(\widehat{V}(t) - 1)\right\} \exp\left\{\frac{a}{2j+1}(\widehat{V}(t) - 1)\right\} dt$$

$$\leqslant \sup_{x \geqslant 0} x^j \exp\left\{-\frac{aj}{j + 1/2} x\right\} \frac{1}{2\pi} \int_{-\pi}^{\pi} \exp\left\{\frac{a}{2j+1}(\widehat{V}(t) - 1)\right\} dt$$

$$\leqslant 2\left(\frac{j + 1/2}{ae}\right)^{j+1/2}.$$

$\square$

3.8   The Method of Convolutions for a Local Metric

In principle, the method of convolutions for $\| \cdot \|_\infty$ is the same, as discussed in previous chapter. We apply (2.4) and a local version of (2.17): if $M_1, M_2 \in \mathcal{M}_Z$ then

$$\| \exp\{M_1\} - \exp\{M_2\} \|_\infty \leqslant \sup_{0 \leqslant \tau \leqslant 1} \| (M_1 - M_2) \exp\{\tau M_1 + (1-\tau)M_2\} \|_\infty. \qquad (3.27)$$

Observe also that, if $F\{0\} > 0$, then, similarly to (3.19),

$$F = F\{0\}I + (1 - F\{0\})V, \qquad V = \sum_{\substack{k \in \mathbb{Z} \\ k \neq 0}} \frac{F\{k\}}{1 - F\{0\}} I_k.$$

We illustrate the method of convolutions for local estimates by considering an approximation to a symmetric distribution. Let

$$H = (1 - 2p)I + pI_1 + pI_{-1}.$$

Theorem 3.5 *Let* $p \leqslant 1/8$, $n \in \mathbb{N}$. *Then*

$$\| H^n - \exp\{n(H - I)\} \|_\infty \leqslant \frac{C}{n\sqrt{np}}. \tag{3.28}$$

Proof Observe that

$$H = (1 - 2p)I + 2pV, \qquad V = \frac{1}{2}I_1 + \frac{1}{2}I_{-1}.$$

Let $M_1 = n \ln H$, $M_2 = n(H - I)$. Then, from Example 2.4 with $s = n = 1$, (see also the proof of Theorem 2.4), we have

$$M_1 = n2p(V - I) + n\frac{(2p)^2(V - I)^2}{2(1 - 4p)}\Theta = 2np(V - I) + \frac{1}{4}2np(V - I)^2\Theta.$$

Therefore

$$M_1 - M_2 = Cnp^2(V - I)^2\Theta.$$

Applying (2.18) we get

$$\exp\{\tau M_1 + (1 - \tau)M_2\} = \exp\left\{\frac{7\tau np}{4}(V - I) + \frac{2}{7}\frac{7\tau np}{4}(V - I)^2\Theta\right\} \exp\left\{\frac{np}{4}(V - I)\right\}$$

$$\times \exp\left\{\frac{7(1 - \tau)np}{4}(V - I)\right\} = \exp\left\{\frac{np}{4}(V - I)\right\}C\Theta.$$

From these estimates, (3.27) and (3.24) it follows that

$$\| H^n - \exp\{n(H - I)\} \|_\infty = \| \exp\{M_1\} - \exp\{M_2\} \|_\infty$$

$$\leqslant \sup_{0 \leqslant \tau \leqslant 1} \| (M_1 - M_2)\exp\{\tau M_1 + (1 - \tau)M_2\} \|_\infty$$

$$\leqslant C \left\| (M_1 - M_2) \exp\left\{ \frac{np}{4}(V - I) \right\} \right\|_\infty \|\Theta\|$$

$$\leqslant C n p^2 \left\| (V - I)^2 \exp\left\{ \frac{np}{4}(V - I) \right\} \right\|_\infty$$

$$\leqslant C \frac{np^2}{(np)^{5/2}} = \frac{C}{n\sqrt{np}}.$$

$\square$

3.9 Problems

3.1 Let F and G be concentrated on nonnegative integers with the same $s - 1$ factorial moment and finite factorial moments of the s-th order. Prove that

$$\| F^n - G^n \|_\infty \leqslant C(s) n [v_s(F) + v_s(G)].$$

3.2 Let F be defined by (3.17) with $q_0 = 1/5$. Prove that

$$\| (F - I)^k \exp\{n(F - I)\} \|_\infty \leqslant C(k) n^{-(k+1)/2}.$$

3.3 Let $F \in \mathcal{F}_+$ be defined by (3.17) with $q_0 > 0$. Prove that

$$\| (F - I)^k F^n \|_\infty \leqslant \frac{C(k)}{n^{k+1/2}\sqrt{1 - q_0}}.$$

3.4 Let $n \in \mathbb{N}$, F be defined by (3.17) with $q_0 = 0.7$,

$$H = \sum_{m=0}^\infty \left(\frac{1}{2}\right)^{m+1} F^m.$$

Prove that

$$\| H^n - F^n \|_\infty \leqslant \frac{C}{\sqrt{n}}.$$

3.5 Let $F \in \mathcal{F}_Z$, $F\{0\} < 1$ and $a > 0, b > 0$. Prove that

$$\| \exp\{a(F - I)\} - \exp\{b(F - I)\} \|_\infty \leqslant C \frac{|a - b|}{b\sqrt{1 - F\{0\}}}.$$

3.6 Let $n \in \mathbb{N}$, F be defined by (3.17) with $q_0 = 0.5$. Prove that

$$\| F^n - F^{n+1} \|_\infty \leq \frac{C}{\sqrt{n}}.$$

3.7 Let $F \in \mathcal{F}_s$ be concentrated on $\pm 1, \pm 2, \ldots, b > a > 0$. Prove that

$$\| \exp\{a(F - I)\} - \exp\{b(F - I)\} \|_\infty \leq \frac{C(b - a)}{a\sqrt{a}}.$$

Bibliographical Notes

The inversion formula (3.1) is well known and can be found in numerous papers. The characteristic function method for lattice local estimates under Franken's condition was applied in [63, 89, 137]. Relation (3.23) was noted in [159]. Lemma 3.4 was proved in [42]. The shifted Poisson approximation is also called the translated Poisson approximation, see [131]. In [78], as an alternative to (3.1), a double complex integral is used. Methods of complex analysis are employed in [77].

Chapter 4
Uniform Lattice Estimates

4.1 The Tsaregradskii Inequality

One of the most popular methods for the uniform estimation of $M \in \mathcal{M}_Z$ is the so-called Tsaregradskii inequality, which in fact is a special case of the characteristic function method. It can be written in the following way.

Lemma 4.1 *Let $M \in \mathcal{M}_Z$, then*

$$|M|_K \leqslant \frac{1}{4\pi} \int_{-\pi}^{\pi} \frac{|\widehat{M}(t)|}{|\sin \frac{t}{2}|} \, dt \tag{4.1}$$

$$\leqslant \frac{1}{4} \int_{-\pi}^{\pi} \frac{|\widehat{M}(t)|}{|t|} \, dt. \tag{4.2}$$

Proof The second inequality follows from (4.1) and (1.50). The estimate (4.1) is trivial if the right-hand of (4.1) is infinite. Therefore we shall assume that it is finite. For two integers $s < m$, summing the inversion formula (3.1) and applying (1.26), we obtain

$$\sum_{k=s}^{m} M\{k\} = \sum_{k=s}^{m} \frac{1}{2\pi} \int_{-\pi}^{\pi} e^{-itk} \widehat{M}(t) \, dt = \frac{1}{2\pi} \int_{-\pi}^{\pi} \widehat{M}(t) \sum_{k=s}^{m} e^{-itk} \, dt$$

$$= \frac{1}{2\pi} \int_{-\pi}^{\pi} \widehat{M}(t) \frac{e^{-its} - e^{-it(m+1)}}{1 - e^{-it}} \, dt. \tag{4.3}$$

© Springer International Publishing Switzerland 2016
V. Čekanavičius, *Approximation Methods in Probability Theory*, Universitext,
DOI 10.1007/978-3-319-34072-2_4

The Riemann-Lebesgue theorem states that, for an absolutely integrable function $g(t)$,

$$\lim_{y \to \pm\infty} \int_{-\infty}^{\infty} e^{ity} g(t) \, dt = 0.$$

Therefore the limit of (4.3) when $s \to -\infty$ gives the following inversion formula

$$M\{(-\infty, m]\} = \frac{1}{2\pi} \int_{-\pi}^{\pi} \widehat{M}(t) \frac{e^{-it(m+1)}}{e^{-it} - 1} \, dt. \tag{4.4}$$

Consequently,

$$|M\{(-\infty, m]\}| \leq \frac{1}{2\pi} \int_{-\pi}^{\pi} \frac{|\widehat{M}(t)|}{|\sin \frac{t}{2}|} \, dt.$$

The Tsaregradskii inequality is obtained by taking the supremum over all $m \in \mathbb{Z}$. $\square$

Version (4.2) is more convenient if combined with an expansion in powers of (it), that is, with (1.12). Meanwhile, (4.1) is more convenient for an expansion in powers of $(e^{it} - 1)$, that is, with (1.16).

Example 4.1 Without any difficulty we can get an integral estimate for the difference of two distributions satisfying Franken's condition (3.13). Let us assume that $F, G \in \mathcal{F}$ are concentrated on nonnegative integers and have s finite moments ($s > 1$ some fixed integer) and let $\nu_k(F) = \nu_k(G)$, ($k = 1, 2, \ldots, s - 1$),

$$\lambda = \min\{\nu_1(F) - \nu_1^2(F) - \nu_2(F), \nu_1(G) - \nu_1^2(G) - \nu_2(G)\} > 0.$$

Then, as demonstrated in the proof of Theorem 3.3

$$|\widehat{F}(t) - \widehat{G}(t)| \leq C(s)[\nu_s(F) + \nu_s(G)] \left| \sin \frac{t}{2} \right|^s,$$

$$\max\{|\widehat{F}(t)|, |\widehat{G}(t)|\} \leq \exp\left\{-2\lambda \sin^2 \frac{t}{2}\right\},$$

$$|\widehat{F}^n(t) - \widehat{G}^n(t)| \leq n \max(|\widehat{F}(t)|^{n-1}, |\widehat{G}(t)|^{n-1})|\widehat{F}(t) - \widehat{G}(t)|.$$

Consequently, applying (1.31) and (4.1) we get

$$|F^n - G^n|_K \leq C(s)n[\nu_s(F) + \nu_s(G)] \int_{-\pi}^{\pi} \left| \sin \frac{t}{2} \right|^{s-1} \exp\left\{-2(n-1)\lambda \sin^2 \frac{t}{2}\right\}$$

$$\leqslant C(s)n[\nu_s(F) + \nu_s(G)] \int\limits_{-\pi}^{\pi} \left| \sin \frac{t}{2} \right|^{s-1} \exp\left\{ -2n\lambda \sin^2 \frac{t}{2} \right\} dt$$

$$\leqslant C(s)\lambda^{-s/2}n^{-(s-2)/2}.$$

Application of the Tsaregradskii inequality is practically the same as that of its local counterpart (3.2).

> **Typical application.** We use the same methods as apply to the local lattice estimates, combining them with (4.1).
>
> **Advantages.** The method is not complicated. It can be applied for distributions of dependent summands.
>
> **Drawbacks.** Usually the same order of accuracy can be proved for the stronger total variation norm.

Note that, for $M \in \mathcal{M}_Z$ we can also switch to other metrics by applying $|M|_K \leqslant \|M\|$ or $|(I_1 - I)M|_K = \|M\|_\infty$.

4.2   The Second Order Poisson Approximation

We demonstrate how asymptotic expansions can be estimated, rewriting them in a form convenient for the application of (4.1). In this section, for the sake of brevity, we use the notation $z := e^{it} - 1$. Formally, we can write the following series:

$$(1 + pz)^n = \exp\{n \ln(1 + pz)\} = \exp\left\{ npz - n\frac{(pz)^2}{2} + n\frac{(pz)^3}{3} - \ldots \right\}$$

$$= e^{npz}\left(1 - n\frac{(pz)^2}{2} + \ldots \right).$$

Therefore it is natural to construct a short asymptotic expansion for the Poisson approximation to the binomial law in the following way:

$$D := \exp\{np(I_1 - I)\}\left(I - \frac{np^2}{2}(I_1 - I)^2 \right).$$

Theorem 4.1 Let $0 \leqslant p \leqslant 1/2, n \in \mathbb{N}$. Then

$$\left| ((1 - p)I + pI_1)^n - D \right|_K \leqslant C \min\{np^3, p^2\}.$$

Proof We apply (4.1). We need an estimate for the difference of Fourier transforms of corresponding measures. The following technical trick might be quite effective, especially when minimizing constants (we will not). Observe that

$$\left| ((1-p) + pe^{it})^n - \widehat{D}(t) \right| = \left| (1+pz)^n - e^{npz}\left(1 - n\frac{(pz)^2}{2}\right) \right|$$

$$= \left| e^{npz} \right| \left| \left((1+p\tau z)^n e^{-np\tau z} - \left(1 - n\frac{(p\tau z)^2}{2}\right) \right) \Big|_0^1 \right|$$

$$= \left| e^{npz} \right| \left| \int_0^1 \frac{\partial}{\partial \tau}\left((1+p\tau z)^n e^{-np\tau z} - \left(1 - n\frac{(p\tau z)^2}{2}\right) \right) d\tau \right|$$

$$= \left| e^{npz} \right| \left| \int_0^1 \left(n(1+p\tau z)^{n-1} pz e^{-np\tau z} - npz(1+p\tau z)^n e^{-np\tau z} + np^2 z^2 \tau \right) d\tau \right|$$

$$\leq \left| e^{npz} \right| \int_0^1 np^2 |z|^2 \tau \left| (1+p\tau z)^{n-1} pz e^{-np\tau z} - 1 \right| d\tau. \tag{4.5}$$

Similarly to the proof of (3.8) we prove that

$$|1+pz\tau| \leq \exp\left\{-2p\tau(1-p\tau)\sin^2\frac{t}{2}\right\}.$$

Due to (3.4)

$$|e^{-pz\tau}| = \exp\{-p\tau(\cos t - 1)\} = \exp\left\{2p\tau \sin^2\frac{t}{2}\right\}.$$

Therefore

$$|1+pz\tau||e^{-pz\tau}| \leq \exp\left\{2p^2\tau^2 \sin^2\frac{t}{2}\right\}$$

and, noting that $0 \leq \tau \leq 1$,

$$|1+pz\tau|^{n-1} |e^{-pz\tau}|^{n-1} \leq \exp\left\{2(n-1)p^2\tau^2 \sin^2\frac{t}{2}\right\}$$

$$\leq \exp\left\{2np^2\tau^2 \sin^2\frac{t}{2}\right\} \exp\{2p^2\tau^2\}$$

$$\leq \sqrt{e}\exp\left\{2np^2\tau^2 \sin^2\frac{t}{2}\right\}. \tag{4.6}$$

Applying Taylor's expansion (see (1.33)) we obtain

$$
\begin{aligned}
(1 + pz\tau)e^{-pz\tau} - 1 &= (1 + pz\tau)(1 - pz\tau + p^2|z|^2\tau^2\theta e^{p|z|\tau}) - 1 \\
&= 1 - p^2z^2\tau^2 + p^2|z|^2\tau^2\theta C + p^3|z|^3\tau^3\theta C - 1 \\
&= -p^2z^2\tau^2 + p^2|z|^2\theta C = p^2|z|^2\theta C.
\end{aligned}
\tag{4.7}
$$

Taking into account (4.6) and (4.7), we prove

$$
\left| (1 + p\tau z)^{n-1}pze^{-np\tau z} - 1 \right|
$$

$$
\leq \left| (1 + p\tau z)^{n-1}pze^{-np\tau z} - (1 + pz\tau)^n e^{-np\tau z} \right| + \left| (1 + p\tau z)^n pze^{-np\tau z} - 1 \right|
$$

$$
\leq |1 + pz\tau|^{n-1}|e^{-np z\tau}||z|p\tau
$$

$$
+ n\max\{1, |1 + pz\tau|^{n-1}|e^{-pz\tau}|^{n-1}\}|(1 + pz\tau)e^{-pz\tau} - 1|
$$

$$
\leq C(|z|p + np^2|z|^2)\exp\left\{2np^2\tau^2\sin^2\frac{t}{2}\right\}
$$

$$
\leq C(|z|p + np^2|z|^2)\exp\left\{2np^2\sin^2\frac{t}{2}\right\}.
$$

Substituting the last estimate into (4.5) we get

$$
\left| ((1-p) + pe^{it})^n - \widehat{D}(t) \right| \leq C(|z|^3np^3 + n^2p^4|z|^4)\exp\left\{2np^2\sin^2\frac{t}{2}\right\}
$$

$$
\times \exp\left\{-2np\sin^2\frac{t}{2}\right\}
$$

$$
= C(|z|^3np^3 + n^2p^4|z|^4)\exp\left\{-2np(1-p)\sin^2\frac{t}{2}\right\}.
\tag{4.8}
$$

Let $np \leq 1$. Then, recalling that $|z| = 2|\sin(t/2)|$, (see (1.28)) we have

$$
\left| ((1-p) + pe^{it})^n - \widehat{D}(t) \right| \leq Cnp^3|\sin(t/2)|(|\sin(t/2)|^2 + np|\sin(t/2)|^2)
$$

$$
\leq Cnp^3|\sin(t/2)|.
$$

It remains to apply (4.1) and (1.31) to the last estimate and (4.8). □

Remark 4.1 There exist other methods for estimation of the difference of Fourier transforms. For example, one can use

$$\left| ((1-p) + pe^{it})^n - \widehat{D}(t) \right| = \left| e^{npz} \int_0^1 \frac{(1-\tau)^2}{2} \frac{\partial^3}{\partial \tau^3} \left(e^{-npz\tau} (1 + pz\tau)^n \right) d\tau \right|$$

or Bergström's expansion (1.37) with $s = 1$ and the triangle inequality.

4.3   Taking into Account Symmetry

When a distribution is concentrated not only on nonnegative integers, it is necessary to take into account its possible symmetry. In this section, we demonstrate how to arrange suitably the difference of characteristic functions for symmetric random variables. Let $F = p_0 I + p_1 I_1 + p_2 I_{-1}$, $G = \exp\{F - I\}$.

Theorem 4.2 *Let* $p_0, p_1, p_2 \in (0, 1)$, $n \in \mathbb{N}$. *Then*

$$|F^n - G^n|_K \leqslant C p_0^{-1} (1 - p_0)^{-1} (p_1 - p_2)^2 + p_0^{-2} (1 - p_0)^{-2} p_2^2 n^{-1}.$$

Proof Observe that by Lemma 3.2

$$\max\{|\widehat{F}(t)|, |\widehat{G}(t)|\} \leqslant \exp\{p_0((Re\,\widehat{F}(t) - 1))\} = \exp\left\{-2p_0(1 - p_0) \sin^2 \frac{t}{2}\right\}.$$

Moreover,

$$|\widehat{F}(t) - \widehat{G}(t)| \leqslant C|\widehat{F}(t) - 1|^2 = C|p_1(e^{it} - 1) + p_2(e^{-it} - 1)|^2.$$

We can use the estimate

$$|p_1(e^{it} - 1) + p_2(e^{-it} - 1)| \leqslant p_1|e^{it} - 1| + p_2|e^{-it} - 1| = 2(1 - p_0) \sin^2 \frac{t}{2}.$$

Regrettably, such an estimate might be too rough, because we have not taken into account any possible symmetry of the distribution F. It is possible to solve this problem by using expansions of e^{it} and e^{-it} in powers of (it). We will take a slightly different approach. From (1.29) we have

$$|p_1(e^{it} - 1) + p_2(e^{-it} - 1)| = |p_1(e^{it} - 1) - p_2(e^{it} - 1) - p_2(e^{it} - 1)(e^{-it} - 1)|$$

$$\leqslant |p_1 - p_2| \left| \sin \frac{t}{2} \right| + p_2 \sin^2 \frac{t}{2},$$

which gives

$$|\widehat{F}^n(t) - \widehat{G}^n(t)| \leqslant Cn \exp\left\{-2np_0(1-p_0)\sin^2\frac{t}{2}\right\}|\widehat{F}(t) - \widehat{G}(t)|$$

$$\leqslant Cn \exp\left\{-2np_0(1-p_0)\sin^2\frac{t}{2}\right\}$$

$$\times C[n(p_1-p_2)^2\sin^2(t/2) + np_2^2\sin^4(t/2)].$$

It remains to apply (4.1) and (1.31).   □

Observe that if $p_1 = p_2$ and $p_0 = const$ then the accuracy of approximation is of the order $O(n^{-1})$.

4.4 Problems

4.1 Let $0 \leqslant p \leqslant 1/2$. Prove that

$$\left|((1-p)I + pI_1)^n - \exp\left\{np(I_1-I) - \frac{np^2}{2}(I_1-I)^2\right\}\right|_K \leqslant C\min\left(np^3, p\sqrt{\frac{p}{n}}\right).$$

4.2 Let F be the binomial distribution with parameters $n \in \mathbb{N}$ and $0 < p < 1$. Prove that

$$|F^n - F^{n+1}|_K \leqslant C\sqrt{\frac{p}{n(1-p)}}.$$

4.3 Let $F, G \in \mathcal{F}_Z$ and $b = \sum_{k\in\mathbb{Z}} k^2 F\{k\} < \infty$, $a = G\{1\} + G\{-1\} > 0$. Prove that, for any $\lambda > 0$, $k \in \mathbb{N}$,

$$|(F-I)^k \exp\{\lambda(G-I)\}|_K \leqslant C\left(\frac{b}{a\lambda}\right)^{k/2}.$$

4.4 Let $F = 0.5(I_1 + I)$, $H = \sum_{j=0}^{\infty} 0.5^{j+1} F^j$, $n \in \mathbb{N}$. Prove that

$$|H^n - \exp\{n(F-I)\}|_K \leqslant Cn^{-1}.$$

4.5 Let F, G be concentrated on $\mathbb{Z}_+$, $v_k(F) = v_k(G)$, $k = 1, 2, 3$, $a = v_4(F) + v_4(G) < \infty$, $b = v_1(F) - v_2(F) > 0$. Prove that, for any $n \in \mathbb{N}$,

$$|\exp\{n(F-I)\} - \exp\{n(G-I)\}|_K \leqslant \frac{Ca}{nb^2}.$$

4.6 Let F be the negative binomial distribution with parameters 0.2 and 16, that is

$$\widehat{F}(t) = \left(\frac{0.8}{1 - 0.2e^{it}} \right)^{16}$$

and let $G = \exp\{3(I_1 - I) + 0.5(I_2 - I)\}$. Prove, that for $n \in \mathbb{N}$

$$| F^n - G^n |_K \leq \frac{C}{\sqrt{n}}.$$

4.7 Let F, G be concentrated on $\mathbb{Z}_+$, $v_k(F) = v_k(G)$, $k = 1, 2, \ldots, s - 1$, $a = v_s(F) + v_s(G) < \infty$, $s \geq 3$, $b = v_1(F) - v_2(F) > 0$. Let $CNB(\gamma, 0.2, F)$ and $CNB(\gamma, 0.2, G)$ be defined by (1.5) with some $\gamma > 0$. Prove that

$$| CNB(\gamma, 0.2, F) - CNB(\gamma, 0.2, G) |_K \leq \frac{C(s)a}{\gamma^{s/2-1}b^{s/2}}.$$

Bibliographical Notes

The Tsaregradkii inequality [153] is considered to be a standard technique. We note only [63, 68, 89] as the main sources for this section's results.

Chapter 5
Total Variation of Lattice Measures

We recall that, for $M \in \mathcal{M}$, $|M|_K \leqslant \|M\|$. One of our aims is to show that many estimates in total variation have the same order of accuracy as for the Kolmogorov norm. In this chapter, we do not consider the Stein method, which is presented in Chap. 11.

5.1 Inversion Inequalities

The characteristic function method for total variation is based on a suitable inversion formula, allowing us to switch from measures to their Fourier transforms.

Lemma 5.1 *Let $M \in \mathcal{M}_{\mathbb{Z}}$, $\sum_{k \in \mathbb{Z}} |k| \, |M\{k\}| < \infty$. Then, for any $a \in \mathbb{R}$, $b > 0$ the following inequality holds*

$$\|M\| \leqslant (1 + b\pi)^{1/2} \left(\frac{1}{2\pi} \int_{-\pi}^{\pi} \left(\left| \widehat{M}(t) \right|^2 + \frac{1}{b^2} \left| \left(e^{-ita} \widehat{M}(t) \right)' \right|^2 \right) dt \right)^{1/2}. \qquad (5.1)$$

Proof We begin with the following identity

$$\sum_{k=-\infty}^{\infty} (k - a)^2 |M\{k\}|^2 = \frac{1}{2\pi} \int_{-\pi}^{\pi} |(e^{-ia} \widehat{M}(t))'|^2 \, dt. \qquad (5.2)$$

Indeed, due to the lemma's assumptions, we can calculate the derivative of $\widehat{M}(t)$:

$$\left(\widehat{M}(t) e^{-ita} \right)' = \sum_{k \in \mathbb{Z}} M\{k\} \left(e^{it(k-a)} \right)' = i \sum_{k \in \mathbb{Z}} M\{k\} (k-a) e^{it(k-a)} = i e^{-ita} \sum_{k \in \mathbb{Z}} M\{k\} (k-a) e^{itk}.$$

© Springer International Publishing Switzerland 2016
V. Čekanavičius, *Approximation Methods in Probability Theory*, Universitext,
DOI 10.1007/978-3-319-34072-2_5

Consequently, defining $V\{k\} = (k - a)M\{k\}$, we have

$$\widehat{V}(t) = \sum_{k \in \mathbb{Z}} (k - a)M\{k\}e^{itk} = -ie^{ita}\left(\widehat{M}(t)e^{-ita}\right)'$$

and can apply (3.25) to the measure V.

Applying Hölder's inequality, (3.25), (5.2) and (1.46) we then obtain

$$\| M \|^2 = \left(\sum_{k \in \mathbb{Z}} |M\{k\}| \left(1 + \left(\frac{k-a}{b}\right)^2\right)^{1/2} \left(1 + \left(\frac{k-a}{b}\right)^2\right)^{-1/2}\right)^2$$

$$\leqslant \sum_{k \in \mathbb{Z}} |M\{k\}|^2 \left(1 + \left(\frac{k-a}{b}\right)^2\right) \sum_{k \in \mathbb{Z}} \left(1 + \left(\frac{k-a}{b}\right)^2\right)^{-1}$$

$$\leqslant (1 + b\pi)\frac{1}{2\pi}\int_{-\pi}^{\pi}\left(\left|\widehat{M}(t)\right|^2 + \frac{1}{b^2}\left|\left(e^{-ita}\widehat{M}(t)\right)'\right|^2\right) dt.$$

$\square$

What role do a and b play in (5.1)? We give some heuristic answers. The characteristic function method for lattice measures means that, estimating the Fourier transform of the measure, we get a rough impression about the uniform estimate. Moreover, in quite a lot of cases, we encounter the following principal scheme: if $|\widehat{M}(t)| \approx A\lambda^{-n}$, then $\| M \| \approx A\lambda^{-n}$ and

$$\int_{-\pi}^{\pi} |\widehat{M}(t)| \, dt \approx A\lambda^{-n-1/2} \quad \text{and} \quad \int_{-\pi}^{\pi} |\widehat{M}(t)|/|t| \, dt \approx A\lambda^{-n}.$$

The estimate in total variation cannot be better than the uniform estimate. Considering (5.1) we see that the estimate contains two parts: an integral of the Fourier transform and an integral of its derivative. In both integrals b acts differently. Taking $b = \lambda^{1/2}$ we will preserve the order λ^{-n} for the first integral and improve the order of the second integral.

The role of a can be explained in the following way. Any additional factor $|\sin(t/2)|$ improves the accuracy of approximation by $\lambda^{-1/2}$. As can be seen from the following example, suitable centering can radically improve the accuracy. Indeed,

$$\left|\left(e^{p(e^{it}-1)}\right)'\right| = \left|e^{p(e^{it}-1)}ipe^{it}\right| = pe^{-2p\sin^2\frac{t}{2}}$$

and we get no additional improvement. On the other hand,

$$\left|\left(e^{-itp}e^{p(e^{it}-1)}\right)'\right| = \left|e^{p(e^{it}-1-it)}ip(e^{it}-1)\right| = 2p\left|\sin^2\frac{t}{2}\right|e^{-2p\sin^2\frac{t}{2}}$$

and we get an improvement of the order $\lambda^{-1/2}$. As a rule, a is equal to the mean of the approximated distribution.

The estimate given in Lemma 5.1 is not unique. The idea of derivatives can be carried further.

Lemma 5.2 *Let $M \in \mathcal{M}_Z$, $\sum_{k\in\mathbb{Z}} k^2 |M\{k\}| < \infty$, $a \in \mathbb{R}$, $b > 0$. Then*

$$\| M \| \leqslant \frac{1 + b\pi}{2b\pi} \int_{|t| \leqslant b\pi} \left| v(t) - v''(t) \right| dt.$$

Here $v(t) = \exp\{-ita/b\}\widehat{M}(t/b)$.

Proof A simple calculation shows that

$$v''(t) = \left(\exp\{-ita/b\} \sum_{k\in\mathbb{Z}} M\{k\} \exp\{kit/b\} \right)'' = -\frac{(k-a)^2}{b^2} \sum_{k\in\mathbb{Z}} M\{k\} \exp\{it(k-a)/b\}.$$

Therefore

$$v(t) - v''(t) = \sum_{k\in\mathbb{Z}} \left(1 + \frac{(k-a)^2}{b^2} \right) M\{k\} \exp\{it(k-a)/b\}.$$

Multiplying both sides by $\exp\{-it(j-a)/b\}$ and integrating with respect to t in the interval $[-b\pi, b\pi]$ similar to the proof of (3.1) we get

$$\left(1 + \frac{(j-a)^2}{b^2} \right) M\{j\} = \frac{1}{2b\pi} \int_{|t| \leqslant b\pi} (v(t) - v''(t)) \exp\{-it(j-a)/b\} dt.$$

Therefore

$$|M\{j\}| \leqslant \left(1 + \frac{(j-a)^2}{b^2} \right)^{-1} \frac{1}{2b\pi} \int_{|t| \leqslant b\pi} |v(t) - v''(t)| dt.$$

Summing the last estimate by j and applying (1.46) we complete the proof of Lemma 5.2. $\qquad\square$

5.2 Examples of Applications

First we discuss a typical approach to the application of Lemma 5.1. Let us assume that for $j = 1, \ldots, n$, some $s \in \mathbb{N}$ and all $t \in \mathbb{R}$:

(a) $F_j, G_j \in \mathcal{F}_Z$,
(b) $\max(|\widehat{F}_j(t)|, |\widehat{G}_j(t)|) \leqslant \exp\{-\lambda_j \sin^2(t/2)\}$, $0 < \lambda_j < C_1$.

(c) $|\widehat{F}_j(t) - \widehat{G}_j(t)| \leqslant C(s)\beta_j| \sin(t/2)|^{s+1}$.

(d) $|\widehat{F}'_j(t) - \widehat{G}'_j(t)| \leqslant C(s)\beta_j| \sin(t/2)|^{s}$.

(d) $\sum_{k\in\mathbb{Z}} kF_j\{k\} = \sum_{k\in\mathbb{Z}} kG_j\{k\}$.

Observe that, in conditions (b)–(c), expression $|\sin(t/2)|$ can be replaced by $|t|$ if $|t| \leqslant \pi$. Indeed, then $|t|/\pi \leqslant |\sin(t/2)| \leqslant |t|/2$. Note also that all λ_j are bounded by *the same* absolute constant C_1.

To make the notation shorter let

$$\mu_j = \sum_{k\in\mathbb{Z}} kF_j\{k\}, \quad \lambda = \sum_{j=1}^{n} \lambda_j,$$

$$\sigma_j^2 = \max\left(\sum_{k\in\mathbb{Z}}(k - \mu_j)^2 F_j\{k\}, \sum_{k\in\mathbb{Z}}(k - \mu_j)^2 G_j\{k\}\right), \quad \sigma^2 = \sum_{j=1}^{n} \sigma_j^2.$$

The following proposition shows how Lemma 5.1 can be used for the estimation of n-fold convolutions of distributions.

Proposition 5.1 *Let assumptions (a)–(d) hold and let $\lambda \geqslant 1$. Then*

$$\left\| \prod_{j=1}^{n} F_j - \prod_{j=1}^{n} G_j \right\| \leqslant C(s)\frac{1}{\lambda^{(s+1)/2}} \sum_{j=1}^{n} \beta_j\left(1 + \frac{\max_{1\leqslant j\leqslant n}|\mu_j|}{\sqrt{\lambda}} + \frac{\sigma^2}{\lambda}\right). \qquad (5.3)$$

Proof Further on we can assume that all $\sigma_j^2 < \infty$, since otherwise (5.3) is trivial. Later in the proof we assume that $|t| \leqslant \pi$. Let, for the sake of brevity,

$$f_j = \widehat{F}_j(t) \exp\{-it\mu_j\}, \quad g_j = \widehat{G}_j(t) \exp\{-it\mu_j\}, \quad \widehat{M}(t) = \prod_{j=1}^{n} f_j - \prod_{j=1}^{n} g_j.$$

Note that $|f_j| = |\widehat{F}_j(t)|$ and $|g_j| = |\widehat{G}_j(t)|$ and $|f_j - g_j| = |\widehat{F}_j(t) - \widehat{G}_j(t)|$. Taking into account (a)–(d) and (1.48), we prove

$$|\widehat{M}(t)| \leqslant \sum_{j=1}^{n} |f_j - g_j| \prod_{k=1}^{j-1} |f_k| \prod_{k=j+1}^{n} |g_k|$$

$$\leqslant C(s) \sum_{j=1}^{n} \beta_j| \sin(t/2)|^{s+1} \prod_{l\neq j} \exp\{-\lambda_l \sin^2(t/2)\}$$

$$\leqslant C(s) \sum_{j=1}^{n} \beta_j| \sin(t/2)|^{s+1} \exp\{-\lambda \sin^2(t/2)\} \exp\{\lambda_j \sin^2(t/2)\}$$

$$\leq C(s)\exp\{C_1\}\sum_{j=1}^{n}\beta_j|\sin(t/2)|^{s+1}\exp\left\{-\frac{\lambda}{2}\sin^2\frac{t}{2}\right\}\exp\left\{-\frac{\lambda}{2}\sin^2\frac{t}{2}\right\}$$

$$\leq C(s)\sum_{j=1}^{n}\beta_j\lambda^{-(s+1)/2}\exp\left\{-\frac{\lambda}{2}\sin^2\frac{t}{2}\right\}. \tag{5.4}$$

Similarly,

$$|\widehat{M}'(t)|\leq\sum_{j=1}^{n}|f_j'-g_j'|\prod_{k=1}^{j-1}|f_k|\prod_{k=j+1}^{n}|g_k|+\sum_{j=1}^{n}|f_j-g_j|\sum_{k=1}^{j-1}|f_k'|\prod_{l\neq k}^{j-1}|f_l|\prod_{k=j+1}^{n}|g_k|$$

$$+\sum_{j=1}^{n}|f_j-g_j|\prod_{k=1}^{j-1}|f_k|\sum_{k=j+1}^{n}|g_k'|\prod_{l\neq k,j+1}^{n}|g_l|$$

$$\leq C\exp\{-\lambda\sin^2(t/2)\}\Bigg[\exp\{C_1\}\sum_{j=1}^{n}|f_j'-g_j'|$$

$$+\exp\{2C_1\}\sum_{j=1}^{n}|f_j-g_j|\left(\sum_{k=1}^{j-1}|f_k'|+\sum_{k=j+1}^{n}|g_k'|\right)\Bigg]$$

$$\leq C\exp\{-\lambda\sin^2(t/2)\}\Bigg[\sum_{j=1}^{n}|f_j'-g_j'|+\sum_{j=1}^{n}|f_j-g_j|\sum_{k=1}^{n}(|f_k'|+|g_k'|)\Bigg].$$

Next observe that, due to (1.32) and (1.50),

$$|\exp\{it(k-\mu_j)\}-1|\leq|k-\mu_j||t|\leq|k-\mu_j|\pi|\sin(t/2)|.$$

Therefore, due to (d),

$$|f_j'|=\left|\left(\sum_{k\in\mathbb{Z}}F_j\{k\}\exp\{it(k-\mu_j)\}\right)'\right|=\left|i\sum_{k\in\mathbb{Z}}(k-\mu_j)F\{k\}\exp\{it(k-\mu_j)\}\right|$$

$$=\left|\sum_{k\in\mathbb{Z}}(k-\mu_j)F_j\{k\}+\sum_{k\in\mathbb{Z}}(k-\mu_j)F_j\{k\}(\exp\{it(k-\mu_j)\}-1)\right|$$

$$=\left|\sum_{k\in\mathbb{Z}}(k-\mu_j)F_j\{k\}(\exp\{it(k-\mu_j)\}-1)\right|$$

$$\leq\pi\sum_{k\in\mathbb{Z}}(k-\mu_j)^2F_j\{k\}|\sin(t/2)|\leq\pi\sigma_j^2|\sin(t/2)|.$$

Similarly, $|g_j'| \leq \pi\sigma_j^2 |\sin(t/2)|$. Due to (c) and (d),

$$
\begin{aligned}
|f_j' - g_j'| &= \left| -i\mu_j \exp\{-it\mu_j\}(\widehat{F}_j(t) - \widehat{G}_j(t)) + \exp\{-it\mu_j\}(\widehat{F}_j'(t) - \widehat{G}_j'(t)) \right| \\
&\leq |\mu_j| |\widehat{F}_j(t) - \widehat{G}_j(t)| + |\widehat{F}_j'(t) - \widehat{G}_j'(t)| \\
&\leq C(s)(\beta_j |\mu_j| |\sin(t/2)|^{s+1} + \beta_j |\sin(t/2)|^s).
\end{aligned}
$$

Combining the above estimates we obtain

$$
\begin{aligned}
|\widehat{M}'(t)| &\leq C(s) \exp\{-\lambda \sin^2(t/2)\} \Bigg[|\sin(t/2)|^{s+1} \sum_{j=1}^{n} |\mu_j|\beta_j + |\sin(t/2)|^s \sum_{j=1}^{n} \beta_j \\
&\quad + |\sin(t/2)|^{s+2}\sigma^2 \sum_{j=1}^{n} \beta_j \Bigg] \leq C(s) \exp\left\{ -\frac{\lambda}{2} \sin^2(t/2) \right\} \\
&\quad \times \Bigg[\lambda^{-(s+1)/2} \max_{1 \leq j \leq n} |\mu_j| \sum_{j=1}^{n} \beta_j + \lambda^{-s/2} \sum_{j=1}^{n} \beta_j + \lambda^{-(s+2)/2}\sigma^2 \sum_{j=1}^{n} \beta_j \Bigg] \\
&\leq C(s) \exp\left\{ -\frac{\lambda}{2} \sin^2(t/2) \right\} \lambda^{-s/2} \sum_{j=1}^{n} \beta_j \left(1 + \frac{\max_{1 \leq j \leq n} |\mu_j|}{\sqrt{\lambda}} + \frac{\sigma^2}{\lambda} \right). \quad (5.5)
\end{aligned}
$$

Let

$$
A := \lambda^{-(s+1)/2} \sum_{j=1}^{n} \beta_j \left(1 + \frac{\max_{1 \leq j \leq n} |\mu_j|}{\sqrt{\lambda}} + \frac{\sigma^2}{\lambda} \right).
$$

Then from (5.4) and (5.5) it follows that

$$
|\widehat{M}(t)| \leq C(s)A \exp\left\{ -\frac{\lambda}{2} \sin^2 \frac{t}{2} \right\}, \qquad |\widehat{M}'(t)| \leq \sqrt{\lambda} C(s)A \exp\left\{ -\frac{\lambda}{2} \sin^2 \frac{t}{2} \right\}.
$$

We apply Lemma 5.1 with $a = \sum_{j=1}^{n} \mu_j$ and $b = \sqrt{\lambda}$. Then, taking into account (1.31), we prove

$$
\left\| \prod_{j=1}^{n} F_j - \prod_{j=1}^{n} G_j \right\|^2 \leq C(s)(1 + \sqrt{\lambda}\pi) \int_{-\pi}^{\pi} \left(|\widehat{M}(t)|^2 + \frac{1}{\lambda} |\widehat{M}'(t)|^2 \right) dt
$$

$$
\leq C(s)\sqrt{\lambda}A^2 \int_{-\pi}^{\pi} \exp\{-\lambda \sin^2(t/2)\}\, dt \leq C(s)A^2.
$$

$\square$

Typical application. We use (5.3) or other estimates for characteristic functions, combining them with (5.1) or Lemma 5.2.
Advantages. The assumptions are less restrictive in comparison to the ones needed for the method of convolutions. The method can be applied for measures concentrated on the whole of $\mathbb{Z}$, not only on $\mathbb{Z}_+$. The method can be applied for distributions of dependent summands with explicit characteristic functions.
Drawbacks. The constants of the estimates are not small. Requires lengthy routine calculations when estimating the derivatives of characteristic functions.

Example 5.1 We estimate the closeness of distributions under Franken's condition. Let F and G be concentrated on $\mathbb{Z}_+$, $s \in \mathbb{N}$, $\beta_{s+1} := \nu_{s+1}(F) + \nu_{s+1}(G) < \infty$ and $\nu_k(F) = \nu_k(G)$, for $k = 1, 2, \ldots, s$ and let

$$\lambda(F, G) = \min\{\nu_1(F) - \nu_1^2(F) - \nu_2(F), \nu_1(G) - \nu_1^2(G) - \nu_2(G)\} > 0. \qquad (5.6)$$

Here $\nu_k(F)$ is k-th factorial moment, see (1.15). Then, from (1.16) and (1.17) it follows that

$$|\widehat{F}(t) - \widehat{G}(t)| \leqslant C(s)\beta_{s+1}\left| \sin \frac{t}{2} \right|^{s+1}, \quad |\widehat{F}'(t) - \widehat{G}'(t)| \leqslant C(s)\beta_{s+1}\left| \sin \frac{t}{2} \right|^{s}.$$

Moreover, by (3.14) we have $\max(|\widehat{F}(t)|, |\widehat{G}(t)|) \leqslant \exp\{-2\lambda(F, G) \sin^2(t/2)\}$.

Let $n\lambda > 1$. We apply Proposition 5.1. Observe that, due to (5.6), $\nu_1(F) = \nu_1(G) < 1$ and $\nu_2(F) \leqslant \nu_1(F)$. Therefore $\nu_1(F)/\sqrt{n\lambda} \leqslant C$ and

$$\sum_{k=0}^{\infty}(k - \nu_1(F))^2 F\{k\} = \sum_{k=0}^{\infty} k^2 F\{k\} - \nu_1^2(F) = \nu_2(F) + \nu_1(F) - \nu_1^2(F) \leqslant 2\nu_1(F).$$

Applying (5.3) and noting that $\sigma^2 \leqslant 2n\nu_1(F)$, we obtain

$$\| F^n - G^n \| \leqslant \frac{C(s)\beta_{s+1}}{n^{(s-1)/2}\lambda(F, G)^{(s+1)/2}}\left(1 + \frac{\nu_1(F)}{\lambda(F, G)}\right).$$

Particularly,

$$\| ((1 - p)I + pI_1)^n - \exp\{np(I_1 - I)\} \| \leqslant Cp. \qquad (5.7)$$

Indeed, it suffices to prove (5.7) for $p \leqslant 1/4$ only, see the discussion after (2.19). Corresponding factorial moments are already given in Example 3.1.

Proposition 5.1 is not the only way to estimate total variation. Similarly to Sect. 3.5, properties of the concentration function can be also applied.

Theorem 5.1 *Let $F \in \mathcal{F}_{\mathbb{Z}} \cap \mathcal{F}_+$, $F\{0\} > 0$, and let F have a finite second moment. Then*

$$\| F^n - \exp\{n(F - I)\} \| \leqslant C(F)n^{-1}. \tag{5.8}$$

Proof For the proof of (5.8) noting that $\widehat{F}(t) \leqslant \exp\{\widehat{F}(t) - 1\}$ and, applying (1.48), we obtain

$$
\begin{aligned}
|\widehat{F}^n(t) - \exp\{n(\widehat{F}(t) - 1)\}| &\leqslant n\exp\{(n-1)(\widehat{F}(t) - 1)\}|\widehat{F}(t) - \exp\{\widehat{F}(t) - 1\}| \\
&\leqslant Cn\exp\{(n-1)(\widehat{F}(t)-1)\}(\widehat{F}(t)-1)^2 \leqslant e^2 Cn\exp\{n(\widehat{F}(t) - 1)\}(\widehat{F}(t) - 1)^2 \\
&\leqslant Cn^{-1}\exp\{n(\widehat{F}(t) - 1)/2\}.
\end{aligned}
$$

Set $p_k = 2F\{k\}$. Then $\widehat{F}(t) = \sum_j p_j \cos(tj)$ and

$$|\widehat{F}'(t)| = \Big| \sum_{j=0}^{\infty} jp_j \sin(tj) \Big| \leqslant \Big(\sum_{j=0}^{\infty} jp_j \Big)^{1/2} \Big(\sum_{k=0}^{\infty} p_k \sin^2(tk) \Big)^{1/2}$$

$$= \Big(\sum_{j=0}^{\infty} jp_j \Big)^{1/2} \Big(4\sum_{k=0}^{\infty} p_k \sin^2 \frac{tk}{2} \cos^2 \frac{tk}{2} \Big)^{1/2}$$

$$\leqslant 2\Big(\sum_{j=0}^{\infty} jp_j \Big)^{1/2} \Big(\sum_{k=0}^{\infty} p_k \sin^2 \frac{tk}{2} \Big)^{1/2} = C(F)(1 - \widehat{F}(t))^{1/2}.$$

Therefore

$$
\begin{aligned}
|(\widehat{F}^n(t) - \exp\{n(\widehat{F}(t) - 1)\})'| &\leqslant |n\widehat{F}^{n-1}(t)\widehat{F}'(t) - n\widehat{F}'(t)\exp\{n(\widehat{F}(t) - 1)\}| \\
&= n|\widehat{F}'(t)||\widehat{F}^{n-1}(t) - \widehat{F}^n(t) + \widehat{F}^n(t) - \exp\{n(\widehat{F}(t) - 1)\}| \\
&\leqslant n|\widehat{F}'(t)|[|\widehat{F}(t)|^{n-1}|1 - \widehat{F}(t)| + |\widehat{F}^n(t) - \exp\{n(\widehat{F}(t) - 1)\}|] \\
&\leqslant n(1 - \widehat{F}(t))^{1/2}\exp\{(n-1)(\widehat{F}(t) - 1)\}\Big((1 - \widehat{F}(t)) + (1 - \widehat{F}(t))^2\Big) \\
&\leqslant C(F)n^{-1/2}\exp\{n(\widehat{F}(t) - 1)/2\}.
\end{aligned}
$$

Taking $b = \sqrt{n}$ in Lemma 5.1 we prove

$$\| F^n - \exp\{n(F - I)\} \|^2 \leqslant C(F)n^{-2}\sqrt{n} \int_{-\pi}^{\pi} \exp\{n(\widehat{F}(t) - 1)\}\, dt.$$

To complete the proof of (5.8) one should note that by Lemma 3.3 we have

$$\int\limits_{-\pi}^{\pi} \exp\{n(\widehat{F}(t) - 1)\}\, dt \leqslant C(F)n^{-1/2}.$$

□

In (5.8), the dependence of $C(F)$ on characteristics of F is not given explicitly. Therefore (5.8) is meaningful only in the case when all $F\{k\}$ do not depend on n, i.e. when dealing with the scheme of sequences.

5.3   Smoothing Estimates for Symmetric Distributions

Instead of Lemma 5.1 one can apply the method of convolutions and smoothing estimates. Obviously, Lemma 2.3 holds. Therefore all approaches described in Chap. 2 can be used. Stronger smoothing results hold for symmetric distributions.

Lemma 5.3 *Let $j, n \in \mathbb{N}$, $p = 1 - q \in (0, 1)$, $\lambda > 0$ and $V \in \mathcal{F}_s$ be concentrated on $\{\pm 1, \pm 2, \dots\}$ with finite variance σ^2. Then*

$$\| (V - I)^j \exp\{\lambda(V - I)\} \| \leqslant 3.6 j^{1/4} \sqrt{1 + \sigma} \left(\frac{j}{te}\right)^j, \tag{5.9}$$

$$\| (V - I)^j (qI + pV)^n \| \leqslant 6.73 \sqrt{\sigma}\, \frac{j}{q^{1/4}} \left(\frac{j}{enpq}\right)^j. \tag{5.10}$$

If $V = 0.5I_{-1} + 0.5I_1$, then

$$\| (V - I)^j \exp\{\lambda(V - I)\} \| \leqslant \frac{j!}{\lambda^j}, \tag{5.11}$$

$$\| (V - I)^j (qI + pV)^n \| \leqslant \frac{j!}{((n + 1)pq)^j}. \tag{5.12}$$

Typical application. Methods of Sect. 2.5 are combined with Lemma 5.3.
Advantages. One can avoid estimation of characteristic functions and their derivatives.
Drawbacks. The method is not applicable for dependent random variables. Usually stronger assumptions on the parameters of distributions are needed than for estimation of their Fourier transforms, especially when Lemma 2.5 is applied.

We illustrate the method of convolutions by considering an accompanying approximation to a symmetric distribution. Let

$$H = (1 - 2p)I + pI_1 + pI_{-1}.$$

Theorem 5.2 *Let* $p \leqslant 1/8$, $n \in \mathbb{N}$. *Then*

$$\| H^n - \exp\{n(H - I)\} \| \leqslant \frac{C}{n}.$$

Proof Similarly to the proof of (3.28), we observe that

$$H = (1 - 2p)I + 2pV, \qquad V = \frac{1}{2}I_1 + \frac{1}{2}I_{-1}.$$

Repeating all the steps of the proof of (3.28) for the total variation norm, we obtain

$$\| H^n - \exp\{n(H - I)\} \| \leqslant Cnp^2 \left\| (V - I)^2 \exp\left\{ \frac{np}{4}(V - I) \right\} \right\|.$$

It remains to apply (5.11).   □

5.4 The Barbour-Xia Inequality

The smoothing inequalities of the previous section depend on the exponential or compound binomial structure of the distribution. As it turns out, a similar smoothing estimate holds for an arbitrary lattice distribution.

Lemma 5.4 *Let* $F_j \in \mathcal{F}_{\mathbb{Z}}$, $j = 1, 2, \ldots, n$. *Then*

$$\left\| (I_1 - I) \prod_{j=1}^{n} F_j \right\| \leqslant 2\sqrt{\frac{2}{\pi}} \left(\sum_{j=1}^{n} u_j \right)^{-1/2}. \tag{5.13}$$

Here

$$u_j = 1 - \frac{1}{2} \| (I_1 - I)F_j \|.$$

The proof of (5.13) employs some facts from the theory of random walks and is beyond the scope of our book. Estimate (5.13) is also known as the Mineka coupling inequality. It can also be formulated in terms of random variables. Let $S = \xi_1 + \xi_2 + \cdots + \xi_n$ be the sum of independent integer-valued random variables and let $\mathcal{L}(S_n)$

denote its distribution. Then

$$\left\| (I_1 - I) \prod_{j=1}^{n} F_j \right\| = \| \mathcal{L}(S_n + 1) - \mathcal{L}(S_n) \|, \quad \| (I_1 - I) F_j \| = \| \mathcal{L}(\xi_j + 1) - \mathcal{L}(\xi_j) \|.$$

It is not difficult to extend (5.13) to k-fold smoothing inequalities. Let $n, k \in \mathbb{N}$, $F \in \mathcal{F}_Z$. Then

$$\| (I_1 - I)^k F^n \| \leq C(k)(nu(F))^{-k/2}, \quad u(F) = 1 - \frac{1}{2} \| (I_1 - I) F \|. \tag{5.14}$$

Indeed, it suffices to assume $n/k > 2$. Let $m = \lfloor n/k \rfloor$ be the integer part of n/k. Then

$$\| (I_1 - I)^k F^n \| \leq \| (I_1 - I) F^m \|^k \leq C(mu(F))^{-k/2} \leq C(k)(nu(F))^{-k/2}.$$

Remark 5.1 Note that for a unimodal distribution F, we can use the following estimate:

$$u(F) \geq 1 - \max_k F\{k\}.$$

Indeed, let $F\{k\} \leq F\{k + 1\}$, for $k \leq m$; and let $F\{k\} \geq F\{k + 1\}$, for $k > m$. Then

$$\| F(I_1 - I) \| = \sum_{k=-\infty}^{\infty} | F\{k\} - F\{k - 1\} | \leq 2F\{m\} \leq 2 \max_k F\{k\}.$$

Remark 5.2 Combining the definition of $u(F)$ with (1.9) we obtain

$$u(F) = 1 - \frac{1}{2} \left(2 - 2 \sum_{k=-\infty}^{\infty} \min(F\{k\}, F\{k - 1\}) \right) = \sum_{k=-\infty}^{\infty} \min(F\{k\}, F\{k - 1\}). \tag{5.15}$$

The Barbour-Xia inequality allows us to formulate a very general estimate of the closeness of two lattice distributions.

Theorem 5.3 *Let F and G be distributions concentrated on nonnegative integers and let, for $s \geq 2$, $v_k(F) = v_k(G)$, $k = 1, \ldots, s - 1$, $v_s(F) + v_s(G) < \infty$ and $n \in \mathbb{N}$. Then*

$$\| F^n - G^n \| \leq C(s) n^{-(s-2)/2} [v_s(F) + v_s(G)]$$

$$\times \left(\sum_{k=-\infty}^{\infty} \min(F\{k\}, F\{k - 1\}) + \sum_{k=-\infty}^{\infty} \min(G\{k\}, G\{k - 1\}) \right)^{-s/2}.$$

Proof Without loss of generality we assume that $n > 2$. Indeed, otherwise, we make use of the fact that $\| F^n - G^n \| \leqslant \| F^n \| + \| G^n \| = 2$. From the equality of factorial moments we have

$$F - G = \Theta(I_1 - I)^s(v_s(F) + v_s(G))/s!,$$

see (2.2). Let $m = \lfloor n/2 \rfloor$. Then, applying (5.15), we obtain

$$\| F^n - G^n \| \leqslant \sum_{k=1}^{n} F^{k-1} G^{(n-k)}(F - G)$$

$$\leqslant C(s)[v_s(F) + v_s(G)] \sum_{k=1}^{n} \| F^{k-1} G^{(n-k)}(I_1 - I) \|$$

$$\leqslant C(s)[v_s(F) + v_s(G)]n[\| F^m (I_1 - I)^2 \| + \| G^m (I_1 - I)^s \|]$$

$$\leqslant C(s)[v_s(F) + v_s(G)]n$$

$$\times \left(n \sum_{k=-\infty}^{\infty} \min(F\{k\}, F\{k-1\}) + n \sum_{k=-\infty}^{\infty} \min(G\{k\}, G\{k-1\}) \right)^{-s/2}.$$

$$\square$$

Example 5.2 Let ξ be concentrated at 4 points and have the following distribution:

$$P(\xi = 0) = 20/45, \quad P(\xi = 1) = 18/45, \quad P(\xi = 3) = 5/45, \quad P(\xi = 6) = 2/45.$$

Then

$$\widehat{F}(t) = 1 + (e^{it} - 1) + (e^{it} - 1)^2 + (e^{it} - 1)^3 + \theta| e^{it} - 1|^4/3.$$

It is easy to check that $v_1(F) = v_2(F)/2 = v_2(F)/6 = 1$ and $v_4(F) < C$. Consequently, Franken's condition (3.13) is not satisfied. On the other hand,

$$\sum_{k \in \mathbb{Z}} \min(F\{k\}, F\{k - 1\}) = \min(F\{1\}, F\{0\}) = \frac{18}{45}.$$

Let G have a geometric distribution with parameter $p = 1/2$, i.e.

$$\widehat{G}(t) = \frac{1}{2 - e^{it}} = 1 + (e^{it} - 1) + (e^{it} - 1)^2 + (e^{it} - 1)^3 + \frac{(e^{it} - 1)^4}{2 - e^{it}}.$$

Taking into account Example 2.2 we observe that $v_1(G) = v_2(G)/2 = v_3/6(G) = v_4(G) = 1$. Moreover, $G\{k\} = 2^{-(k+1)} < G\{k-1\}, k = 1, 2, \ldots$. Therefore

$$\sum_{k=-\infty}^{\infty} \min(G\{k\}, G\{k-1\}) = \min(2^{-1}, 0) + \sum_{k=1}^{\infty} \min(2^{-(k+1)}, 2^{-k}) = \sum_{k=1}^{\infty} \frac{1}{2^{k+1}} = \frac{1}{2}.$$

Applying Theorem 5.3 with $s = 4$ we prove that $\| F^n - G^n \| \leqslant Cn^{-1}$.

5.5 Application to the Wasserstein Norm

Observe that

$$\| M(I_1 - I) \|_W = \| M \|. \tag{5.16}$$

Therefore we can apply the method of convolutions, rewriting smoothing estimates of Lemma 2.3 in the following way

Lemma 5.5 *Let $a \in (0, \infty)$, $p = 1 - q \in (0, 1)$, and $k, n \in \mathbb{N}$. Then*

$$\| (I_1 - I)^k \exp\{a(I_1 - I)\} \|_W \leqslant \frac{\sqrt{k!}}{a^{(k-1)/2}},$$

$$\| (I_1 - I)^k (qI + pI_1)^n \|_W \leqslant \frac{C(k)}{(npq)^{(k-1)/2}}.$$

Moreover, from (5.16) it follows that, if $M \in \mathcal{M}_Z$ can be expressed as $M = (I_1 - I)W$, for some $W \in \mathcal{M}_Z$, such that $\sum_{k \in \mathbb{Z}} |k| |W\{k\}| < \infty$, then Lemmas 5.1 and 5.2 can be applied to estimate $\| M \|_W$ with $\widehat{M}(t)$ replaced by $\widehat{M}(t)/(e^{it} - 1)$, that is, for $a \in \mathbb{R}, b > 0$,

$$\| M \|_W \leqslant (1 + b\pi)^{1/2} \left(\frac{1}{2\pi} \int_{-\pi}^{\pi} \left(\frac{|\widehat{M}(t)|^2}{|e^{it} - 1|^2} + \frac{1}{b^2} \left| \left(e^{-ita} \frac{\widehat{M}(t)}{e^{it} - 1} \right)' \right|^2 \right) dt \right)^{1/2}. \tag{5.17}$$

Example 5.3 Let $F = (1 - p)I + pI_1$, $G = \exp\{p(I_1 - I)\}$, $n \in \mathbb{N}, p \leqslant 1/2, np \geqslant 1$. Then

$$F^n - G^n = (F - G) \sum_{j=1}^{n} F^{j-1} G^{n-j} = (I_1 - I) \sum_{k=1}^{\infty} \frac{p^k (I_1 - I)^{k-1}}{k!} \sum_{j=1}^{n} F^{j-1} G^{n-j}.$$

Therefore we can apply (5.17). From (1.16) and (1.17) and (3.8) it follows that

$$|\widehat{F}(t) - \widehat{G}(t)| \leqslant Cp^2 \sin^2(t/2), \quad |\widehat{F}'(t) - \widehat{G}'(t)| \leqslant Cp^2| \sin(t/2)|,$$

and $\max(|\widehat{F}(t)|, |\widehat{G}(t)|) \leqslant \exp\{-p \sin^2(t/2)\}$. Setting

$$\widehat{U}(t) = \exp\{-itnp\}[(1 + p(e^{it} - 1))^n - \exp\{np(e^{it} - 1)\}],$$

and arguing similarly to the proof of (5.4) and (5.5), we prove that

$$|\widehat{U}(t)| \leqslant Cnp^2 \sin^2(t/2) \exp\{-np \sin^2(t/2)\}, \tag{5.18}$$

$$|\widehat{U}'(t)| \leqslant Cnp^2| \sin(t/2)| \exp\{-Cnp \sin^2(t/2)\}. \tag{5.19}$$

Next, applying (1.48), we obtain

$$\frac{|\widehat{U}(t)|}{|\sin(t/2)|} \leqslant Cnp^2| \sin(t/2)| \exp\{-np \sin^2(t/2)\} \leqslant C(p\sqrt{np}) \exp\{-Cnp \sin^2(t/2)\},$$

$$\left|\left(\frac{\widehat{U}(t)}{e^{it} - 1}\right)'\right| \leqslant \frac{|\widehat{U}'(t)||e^{it} - 1| + |\widehat{U}(t)||ie^{it}|}{|e^{it} - 1|^2} \leqslant \frac{Cnp^2 \sin^2(t/2) \exp\{-Cnp \sin^2(t/2)\}}{4 \sin^2(t/2)}$$

$$\leqslant C(p\sqrt{np}) \sqrt{np} \exp\{-Cnp \sin^2(t/2)\}.$$

Substituting the last estimates into (5.17) with $b = \sqrt{np}$, $a = np$ and applying (1.31) we, finally, arrive at

$$\| F^n - G^n \| \leqslant Cp\sqrt{np}.$$

It must be noted that assumption $M = (I_1 - I)W$ is not very restrictive.

Lemma 5.6 *Let $F, G \in \mathcal{F}_Z$ have finite first absolute moments, $n \in \mathbb{N}$. Then $F^n - G^n = (I_1 - I)W$, $W \in \mathcal{M}_Z$.*

Proof Observe that, for any $k \in \mathbb{N}$,

$$I_k = I + (I_1 - I) \sum_{j=0}^{k-1} I_j, \quad I_{-k} = I + (I_{-1} - I) \sum_{j=0}^{k-1} I_{-j}.$$

Let

$$V = \sum_{j=1}^{n} F^{j-1} G^{n-j}.$$

Then

$$F^n - G^n = (F - G)V = V \sum_{k \in \mathbb{Z}} (F\{k\} - G\{k\})I_k$$

$$= V \left[(F\{0\} - G\{0\})I + \sum_{k=1}^{\infty} (F\{k\} - G\{k\})I_k + \sum_{k=1}^{\infty} (F\{-k\} - G\{-k\})I_{-k} \right]$$

$$= V \left[(F\{0\} - G\{0\})I + \sum_{k=1}^{\infty} (F\{k\} - G\{k\})I + \sum_{k=1}^{\infty} (F\{-k\} - G\{-k\})I \right]$$

$$+ V \left[(I_1 - I) \sum_{k=1}^{\infty} (F\{k\} - G\{k\}) \sum_{j=0}^{k-1} I_j \right.$$

$$+ (I_{-1} - I) \sum_{k=1}^{\infty} (F\{-k\} - G\{-k\}) \sum_{j=0}^{k-1} I_{-j} \right] = V[1 - 1]$$

$$+ V(I_1 - I) \left[\sum_{k=1}^{\infty} (F\{k\} - G\{k\}) \sum_{j=0}^{k-1} I_j - I_{-1} \sum_{k=1}^{\infty} (F\{-k\} - G\{-k\}) \sum_{j=0}^{k-1} I_{-j} \right]$$

$$= (I_1 - I)W = (I_1 - I)n \sum_{k \in \mathbb{Z}} |k|(F\{k\} + G\{k\})\Theta.$$

Thus, we see that W is concentrated on $\mathbb{Z}$ and has bounded variation.   □

5.6 Problems

5.1 Let $\lambda > 0, k \in \{1, 2 \ldots\}$. Applying Lemma 5.1 prove that

$$\| (I_1 - I)^k \exp\{\lambda(I_1 - I)\} \|' \leq C(k)\lambda^{-k/2}.$$

5.2 Let F be the binomial distribution with parameters n and $p < 1/2$. Prove that

$$\| F^n - F^{n+1} \| \leq C \min\left(p, \sqrt{\frac{p}{n}}\right).$$

5.3 Let $0 \leq p \leq 1/2$. Prove that

$$\left\| ((1-p)I + pI_1)^n - \exp\left\{ np(I_1 - I) - \frac{np^2}{2}(I_1 - I)^2 \right\} \right\| \leq \left(np^3, p\sqrt{\frac{p}{n}} \right).$$

5.4 Let $F = qI + pV$, $G = \exp\{p(V - I)\}$. Here $V \in \mathcal{F}_s$ is concentrated on $\{\pm 1, \pm 2, \dots\}$ with finite variance σ^2. Prove that, for any $n \in \mathbb{N}$,

$$\| F^n - G^n - nG^{n-1}(F - G) \| \leqslant \frac{C\sqrt{\sigma + 1}}{n^2}.$$

5.5 Let $p < 1/5$. Prove that

$$\| ((1 - 2p)I + pI_1 + pI_{-1})^n (I_1 - I)^2 (I_{-1} - I) \| \leqslant \frac{C}{np\sqrt{np}}.$$

5.6 Prove (5.16).

5.7 Prove (5.18) and (5.19).

Bibliographical Notes

The proofs of Lemmas 5.1 and 5.2 can be found in [111] and [112]. The proofs of (5.9), (5.10), (5.11), and (5.12) are based on Lemma 5.1 and can be found in [42], Proposition 3.2 and in [43], Lemma 4.6. Estimate (5.13) follows from Corollary 1.6 in [98], which is a constant-improved version of Proposition 4.6 from the paper of Barbour and Xia [14]. The Barbour-Xia inequality was further investigated in [108] and [12]. It is worth noting that a similar shifted measure as in the Barbour-Xia inequality was considered by Gamkrelidze, see [64]. Prohorov [113] was the first to prove the estimate (5.7). He used a direct asymptotic expansion of the binomial probability. The improvement of the constant in (5.7) is one of the most comprehensively studied problems in limit theorems. Hwang in [77] applied the inversion formula for the moment generating function via a Cauchy integral and used asymptotic analysis of complex functions, proving estimates of the form $o(1)$. Lemma 5.1 was applied for sums of 1-dependent random variables in [45].

Chapter 6
Non-uniform Estimates for Lattice Measures

In this chapter, we demonstrate that for lattice distributions with a sufficient number of finite moments the non-uniform estimates can be proved via a somewhat modified Tsaregradskii inequality.

6.1 Non-uniform Local Estimates

In many respects non-uniform estimates for $M \in \mathcal{M}_Z$ can be obtained very similarly to the estimates in total variation. Let $M \in \mathcal{M}_Z$ and let $\sum_{k \in \mathbb{Z}} k^2 |M\{k\}| < \infty$. Note that, since $\sin(k\pi) = 0$,

$$\widehat{M}\{\pi\} = \sum_{k \in \mathbb{Z}} M\{k\} e^{ik\pi} = \sum_{k \in \mathbb{Z}} M\{k\} \cos(k\pi) = \sum_{k \in \mathbb{Z}} M\{k\} \cos(-k\pi) = \widehat{M}(-\pi).$$

Similarly,

$$M'\{\pi\} = i \sum_{k \in \mathbb{Z}} k M\{k\} e^{ik\pi} = i \sum_{k \in \mathbb{Z}} k M\{k\} \cos(k\pi) = M'\{-\pi\}.$$

Integrating by parts the inversion formula (3.1), for $a \neq k$, we get

$$M\{k\} = \frac{1}{2\pi} \int_{-\pi}^{\pi} \widehat{M}(t) e^{-itk} \, dt = \frac{1}{2\pi} \int_{-\pi}^{\pi} \left(\widehat{M}(t) e^{-ita} \right) e^{it(a-k)} \, dt$$

$$= \frac{1}{i(a-k)} \frac{1}{2\pi} \int_{-\pi}^{\pi} \left(\widehat{M}(t) e^{-ita} \right) \left(e^{it(a-k)} \right)' \, dt$$

© Springer International Publishing Switzerland 2016
V. Čekanavičius, *Approximation Methods in Probability Theory*, Universitext,
DOI 10.1007/978-3-319-34072-2_6

$$= \frac{-1}{i(a-k)} \frac{1}{2\pi} \int\limits_{-\pi}^{\pi} \left(\widehat{M}(t)e^{-ita}\right)' e^{it(a-k)} \, dt$$

$$= \frac{1}{(a-k)^2} \frac{1}{2\pi} \int\limits_{-\pi}^{\pi} \left(\widehat{M}(t)e^{-ita}\right)' \left(e^{it(a-k)}\right)' \, dt$$

$$= -\frac{1}{(a-k)^2} \frac{1}{2\pi} \int\limits_{-\pi}^{\pi} \left(\widehat{M}(t)e^{-ita}\right)'' e^{it(a-k)} \, dt.$$

Therefore, for any $a \in \mathbb{R}$,

$$|k-a|\,|M\{k\}| \leqslant \frac{1}{2\pi} \int\limits_{-\pi}^{\pi} \left|\left(\widehat{M}(t)e^{-ita}\right)'\right| dt, \tag{6.1}$$

$$(k-a)^2 \,|M\{k\}| \leqslant \frac{1}{2\pi} \int\limits_{-\pi}^{\pi} \left|\left(\widehat{M}(t)e^{-ita}\right)''\right| dt. \tag{6.2}$$

Observe that, for $a = k$, estimates (6.1) and (6.2) are trivial. In general, the process can be continued for estimates with a higher power of $(k - a)$. However, additional derivatives usually result in reduced accuracy.

> **Typical application.** The same problems as considered in Chap. 3 combined with (6.1) or (6.2).
> **Advantages.** Due to the very similar estimating technique, non-uniform local estimates can easily be obtained jointly with estimates for the local norm.
> **Drawbacks.** Quite long routine estimations of derivatives are needed. Constants are not small.

Example 6.1 Let $F = (1-p)I + pI_1$, $G = \exp\{p(I_1 - I)\}$, $n \in \mathbb{N}, p \leqslant 1/2, np \geqslant 1$. Let

$$\widehat{U}(t) = \exp\{-itnp\}[(1 + p(e^{it} - 1))^n - \exp\{np(e^{it} - 1)\}].$$

Then, as noted in Example 5.3

$$|\widehat{U}(t)| \leqslant Cnp^2 \sin^2(t/2) \exp\{-np \sin^2(t/2)\}, \tag{6.3}$$

$$|\widehat{U}'(t)| \leqslant Cnp^2 |\sin(t/2)| \exp\{-Cnp \sin^2(t/2)\}. \tag{6.4}$$

Therefore, applying (6.1) and (1.31), we obtain

$$| k - np | | F^n\{k\} - G^n\{k\} | \leqslant Cp.$$

As a rule, non-uniform estimates are combined with a local norm estimate. In our case, from (3.2) and (6.3) we get

$$| F^n\{k\} - G^n\{k\} | \leqslant \| F^n - G^n \|_\infty \leqslant C \sqrt{\frac{p}{n}}.$$

Therefore we can write the following non-uniform estimate

$$\left(1 + \frac{| k - np |}{\sqrt{np}} \right) | F^n\{k\} - G^n\{k\} | \leqslant C \sqrt{\frac{p}{n}}.$$

In principle, the estimates with $(k - np)^2$ can also be proved.

6.2   Non-uniform Estimates for Distribution Functions

The idea behind the inversion formula for the distributional non-uniform estimates is the same as for the point estimates. Let $M \in \mathcal{M}_Z$,

$$u(t) := \frac{\widehat{M}(t)}{e^{-it} - 1}.$$

We shall assume that $u(t)$ has two continuous derivatives. Then by the inversion formula (4.4) we get

$$M(k) = \frac{1}{2\pi} \int_{-\pi}^{\pi} u(t) e^{-itk} \, dt.$$

Henceforth $M(m) = M\{(-\infty, m)\}$. Observe that $u(\pi) = u(-\pi)$. Therefore, applying the same argument as in the previous section, we prove that, for any $a \in \mathbb{R}$ and $k \in \mathbb{Z}$,

$$| k - a | | M(k) | \leqslant \frac{1}{2\pi} \int_{-\pi}^{\pi} \left| \left(u(t) e^{-ita} \right)' \right| dt, \tag{6.5}$$

$$(k - a)^2 | M(k) | \leqslant \frac{1}{2\pi} \int_{-\pi}^{\pi} \left| \left(u(t) e^{-ita} \right)'' \right| dt. \tag{6.6}$$

Now the estimates can be obtained similarly to the local case. We illustrate this approach by considering symmetric distributions with nonnegative characteristic functions. Let $F \in \mathcal{F}_+ \cap \mathcal{F}_\mathbb{Z}$, $G = \exp\{F - I\}$. We denote by σ^2 the variance of F.

Theorem 6.1 *Let $n \geqslant 3$, $k \in \mathbb{Z}$, $F \in \mathcal{F}_+ \cap \mathcal{F}_\mathbb{Z}$. Then*

$$\left(1 + \frac{k^2}{\max\{1, n\sigma^2\}}\right)\left| F^n(k) - \exp\{n(F - I)\}(k)\right| \leqslant \frac{C\sigma}{n\sqrt{1 - F\{0\}}}.$$

Proof Let $G = \exp\{F - I\}$, $p_k = 2F\{k\}$, $r(t) = \sum_{k=1}^{\infty} p_k \sin^2(tk/2)$. Note that $\sigma^2 = \sum_{k=1}^{\infty} k^2 p_k$.

By conditioning $F\{-k\} = F\{k\}$ and $\widehat{F}(t) = \mathrm{Re}\widehat{F}(t) = \sum_{k \in \mathbb{Z}} F\{k\} \cos(tk)$. Note also that $\cos(tk) = \cos(-tk)$. Therefore

$$\widehat{F}(t) = F\{0\} + \sum_{k=1}^{\infty} (F\{k\} + F\{-k\}) \cos(tk) = 1 + \sum_{k=1}^{\infty} p_k(\cos tk - 1) \leqslant \exp\{-2r(t)\} = \widehat{G}(t).$$

Using Hölder's inequality we prove

$$|\widehat{F}'(t)| = \left| \sum_{k=1}^{\infty} k p_k \sin tk \right| \leqslant \sum_{k=1}^{\infty} (k\sqrt{p_k})(\sqrt{p_k}\, |\sin tk|) \leqslant C\sigma \sqrt{r(t)}.$$

Similarly, applying (1.33) and Example 1.9, we obtain

$$|\widehat{F}''(t)| = \left| \sum_{k=1}^{\infty} k^2 p_k \cos tk \right| \leqslant \sigma^2, \quad |\widehat{F}(t) - 1| = 2r(t),$$

$$|\widehat{G}'(t)| = |\widehat{G}(t)\widehat{F}'(t)| \leqslant C\sigma \sqrt{r(t)},$$

$$|\widehat{G}''(t)| \leqslant |\widehat{G}(t)||\widehat{F}'(t)|^2 + |\widehat{G}(t)\widehat{F}''(t)| \leqslant C\sigma^2,$$

$$|r(t)| \leqslant \frac{1}{4}\sigma^2 t^2, \quad e^{2r(t)} \leqslant e^2,$$

$$|\widehat{F}(t) - \widehat{G}(t)| \leqslant C|\widehat{F}(t) - 1|^2 \leqslant Cr^2(t),$$

$$|\widehat{F}'(t) - \widehat{G}'(t)| \leqslant |\widehat{F}'(t)||\widehat{F}(t) - 1| \leqslant C\sigma r^{3/2}(t),$$

$$|\widehat{F}''(t) - \widehat{G}''(t)| \leqslant C\left\{|\widehat{F}''(t)||\widehat{F}(t) - 1| + |\widehat{F}'(t)|^2\right\} \leqslant C\sigma^2 r(t).$$

Using these estimates and (1.48) and, omitting for the sake of brevity the dependence on t, we prove that

$$|\widehat{F}^n - \widehat{G}^n| \leqslant n|\widehat{F} - \widehat{G}| \max(\widehat{F}^{n-1}, \exp\{(n-1)(\widehat{F} - 1)\})$$
$$\leqslant Cn|\widehat{F} - \widehat{G}|e^{-2(n-1)r(t)} \leqslant Cnr^2(t)e^{-2nr(t)}$$
$$\leqslant Cn\sqrt{r(t)}(\sigma^2 t^2)^{3/2}e^{-2nr(t)} \leqslant C\sigma^3\sqrt{n}|t|^3 e^{-nr(t)},$$

$$|(\widehat{F}^n - \widehat{G}^n)'| \leqslant n|\widehat{F}^{n-1} - \widehat{G}^{n-1}| \, |\widehat{F}'| + n|\widehat{G}^{n-1}| \, |\widehat{F}' - \widehat{G}'|$$
$$\leqslant Cn^2\sigma\sqrt{r(t)}r^2(t)\exp\{-2nr(t)\} + Cn\sigma r^{3/2}(t)\exp\{-2nr(t)\}$$
$$\leqslant Cn\sigma r^{3/2}(t)\exp\{-1.5nr(t)\} \leqslant Cn\sigma\sqrt{r(t)}\sigma^2 t^2\exp\{-1.5nr(t)\}$$
$$\leqslant C\sigma^3\sqrt{n}t^2 e^{-nr(t)},$$

$$|(\widehat{F}^n - \widehat{G}^n)''| = |n(n-1)\widehat{F}^{n-2}(\widehat{F}')^2 + n\widehat{F}^{n-1}\widehat{F}'' - n(n-1)\widehat{G}^{n-2}(\widehat{G}')^2 - n\widehat{G}^{n-1}\widehat{G}''|$$
$$\leqslant n(n-1)|\widehat{F}^{n-2} - \widehat{G}^{n-2}| \, |\widehat{F}'|^2 + n(n-1)|\widehat{G}^{n-2}| \, |\widehat{F}' - \widehat{G}'| \, |\widehat{F}' + \widehat{G}'|$$
$$+n|\widehat{F}^{n-1} - \widehat{G}^{n-1}| \, |\widehat{F}''| + n|\widehat{G}^{n-1}| \, |\widehat{F}'' - \widehat{G}''|$$
$$\leqslant C\exp\{-2nr(t)\}[n^3 r^3(t)\sigma^2 + n^2 r^2(t)\sigma^2 + nr(t)\sigma^2]$$
$$\leqslant Ce^{-2nr(t)}\sigma^3|t|[n^3 r^{5/2}(t) + n^2 r^{3/2}(t) + nr^{1/2}(t)] \leqslant C\sigma^3\sqrt{n}|t|e^{-nr(t)}.$$

Let $M = F^n - \exp\{n(F - I)\}$. By using (1.38) (or an expansion of $\widehat{F}(t) - 1$ in powers of t) we can prove that $u(t)$ and its first and second derivatives are continuous. Noting that $|e^{-it} - 1| \geqslant C|t|$ for $|t| \leqslant \pi$ we prove

$$|u''(t)| \leqslant C\left(\frac{|\widehat{M}''(t)|}{|t|} + \frac{|\widehat{M}'(t)|}{t^2} + \frac{|\widehat{M}(t)|}{|t|^3}\right) \leqslant Ce^{-nr(t)}\sigma^3\sqrt{n}. \tag{6.7}$$

Observe that by (3.20)

$$\int_{-\pi}^{\pi} e^{-nr(t)}\,dt \leqslant Cn^{-1/2}(1 - F\{0\})^{-1/2}. \tag{6.8}$$

We also have the following estimate

$$|\widehat{F}^n(t) - \widehat{G}^n(t)| \leqslant Cnr^2(t)e^{-2nr(t)} \leqslant \sigma|t|r^{3/2}(t)e^{-2nr(t)} \leqslant C\sigma|t|n^{-1/2}e^{-nr(t)}.$$

Consequently, applying (4.1) we prove that

$$|F^n(k) - G^n(k)| \leqslant |F^n - G^n|_K \leqslant \frac{C\sigma}{n\sqrt{1 - F\{0\}}}.$$

Combining all estimates with (6.6) we get

$$\left(1 + \frac{k^2}{\max\{1, n\sigma^2\}}\right)\left| F^n(k) - G^n(k) \right| \leqslant C\sigma(1 - F\{0\})^{-1/2}n^{-1}.$$

□

Note that, for the proof of (6.2) we have not applied any centering. Indeed, the mean of F already equals zero. It is obvious that non-uniform estimates for distributions can be used to prove estimates in the Wasserstein norm.

Example 6.2 Let $F \in \mathcal{F}_+ \cap \mathcal{F}_Z$, $\sigma^2 = \sum_{k=-\infty}^{\infty} k^2 F\{k\}$ and let $n\sigma > 1$. Then from (6.2) and (1.46) it follows that

$$\| F^n - \exp\{n(F - I)\} \|_W \leqslant C\sigma^2(1 - F\{0\})^{-1/2}n^{-1/2}.$$

One should bear in mind that, if non-uniform estimates are used to prove results for the Wasserstein norm, then the absolute constants in the estimates are not small.

6.3 Applying Taylor Series

Usually it is simpler to write the second derivative as a sum of a few fractions and to estimate each fraction separately, as was demonstrated in (6.7). However, in some cases, this approach might be inconvenient. We show how (1.33) can simplify the problem. Let $F = (1 - p)I + pI_1$, $G = \exp\{p(I_1 - I)\}$.

Theorem 6.2 *Let $n \in \mathbb{N}$, $p \leqslant 1/2$, $np \geqslant 1$. Then for any $k \in \mathbb{N}$ the following estimate holds*

$$\left(1 + \frac{(k - np)^2}{np}\right)\left| F^n(k) - G^n(k) \right| \leqslant Cp. \tag{6.9}$$

Proof For the sake of brevity, let $z = e^{it} - 1$, $M = F^n - G^n$. Then by (1.40) and (1.33) we get

$$\widehat{M}(t) = (1 + pz)^n - e^{npz} = (1 + pz - e^{pz})\sum_{j=1}^{n}(1 + pz)^{j-1}\exp\{(n - j)pz\}$$

$$= -(pz)^2\int_0^1(1 - \tau)\exp\{\tau pz\}d\tau\sum_{j=1}^{n}(1 + pz)^{j-1}\exp\{(n - j)pz\}.$$

Therefore

$$u(t) = \frac{\widehat{M}(t)}{e^{-it} - 1} = -\frac{e^{it}\widehat{M}(t)}{z} = p^2 z e^{it}h(t)H(t).$$

Here

$$h(t) = \int_0^1 (1 - \tau) \exp\{\tau pz\} d\tau, \qquad H(t) = \sum_{j=1}^n (1 + pz)^{j-1} \exp\{(n - j)pz\}.$$

Taking into account (3.8) we prove that

$$|ze^{it}| \leq 2|\sin(t/2)|, \quad |(ze^{it})'| \leq 3, \quad |(ze^{it})''| \leq 5,$$

$$|h(t)| \leq C, \quad |(h(t)\exp\{-itp\})'| \leq C, \quad |(h(t)\exp\{-itp\})''| \leq C,$$

$$|(1 + pz)e^{-itp}| \leq \exp\{-2p(1 - p)\sin^2(t/2)\} \leq \exp\{-p\sin^2(t/2)\},$$

$$|((1 + pz)e^{-itp})'| \leq p(1 - p)|e^{-itp}z| \leq Cp|\sin(t/2)|,$$

$$|((1 + pz)e^{-itp})''| \leq p(1 - p)|e^{-itp}(p + (1 - p)e^{it})| \leq Cp,$$

$$|\exp\{pz - pit\}| \leq \exp\{-2p\sin^2(t/2)\} \leq \exp\{-p\sin^2(t/2)\},$$

$$|(\exp\{pz - pit\})'| \leq Cp|\sin(t/2)|, \quad |(\exp\{pz - pit\})''| \leq Cp$$

$$|H(t)| \leq Cn\exp\{-2(n - 1)p(1 - p)\sin^2(t/2)\} \leq Cn\exp\{-np\sin^2(t/2)\},$$

$$|(H(t)\exp\{-(n - 1)itp\})'| \leq Cn^2p|\sin(t/2)|\exp\{-np\sin^2(t/2)\}$$

$$\leq Cn^{3/2}\sqrt{p}\exp\{-0.5np\sin^2(t/2)\},$$

$$|(H(t)\exp\{-(n - 1)itp\})''| \leq Cn[n^2p^2\sin^2(t/2) + np]\exp\{-np\sin^2(t/2)\}$$

$$\leq Cnp\exp\{-0.5np\sin^2(t/2)\}.$$

Observing that

$$(abc)'' = a''bc + ab''c + abc'' + 2[a'b'c + a'bc' + ab'c'],$$

we obtain

$$|(u(t)\exp\{-nitp\})''| \leq Cnp^2\sqrt{np}\exp\{-0.5np\sin^2(t/2)\}. \tag{6.10}$$

Substituting (6.10) into (6.6) and applying (1.31) we get

$$(k - np)^2|F^n(k) - G^n(k)| \leq Cnp^2.$$

It is not difficult to check that

$$|u(t)| \leq Cnp^2|\sin(t/2)|\exp\{-np\sin^2(t/2)\}.$$

Combining the last two estimates and the Tsaregradskii inequality (4.2) we complete the proof of (6.9). □

6.4   Problems

6.1 Let the assumptions of Theorem 6.1 be satisfied. Let $M = F^n - \exp\{n(F-I)\}$. Prove that, for all $k \in \mathbb{Z}$,

$$\left(1 + \frac{k^2}{\max(1, n\sigma^2)}\right)|M\{k\}| \leq \frac{C}{n\sqrt{n(1-F\{0\})}}.$$

6.2 Let the assumptions of Theorem 6.1 be satisfied. Prove that

$$k^2|F^{n+1}\{k\} - F^n\{k\}| \leq \frac{C\sigma^2}{\sqrt{n(1-F\{0\})}}.$$

6.3 Prove that

$$\| ((1-p)I + pI_1)^n - \exp\{np(I_1 - I)\} \|_W \leq Cp\sqrt{np}.$$

6.4 Let $p \leq 1/2, q = 1-p, np > 1, F = qI + pI_1, G = \exp\{p(I_1-I) - (p^2/2)(I_1 - I)^2\}$. Prove that for $k \in \mathbb{N}$

$$\left(1 + \frac{|k - np|}{\sqrt{np}}\right)|F^n(k) - G^n(k)| \leq Cp\sqrt{\frac{p}{n}}.$$

6.5 Let the assumptions of the previous problem hold. Prove that, for $k \in \mathbb{N}$,

$$\left(1 + \frac{(k - np)^2}{np}\right)|F^n(k) - G^n(k)| \leq Cp\sqrt{\frac{p}{n}}.$$

Bibliographical Notes

The main body of research in non-uniform estimates is related to normal approximation, see, for example, [105], Chapter 6 or [21, 22]. Non-uniform estimates for lattice approximations have attracted little attention. In this section we presented slightly simplified results from [32]. Note also that Karymov [83] used conjugate distributions to prove non-uniform estimates for the Poisson approximation to the binomial law, when $p = O(n^{-1})$.

Chapter 7
Discrete Non-lattice Approximations

7.1 Arak's Lemma

Not all discrete distributions are lattice distributions. A lattice distribution F, for some $h > 0$, is concentrated on a set $K = \{a + hk,\ k \in \mathbb{Z}\}$. For convenience, it is usually assumed that h is the maximal common divisor of the lattice points. It is obvious that, suitably shifted and normed by h, any lattice measure can be reduced to the case of a measure concentrated on $\mathbb{Z}$. Then, the methods of the previous sections apply. Now let us consider F concentrated at 0, 1 and $\sqrt{2}$. For such a distribution there is no simple analogue of (3.1). To some extent one can avoid the problem, since any discrete distribution admits the representation

$$(1 - p)I + pV, \qquad V \in \mathcal{F}$$

and results from Sect. 7.2 apply. However, such a simplification might lead to very inaccurate estimates, especially if some symmetry of distribution is involved.

In this section, we use Arak's inequality. It allows estimation of discrete distributions with supports that are one-dimensional projections of some multi-dimensional lattice, that is, have quite a regular structure. Note that another inversion formula for discrete approximations is given in Theorem 9.5, Case 1. Let $\mathbf{u} = (u_1, u_2, \ldots, u_N) \in \mathbb{R}^N$. Set

$$K_m(\mathbf{u}) = \left\{ \sum_{i=1}^{N} j_i u_i : j_i \in \{-m, -m+1, \ldots, m\};\ i = 1, \ldots, N \right\}.$$

© Springer International Publishing Switzerland 2016
V. Čekanavičius, *Approximation Methods in Probability Theory*, Universitext,
DOI 10.1007/978-3-319-34072-2_7

For any finite signed measure M the Jordan-Hahn decomposition $M = M^+ - M^-$ allows us to express M as the difference of two nonnegative measures with different supports. Let

$$\delta(M, m, \mathbf{u}) = M^+\{\mathbb{R} \setminus K_m(\mathbf{u})\} + M^-\{\mathbb{R} \setminus K_m(\mathbf{u})\}. \tag{7.1}$$

Arak's lemma allows us to switch from the Kolmogorov norm to Fourier transforms and Lévy's concentration function.

Lemma 7.1 *Let $M \in \mathcal{M}$, $M\{\mathbb{R}\} = 0$, $N, m \in \mathbb{N}$, $\mathbf{u} \in \mathbb{R}^{N+1}$, $h > 0$ and $U \in \mathcal{F}_+$. Then*

$$|M|_K \leqslant C \int\limits_{|t| \leqslant 1/h} \left| \frac{\widehat{M}(t)}{t} \right| dt + C(N) \ln(m+1) \sup_{t \in \mathbb{R}} \frac{|\widehat{M}(t)|}{\widehat{U}(t)} Q(U, h)$$

$$+ \delta(M, m, \mathbf{u}). \tag{7.2}$$

The proof of (7.2) is beyond the scope of this book. Note that h, N, m and $U \in \mathcal{F}_+$ can be chosen arbitrary. We can see that, unlike the lattice case, (7.2) has three different summands. The first one is quite similar to the Tsaregradskii inequality and deals with the behavior of the Fourier transform in a neighborhood of zero. The second replaces the estimate of measure by an estimate of its Fourier transform. However, some additional distribution $U \in \mathcal{F}_+$ appears. How does this distribution affect the estimate? In principle, its concentration function should neutralize the logarithm factor. The third summand imposes limits on the applicability of (7.2). Indeed, the support of M should not differ much from $K_m(\mathbf{u})$.

What can be said about the structure of $K_m(\mathbf{u})$? The idea can probably be grasped from the following example: $K_1(1, \sqrt{2}) = \{0, \pm\sqrt{2}, -1 \pm \sqrt{2}, 1 \pm \sqrt{2}\}$.

Let *suppM* denote the support of M. Obviously, if $supp M_i \subset K_m(\mathbf{u})$, $(i = 1, \ldots, n)$ then

$$supp M_1 M_2 \cdots M_n \subset K_{nm}(\mathbf{u}). \tag{7.3}$$

For an effective application of (7.2) we must have estimates of $\delta(M, m, \mathbf{u})$. As it turns out, the support of any signed compound Poisson measure is close to some subset of $K_{my}(\mathbf{u})$ provided its compounding measure is concentrated on $K_m(\mathbf{u})$. We shall formulate this result more precisely.

Lemma 7.2 *Let $W, V \in \mathcal{M}$, $\|W\| \leqslant b_1$, $\|V\| \leqslant b_2$, $N, a \in \mathbb{N}$, $\mathbf{u} \in \mathbb{R}^{N+1}$, $supp W \subset K_s(\mathbf{u})$, $supp V \subset K_a(\mathbf{u})$. Then, for any $y \in \mathbb{N}$, the following inequalities hold*

$$\delta(W \exp\{V\}, s + ay, \mathbf{u}) \leqslant b_1 \exp\{3b_2 - y\},$$

$$\delta(\exp\{V\}, ay, \mathbf{u}) \leqslant \exp\{3b_2 - y\}.$$

Proof From (7.3) we obtain

$$\sum_{k=0}^{y} \frac{V^k \{\mathbb{R} \setminus K_{ay}(\mathbf{u})\}}{k!} = 0.$$

Therefore

$$\delta(\exp\{V\}, ay, \mathbf{u}) \leqslant \sum_{k>y}^{\infty} \frac{\| V^k \|}{k!} \leqslant \sum_{k>y}^{\infty} \frac{\| V \|^k}{k!} \leqslant \sum_{k>y}^{\infty} \frac{b_2^k}{k!} \leqslant e^{-y} \sum_{k=0}^{\infty} \frac{(eb_2)^k}{k!}.$$

Similarly,

$$\delta(W \exp\{V\}, s + ay, \mathbf{u}) \leqslant \sum_{k>y} \frac{b_1 b_2^k}{k!} \leqslant b_1 \exp\{eb_2 - y\}.$$

$\square$

Typical application. Arak's lemma is used when Le Cam's trick or general estimates from Lemma 2.6 cannot be applied.
Advantages. Gives a sharper estimate than can be obtained from the general lemma of Sect. 9.1.
Drawbacks. The difference of Fourier transforms must be estimated for all $t \in \mathbb{R}$.

7.2 Application to Symmetric Distributions

We shall demonstrate how Arak's lemma works by considering a quite simple case of symmetric distributions.

Theorem 7.1 *Let $F_j \in \mathcal{F}_+$ be concentrated on a set $\{0, \pm w_j x_{j1}, \pm w_j x_{j2}, \ldots, \pm w_j x_{js}\}$ $(j = 1, 2)$, $n_1, n_2 \in \mathbb{N}$. Suppose s, w_1, w_2 and all x_{ij} do not depend on n and let $F_i\{x_{ij}\}$ be uniformly bounded by some positive constants, $0 < C_1 \leqslant F_i\{x_{ij}\} \leqslant C_2 < 1$. Then*

$$| F_1^{n_1} F_2^{n_2} - \exp\{n_1(F_1 - I) + n_2(F_2 - I)\} |_K$$

$$\leqslant \frac{C}{\sqrt{n_1 + n_2}} \left(\frac{1}{\sqrt{n_1}} + \frac{1}{\sqrt{n_2}} + \ln(n_1 + n_2 + 1)\left(\frac{1}{n_1} + \frac{1}{n_2} \right) \right) + e^{-n}.$$

$$(7.4)$$

Remark 7.1 Direct application of the triangle inequality combined with (2.38) provides the following estimate

$$|F_1^{n_1}F_2^{n_2} - \exp\{n_1(F_1 - I) + n_2(F_2 - I)\}|_K$$

$$\leqslant \| F_1^{n_1} \| | F_2^{n_2} - \exp\{n_2(F_2 - I)\}|_K + \| \exp\{n_2(F_2 - I)\}\| | F_1^{n_1} - \exp\{n_1(F_1 - I)\}|_K$$

$$= |F_2^{n_2} - \exp\{n_2(F_2 - I)\}|_K + |F_1^{n_1} - \exp\{n_1(F_1 - I)\}|_K \leqslant C\Big(\frac{1}{n_1} + \frac{1}{n_2}\Big). \qquad (7.5)$$

The good accuracy of the estimate (7.5) requires both n_1 and n_2 to be large. Meanwhile, (7.4) can be quite accurate even if n_1 is small and n_2 is large.

Proof For the sake of brevity we omit the dependence of Fourier transforms on t whenever it does not lead to ambiguity. We also use the notation $\sigma_i^2 = \int x^2 F_i\{dx\}$, $n = n_1 + n_2$. Note that $\int x dF_i\{x\} = 0$, since $F_i \in \mathcal{F}_+ \subset \mathcal{F}_s$.

First we use Lemma 7.1 with $N = 2$, $M = F_1^{n_1}F_2^{n_2} - \exp\{n_1(F_1 - I) + n_2(F_2 - I)\}$ and $U = \exp\{(n_1/2)(F_1 - I) + (n_2/2)(F_2 - I)\}$. Quantities N, h, m, $\mathbf{u}$ will be chosen later.

It is obvious that $\widehat{F}_i \leqslant \exp\{\widehat{F}_i - 1\}$, since $0 \leqslant \widehat{F}_i \leqslant 1$. Moreover, $\exp\{1 - \widehat{F}_i\} \leqslant e$ and by (1.33)

$$|\widehat{F}_i - \exp\{\widehat{F}_i - 1\}| = |1 + (\widehat{F}_i - 1) - \exp\{\widehat{F}_i - 1\}| \leqslant C(\widehat{F}_i - 1)^2.$$

Therefore, applying (1.40) we prove

$$|\widehat{M}| \leqslant C[n_1(\widehat{F}_1 - 1)^2 \exp\{(n_1 - 1)(\widehat{F}_2 - 1) + n_2(\widehat{F}_2 - 1)\}$$
$$+ n_2(\widehat{F}_2 - 1)^2 \exp\{n_1(\widehat{F}_1 - 1) + (n_2 - 1)(\widehat{F}_2 - 1)\}]$$
$$\leqslant Ce[n_1(\widehat{F}_1 - 1)^2 + n_2(\widehat{F}_2 - 1)^2] \exp\{n_1(\widehat{F}_1 - 1) + n_2(\widehat{F}_2 - 1)\}.$$

By (1.48)

$$|\widehat{M}| \leqslant C\widehat{U} \sum_{i=1}^{2} n_i(\widehat{F}_i - 1)^2 \exp\{(n_i/2)(\widehat{F}_i - 1)\} \leqslant C\widehat{U}\Big(\frac{1}{n_1} + \frac{1}{n_2}\Big).$$

Similarly, applying (1.12) we get $|\widehat{F}_i - 1| \leqslant \sigma^2 t^2/2$ and, therefore,

$$|\widehat{M}| \leqslant C\widehat{U}^2 \sum_{i=1}^{2} n_i|\widehat{F}_i - 1|^{3/2}\sigma_i|t| \leqslant C\widehat{U} \sum_{i=1}^{2} \frac{\sigma_i|t|}{\sqrt{n_i}}.$$

Applying (1.24) we obtain

$$\int_{|t| \leqslant 1/h} \Big|\frac{\widehat{M}(t)}{t}\Big| dt \leqslant C \sum_{i=1}^{2} \frac{\sigma_i}{\sqrt{n_i}} \int_{|t| \leqslant 1/h} |\widehat{U}(t)| dt \leqslant C\frac{1}{h} \sum_{i=1}^{2} \frac{\sigma_i}{\sqrt{n_i}} Q(U, h).$$

Therefore, applying Lemma 7.1 we obtain

$$|M|_K \leq CQ(U,h)\left(\frac{1}{h}\sum_{j=1}^{2}\frac{\sigma_j}{\sqrt{n_j}} + C(N)\ln(m+1)\sum_{j=1}^{2}\frac{1}{n_j}\right) + \delta(M,m,\mathbf{u}). \quad (7.6)$$

Let $h = \min_{i=1,2;j\leq s} x_{ij}/2$. Then applying (1.22) we get

$$Q(U,h) \leq \frac{C(F_1,F_2)}{\sqrt{n}}. \quad (7.7)$$

We still need to estimate $\delta(M,m,\mathbf{u})$ for suitably chosen m and $\mathbf{u}$. We apply Lemma 7.2 with $W = F_1^{n_1}F_2^{n_2}$ and $V = n_1(F_1 - I) + n_2(F_2 - I)$.

Distribution F_i is concentrated on $\{0, \pm w_i x_{i1}, \pm w_i x_{i2}, \ldots, \pm w_i x_{is}\}$. By choosing $\mathbf{u}_i = (0, w_i\mathbf{x}_i) = (0, w_i x_{i1}, \ldots, w_i x_{is})$, we get that $supp\, F_i \subset K_1(\mathbf{u}_i)$ and $supp\, F_i^{n_i} \subset K_{n_i}(\mathbf{u}_i)$. Let $\mathbf{u} = (\mathbf{u}_1, \mathbf{u}_2) \in \mathbb{R}^{2s+2}$. Obviously, $supp\, W = supp\, F_1^{n_1}F_2^{n_2} \subset K_n(\mathbf{u})$.

From the definition of V it follows that $\|V\| \leq n_1(\|F_1\| + \|I\|) + n_2(\|F_2\| + \|I\|) = 2n$ and $supp\, V \subset K_1(\mathbf{u})$. Applying Lemma 7.2 with $s = n$, $a = 1$, $b_1 = 1$, $b_2 = 2n$, $y = 7n$ we get that $\delta(e^V, y, \mathbf{u}) \leq e^{-n}$.

From the definition of δ (see (7.1)), we also see that $\delta(W, n, \mathbf{u}) = 0$. Consequently, $\delta(W + e^V, 7n, \mathbf{u}) \leq e^{-n}$. Finally, $\delta(M, 7n, \mathbf{u}) = \delta(W - e^V, 7n, \mathbf{u}) \leq \delta(W + e^V, 7n, \mathbf{u}) \leq e^{-n}$. Substituting the last estimate and (7.7) into (7.6) and putting $N = 2s + 2$ we complete the proof. $\square$

7.3 Problems

7.1 Let, for some $a > 1$, $n > 2$

$$F = q_0 I + q_1 I_a + q_2 I_{-a} + q_3 I, \quad \epsilon = \frac{(q_1 - q_1)^2}{q_0(q_1 + q_2)} + \frac{q_2^2}{q_0^2(q_1 + q_2)^2 n} + \frac{q_3}{q_0}.$$

Prove that

$$|F^n - \exp\{n(F - I)\}|_K \leq C\epsilon\left(1 + \frac{a\ln n}{\sqrt{q_0(1 - q_0)n}}\right) + e^{-n}.$$

7.2 Let $F \in \mathcal{F}_s$, $suppF = \{0, \pm x_1, \ldots, \pm x_N\}$, $0 < F\{0\} < 1$. Prove that, for any $k \in \mathbb{N}$,

$$|(F - I)^k \exp\{n(F - I)\}|_K \leq C(F,k)n^{-k}.$$

7.3 Let $F \in \mathcal{F}_+$ be concentrated on the set $\{0, \pm x_1, \ldots, \pm x_N\}$, $0 < F\{0\} < 1$. Prove that

$$|F^n - F^{n+1}|_K \leqslant C(F)n^{-1}.$$

7.4 Show that the estimate in Problem 7.1 can be of the order $O(1/n)$.

Bibliographical Notes

There are very few inversion formulas for discrete non-lattice variables. In [1] the Essen type inversion formulas for purely or partially discontinuous distribution functions were obtained, see Theorem 9.5. However, those formulas seem to be of limited use. Arak's inequality was proved in [3, 4], see also Lemma 5.1 on p. 74 in [5]. Remarkably it was just one of many auxiliary results used for the proof of the first uniform Kolmogorov theorem. Lemma 7.2 is Lemma 4.3 from [46]. Theorem 7.1 is a special case of Theorem 1 from [55]. For other results, see [37].

Chapter 8
Absolutely Continuous Approximations

8.1 Inversion Formula

For absolutely continuous distributions it is more convenient to write all results in terms of distribution functions. We recall that the distribution function of F is $F(x) = F\{(-\infty, x]\}$ and

$$F(x) = \int_{-\infty}^{x} f(y)\,dy, \qquad \widehat{F}(t) = \int_{-\infty}^{\infty} e^{itx} f(x)\,dx. \qquad (8.1)$$

Here $f(x)$ is a nonnegative function integrable on the real line, called the density of F.

If F is an absolutely continuous distribution with density $f(x)$ and $G \in \mathcal{F}$, then FG is also an absolutely continuous distribution having density

$$p(y) = \int_{-\infty}^{\infty} f(y - x)\,G\{dx\}.$$

If, in addition, G has density $g(x)$ then FG has density

$$p(y) = \int_{-\infty}^{\infty} f(y - x)g(x)\,dx = \int_{-\infty}^{\infty} g(y - x)f(x)\,dx.$$

© Springer International Publishing Switzerland 2016
V. Čekanavičius, *Approximation Methods in Probability Theory*, Universitext,
DOI 10.1007/978-3-319-34072-2_8

If the characteristic function $\widehat{F}(t)$ is absolutely integrable on $\mathbb{R}$ ($\int_{-\infty}^{\infty} |\widehat{F}(t)| dt < \infty$), then F has density $f(x)$ and the following inversion formula:

$$f(x) = \frac{1}{2\pi} \int_{-\infty}^{\infty} e^{-itx} \widehat{F}(t) \, dt. \tag{8.2}$$

Typical application of (8.2). Local estimates of $F^n - G^n$ and estimates in the Kolmogorov and total variation norms, when F and G are absolutely continuous with integrable characteristic functions. Usually the scheme of sequences is considered.

Advantages. Easy to apply.

Drawbacks. Estimates for integrals of characteristic functions outside a neighborhood of zero are necessary.

The main assumption limiting the usage of (8.2) is the absolute integrability of the characteristic function. As it turns out, bounded density suffices for its fulfilment. If F has bounded density $f(x) \leqslant C_0(F)$, then

$$\int_{-\infty}^{\infty} |\widehat{F}(t)|^2 dt \leqslant C(F). \tag{8.3}$$

If, in addition, F has finite variance σ^2, then

$$|\widehat{F}(t)| \leqslant \exp\left\{ -\frac{t^2}{96 C_0^2 (2\sigma |t| + \pi)^2} \right\}. \tag{8.4}$$

From (8.4) the more rough estimate follows: if F has finite second moment and bounded density, then there exists an $\varepsilon = \varepsilon(F)$ such that

$$|\widehat{F}(t)| \leqslant \begin{cases} \exp\{-C_1(F)t^2\}, & \text{if } |t| \leqslant \varepsilon, \\ \exp\{-C_2(F)\}, & \text{if } |t| > \varepsilon. \end{cases} \tag{8.5}$$

Observe that the sum of independent iid rvs with bounded densities $S_n = \xi_1 + \xi_2 + \cdots + \xi_n$ and $n \geqslant 2$ satisfies (8.2) and (8.3). Indeed, $|\widehat{F}(t)|^n \leqslant |\widehat{F}(t)|^2$. Note that (8.5) can be applied in the scheme of sequences. On the other hand, (8.5) is not very useful in the scheme of triangular arrays since, in that case, $C(F) = C(F, n)$ and the correct order of approximation is unclear.

8.2 Local Estimates for Bounded Densities

If $F, G \in \mathcal{F}$ have densities $f(x)$, $g(x)$ and absolutely integrable characteristic functions, then (8.2) yields the following estimate

$$\sup_{x \in \mathbb{R}} |f(x) - g(x)| \leq \frac{1}{2\pi} \int_{-\infty}^{\infty} |\widehat{F}(t) - \widehat{G}(t)| \, dt. \tag{8.6}$$

We illustrate the application of (8.6) by considering sums of iid rvs with bounded densities and matching moments.

Let F and G be continuous distributions, having bounded densities $f(x)$ and $g(x)$ respectively, $s \geq 2$ be some fixed integer, and let for $k = 1, 2, \ldots, s - 1$

$$\int_{-\infty}^{\infty} x^k (f(x) - g(x)) \, dx = 0, \quad \int_{-\infty}^{\infty} |x|^s (f(x) + g(x)) dx < C_3(F, G, s) < \infty. \tag{8.7}$$

We denote the densities of F^n and G^n by $f_n(x)$ and $g_n(x)$, respectively.

Theorem 8.1 *Let (8.7) be satisfied. Then*

$$\sup_{x \in \mathbb{R}} |f_n(x) - g_n(x)| \leq C_4(F, G, s) n^{-(s-1)/2}.$$

Proof Without loss of generality, we can assume that $n \geq 2$. Indeed, if $n = 1$, then

$$|f(x) - g(x)| \leq f(x) + g(x) \leq C(F, G).$$

From the expansion in moments (1.12) it follows that, for all $t \in \mathbb{R}$,

$$|\widehat{F}(t) - \widehat{G}(t)| \leq C(F, G, s)|t|^s.$$

Moreover, we can find $\varepsilon = \varepsilon(F, G)$ such that estimates (8.5) apply. Then by (8.6) and (1.45)

$$\sup_x |f_n(x) - g_n(x)| \leq C \int_{-\infty}^{\infty} |\widehat{F}^n(t) - \widehat{G}^n(t)| \, dt$$

$$\leq \int_{-\varepsilon}^{\varepsilon} |\widehat{F}^n(t) - \widehat{G}^n(t)| \, dt + \int_{|t| > \varepsilon} (|\widehat{F}(t)|^n + |\widehat{G}(t)|^n) \, dt$$

$$\leq C(F,G) \int_{-\varepsilon}^{\varepsilon} n|\widehat{F}(t) - \widehat{G}(t)| \exp\{-C_1(F,G)(n-1)t^2\} \, dt$$

$$+ C(F,G) \exp\{-C_6(F,G)(n-2)\} \int_{-\infty}^{\infty} (|\widehat{F}(t)|^2 + |\widehat{G}(t)|^2) \, dt$$

$$\leq C(F,G,s) \int_{-\varepsilon}^{\varepsilon} n|t|^s \exp\{-C_1(F,G)nt^2\} \, dt + C(F,G) \exp\{-C_6(F,G)n\}$$

$$\leq C(F,G,s) \int_{0}^{\infty} n|t|^s \exp\{-C_1(F,G)nt^2\} \, dt + C(F,G) \exp\{-C_6(F,G)n\}$$

$$\leq C(F,G,s) n^{-(s-1)/2}.$$

$\square$

8.3   Approximating Probability by Density

It is obvious that one can combine (3.1) and (8.2) to estimation the closeness of probability and density. As a rule such an approach is used solely for the normal density, that is, for

$$\varphi_{\mu,\sigma}(x) = \frac{1}{\sqrt{2\pi}\sigma} \exp\left\{-\frac{(x-\mu)^2}{2\sigma^2}\right\},$$

with the corresponding characteristic function

$$\widehat{\varphi}(t) = \exp\left\{it\mu - \frac{\sigma^2 t^2}{2}\right\}.$$

From (1.34) it follows that

$$\widehat{\varphi}(t)e^{-it\mu} = 1 - \frac{\sigma^2 t^2}{2} + \theta \frac{\sigma^4 t^4}{2}.$$

By (1.12) we have that for a random variable ξ with the third absolute moment and distribution F the following short expansion holds

$$\widehat{F}(t)e^{-\mu it} = 1 - \frac{\sigma^2 t^2}{2} + \theta \frac{\beta_3 |t|^3}{6}.$$

Here

$$\mu = \mathbb{E}\,\xi, \quad \sigma^2 = \text{Var}\,\xi = \mathbb{E}\,(\xi - \mu)^2, \quad \beta_3 = \mathbb{E}\,|\xi - \mu|^3.$$

Therefore

$$|\widehat{F}(t) - \widehat{\varphi}(t)| = |(\widehat{F}(t) - \widehat{\varphi}(t))e^{-\mu i t}| \leq C(\sigma^4 t^4 + \beta_3 |t|^3). \tag{8.8}$$

It is obvious that in a neighborhood of zero

$$|\widehat{F}(t)| \leq \exp\{-C\sigma^2 t^2\}.$$

Therefore we choose small ε and divide the integral into three parts

$$\left| \int_{-\pi}^{\pi} \widehat{F}^n(t)dt - \int_{-\infty}^{\infty} \widehat{\varphi}^n(t)dt \right| \leq \int_{-\varepsilon}^{\varepsilon} |\widehat{F}^n(t) - \widehat{\varphi}^n(t)|\,dt$$

$$+ \int_{\varepsilon < |t| \leq \pi} |\widehat{F}(t)|^n dt + \int_{|t| > \varepsilon} e^{-n\sigma^2 t^2/2} dt.$$

To estimate the first integral we choose ε to be sufficiently small (for example $\varepsilon = \sigma^2/(4\beta_3)$) and apply (8.8):

$$|\widehat{F}^n(t) - \widehat{\varphi}^n(t)| \leq n\exp\{-(n-1)C\sigma^2 t^2\}|\widehat{F}(t) - \varphi(t)| \leq n\exp\{-(n-1)C\sigma^2 t^2\}(\sigma^4 t^4 + \beta_3 |t|^3).$$

The last integral is estimated by $C(\sigma^2)\exp\{-Cn\varepsilon\sigma^2\}$. The estimate of the second integral depends on the properties of the approximated distribution.

We illustrate this approach by considering an approximation of the binomial distribution. Let $S = \xi_1 + \xi_2 + \cdots + \xi_n$, $P(\xi = 1) = p > 0$, $P(\xi = 0) = 1 - p = q$ and assume all ξ_i are independent.

Theorem 8.2 *The following estimate holds*

$$\sup_{m \in \mathbb{N}} \left| \sqrt{npq}P(S = m) - \frac{1}{\sqrt{2\pi}} \exp\left\{ -\frac{(m - np)^2}{2npq} \right\} \right| \leq \frac{C}{\sqrt{npq}}.$$

Proof We have

$$\left| P(S = m) - \frac{1}{\sqrt{2\pi npq}} \exp\left\{ -\frac{(m - np)^2}{2npq} \right\} \right|$$

$$\leq \frac{1}{2\pi} \int_{-\pi}^{\pi} |(q + pe^{it})^n - \exp\{inpt - npqt^2/2\}|\,dt$$

$$+ \frac{1}{2\pi} \int_{|t| > \pi} \exp\{-npqt^2/2\}dt. \tag{8.9}$$

Let $|t| \leq \pi$. Then due to (1.40), (3.8), (8.8) and (1.50)

$$|(q + pe^{it})^n - \exp\{inpt - npqt^2/2\}|$$
$$\leq n\max(\exp\{-2(n-1)pq\sin^2(t/2)\}, \exp\{-(n-1)pqt^2/2\})$$
$$\times(p^2q^2t^4 + pq(p^2 + q^2)|t|^3)$$
$$\leq Cnpq|t|^3\exp\{-0.2npqt^2\}.$$

Therefore

$$\frac{1}{2\pi}\int_{-\pi}^{\pi}|(q + pe^{it})^n - \exp\{inpt - npqt^2/2\}|dt \leq C(npq)^{-1}.$$

Next, observe that

$$\frac{1}{2\pi}\int_{|t|>\pi}\exp\{-npqt^2/2\}dt \leq \frac{1}{\pi}e^{-npq\pi^2/4}\int_0^\infty\exp\{-npqt^2/4\}dt \leq \frac{C}{\sqrt{npq}}e^{-npq\pi^2/4}.$$

Substituting the last two estimates into (8.9) we complete the theorem's proof. $\square$

8.4   Estimates in the Kolmogorov Norm

It is not difficult to derive a complete analogue of the Tsaregradskii inequality. Let the characteristic functions $\widehat{F}(t), \widehat{G}(t)$ be absolutely integrable and F and G have finite first absolute moments. Then integrating the inversion formula for densities (8.2) we get

$$[F(b) - G(b)] - [F(a) - G(a)] = \int_a^b [f(y) - g(y)]dy$$

$$= \frac{1}{2\pi}\int_a^b\int_{-\infty}^{\infty}e^{-ity}[\widehat{F}(t) - \widehat{G}(t)]dtdy = \frac{1}{2\pi}\int_{-\infty}^{\infty}[\widehat{F}(t) - \widehat{G}(t)]\int_a^b e^{-ity}dydt$$

$$= \frac{1}{2\pi}\int_{-\infty}^{\infty}[\widehat{F}(t) - \widehat{G}(t)]\frac{e^{-itb} - e^{-ita}}{-it}dt.$$

Taking $a \to -\infty$ by Lebesgue's theorem we obtain the following inversion formula

$$F(b) - G(b) = \frac{1}{2\pi}\int_{-\infty}^{\infty}\frac{\widehat{F}(t) - \widehat{G}(t)}{-it}e^{-itb}dt.$$

Therefore

$$|F - G|_K \leqslant \frac{1}{2\pi} \int_{-\infty}^{\infty} \frac{|\widehat{F}(t) - \widehat{G}(t)|}{|t|} dt. \tag{8.10}$$

Observe that (8.10) requires absolute integrability of the difference of characteristic functions. In other words, we assume that both distributions have somewhat nice properties. In subsequent chapters, we will deal with more general Esseen type estimates, where integrability of the characteristic functions is unnecessary.

Proposition 8.1 *Let* $\mathcal{L}(\xi), \mathcal{L}(\eta) \in \mathcal{F}$ *have normal distributions with zero means and variances* $0 < a < b$, *respectively. Then*

$$|\mathcal{L}(\xi) - \mathcal{L}(\eta)|_K \leqslant C\frac{|b - a|}{a}.$$

Proof We have

$$\left| e^{-at^2/2} - e^{-bt^2/2} \right| = e^{-at^2/2} \left| 1 - e^{-(b-a)t^2/2} \right| \leqslant Ce^{-at^2/2}|b - a|t^2.$$

It remains to apply (8.10). □

8.5 Estimates in Total Variation

Let $F, G \in \mathcal{F}$ have densities $f(x)$ and $g(x)$. Then

$$\|F - G\| = \int_{-\infty}^{\infty} |f(x) - g(x)| \, dx.$$

The characteristic function method is based on the following estimate.

Lemma 8.1 *Let* $F, G \in \mathcal{F}$ *be absolutely continuous distributions having densities* $f(x)$ *and* $g(x)$, *respectively. Let*

$$\int_{-\infty}^{\infty} |x| |f(x) - g(x)| \, dx < \infty.$$

Then, for any $b > 0$,

$$\|F - G\| \leqslant \frac{1}{\sqrt{2}} \left(\int_{-\infty}^{\infty} \left(b|\widehat{F}(t) - \widehat{G}(t)|^2 + b^{-1}|\widehat{F}'(t) - \widehat{G}'(t)|^2 \right) dt \right)^{1/2}. \tag{8.11}$$

Observe that the total variation norm is invariant with respect to shift, that is, for any $a \in \mathbb{R}$ we have $\| F - G \| = \| I_a F - I_a G \|$. Therefore, in (8.11), $\widehat{F}(t)$ and $\widehat{G}(t)$ can be replaced by $\exp\{-ita\}\widehat{F}(t)$ and $\exp\{-ita\}\widehat{G}(t)$, respectively.

Proof Let $M = F - G$. Observe that due to the lemma's assumption

$$\widehat{M}'(t) = \widehat{F}'(t) - \widehat{G}'(t) = \int_{-\infty}^{\infty} (f(x) - g(x))(e^{itx})' dx = \int_{-\infty}^{\infty} (ix[f(x) - g(x)])e^{itx} dx.$$

Without loss of generality, we can assume that the right-hand side of (8.11) is finite. Therefore by Parseval's identity (1.19) we have, for $f(x) - g(x)$ and $ix(f(x) - g(x))$,

$$\int_{-\infty}^{\infty} (f(x) - g(x))^2 dx = \frac{1}{2\pi} \int_{-\infty}^{\infty} |\widehat{M}(t)|^2 dt,$$

$$\int_{-\infty}^{\infty} x^2 (f(x) - g(x))^2 dx = \frac{1}{2\pi} \int_{-\infty}^{\infty} |\widehat{M}'(t)|^2 dt.$$

Therefore

$$\begin{aligned}
\| M \| &= \int_{-\infty}^{\infty} |f(x) - g(x)| dx \\
&\leq \left(\int_{-\infty}^{\infty} \frac{1}{1 + x^2} dx \right)^{1/2} \left(\int_{-\infty}^{\infty} (1 + x^2)(f(x) - g(x))^2 dx \right)^{1/2} \\
&= \frac{1}{\sqrt{2}} \left(\int_{-\infty}^{\infty} [|\widehat{M}(t)|^2 + |\widehat{M}'(t)|^2] dt \right)^{1/2}.
\end{aligned} \qquad (8.12)$$

The total variation norm is invariant with respect to scaling. Let $F = \mathcal{L}(\xi)$, $G = \mathcal{L}(\eta)$ for some random variables ξ and η. Then

$$\| F - G \| = \| \mathcal{L}(\xi) - \mathcal{L}(\eta) \| = \| \mathcal{L}(\xi/b) - \mathcal{L}(\eta/b) \|$$

and (8.11) follows from (8.12) and the change of argument, since

$$\int_{-\infty}^{\infty} |\widehat{M}(t/b)|^2 dt = b \int_{-\infty}^{\infty} |\widehat{M}(t)|^2 dt, \quad \int_{-\infty}^{\infty} |\widehat{M}'(t/b)|^2 dt = b^{-1} \int_{-\infty}^{\infty} |\widehat{M}'(t)|^2 dt.$$

$\square$

Typically the first integral in (8.12) is of a smaller order than the second one. Therefore b in (8.11) can be chosen to compensate for this difference in accuracy.

Lemma 8.2 *Let the assumptions of Lemma 8.1 be satisfied. Then*

$$\| F - G \| \leq \left(\int\limits_{-\infty}^{\infty} | \widehat{F}(t) - \widehat{G}(t) |^2 \, dt \int\limits_{-\infty}^{\infty} | \widehat{F}'(t) - \widehat{G}'(t) |^2 \, dt \right)^{1/4}.$$

Proof We apply Lemma 8.1 with

$$b^2 = \int_{-\infty}^{\infty} | \widehat{M}'(t) |^2 dt \left(\int_{-\infty}^{\infty} | \widehat{M}(t) |^2 dt \right)^{-1}.$$

□

Theorem 8.3 *Let F and G be continuous distributions with zero means and bounded densities $f(x)$ and $g(x)$. Let, for some $s \geq 2$*

$$\int\limits_{-\infty}^{\infty} x^k (f(x) - g(x)) \, dx = 0, \quad (k = 1, 2, \ldots, s-1),$$

$$\int\limits_{-\infty}^{\infty} | x |^s (f(x) + g(x)) \, dx \leq C(F, G, s) < \infty.$$

Then

$$\| F^n - G^n \| \leq C(F, G, s) n^{-(s-2)/2}.$$

Proof Without loss of generality, we can assume that $n \geq 3$. Indeed, if $n = 1, 2$, then $\| F - G \| \leq \| F \| + \| G \| = 2 \leq C(s) \cdot 2^{-(s-2)/2}$.

We apply (8.13) with $b = \sqrt{n}$. Due to (8.5) there exists an $\varepsilon = \varepsilon(F, G)$ such that, for $| t | \leq \varepsilon$,

$$\max\{| \widehat{F}(t) |, | \widehat{G}(t) |\} \leq \exp\{-C_7(F, G) t^2\}$$

and, for $| t | > \varepsilon$,

$$\max\{| \widehat{F}(t) |, | \widehat{G}(t) |\} \leq \exp\{-C_8(F, G)\}.$$

Let ξ be a random variable with distribution F. We assumed that the mean of ξ is zero. Therefore, on one hand,

$$| \widehat{F}'(t) | = | \mathbb{E} \, e^{it\xi} i\xi | \leq \mathbb{E} | \xi | \leq C(F). \tag{8.13}$$

On the other hand, by (1.13)

$$|\widehat{F}'(t)| \leq |t| \mathbb{E}\,\xi^2 = C(F)|t|.$$

Similar estimates hold for $|\widehat{G}'(t)|$.

Due to the bounded densities $|\widehat{F}(t)|^2$ and $|\widehat{G}(t)|^2$ are integrable. Therefore

$$\int_{|t|>\varepsilon} (|\widehat{F}(t)^n - \widehat{G}(t)^n|^2)\, dt$$

$$\leq C(F,G)\exp\{-C_8(F,G)(n-2)\} \int_{|t|>\varepsilon} (|\widehat{F}(t)|^2 + |\widehat{G}(t)|^2)\, dt$$

$$\leq C(F,G)\exp\{-C(F,G)n\}$$

and

$$\int_{|t|>\varepsilon} (|(\widehat{F}(t)^n - \widehat{G}(t)^n)'|)^2\, dt$$

$$\leq n \int_{|t|>\varepsilon} (|\widehat{F}'(t)|\,|\widehat{F}(t)|^{n-1} + |\widehat{G}'(t)|\,|\widehat{G}(t)|^{n-1})^2\, dt$$

$$\leq nC(F,G)\exp\{-C_8(F,G)(n-3)\} \int_{|t|>\varepsilon} (|\widehat{F}(t)|^2 + |\widehat{G}(t)|^2)^2\, dt$$

$$\leq C(F,G)\exp\{-C(F,G)n\}.$$

Therefore the initial estimate reduces to

$$\|F^n - G^n\| \leq \frac{1}{\sqrt{2}}\left(\int_{-\varepsilon}^{\varepsilon} \left(\sqrt{n}\,|\widehat{F}^n(t) - \widehat{G}^n(t)|^2 + \frac{1}{\sqrt{n}}|(\widehat{F}^n(t) - \widehat{G}^n(t))'|^2\right)\right)^{1/2}$$

$$+C(F,G,s)n^{-(s-2)/2}. \tag{8.14}$$

From (1.12) and (1.13) it follows that

$$|\widehat{F}(t) - \widehat{G}(t)| \leq C(F,G,s)|t|^s, \quad |\widehat{F}'(t) - \widehat{G}'(t)| \leq C(F,G,s)|t|^{s-1}.$$

Consequently, for $|t| \leq \varepsilon$,

$$|\widehat{F}^n(t) - \widehat{G}^n(t)| \leq n\exp\{-C_7(F,G)nt^2\}|\widehat{F}(t) - \widehat{G}(t)| \leq C(F,G)n\exp\{-C_7(F,G)nt^2\}|t|^s.$$

Similarly, for $|t| \leq \varepsilon$,

$$|(\widehat{F}^n(t) - \widehat{G}^n(t))'|$$

$$\leq n|\widehat{F}'(t)| \, |\widehat{F}^{n-1}(t) - \widehat{G}^{n-1}(t)| + n|\widehat{G}(t)|^{n-1}|\widehat{F}'(t) - \widehat{G}'(t)|$$

$$\leq C(F,G)n^2 \exp\{-C_1(F,G)nt^2\}|t|^{s+1} + C(F,G)n \exp\{-C_7(F,G)nt^2\}|t|^{s-1}.$$

Now it suffices to substitute the last estimates into (8.14) and observe that

$$\int_{-\varepsilon}^{\varepsilon} e^{-Cnt^2}|t|^k \, dt \leq \int_{-\infty}^{\infty} e^{-Cnt^2}|t|^k \, dt \leq C(k)n^{-(k+1)/2}.$$

$\square$

8.6 Non-uniform Estimates

First, we formulate one general lemma, where absolute continuity is required just for approximation. In this sense, Lemma 8.3 is closer to the results of Chap. 9. As usual $\mu(x) = \mu\{(-\infty, x]\}$ and the Lebesgue-Stieltjes integral $\int f(x) d\mu(x) := \int f(x)\mu\{dx\}$.

Lemma 8.3 *Let $F(x)$ be a non-decreasing function and let $G(x)$ be a differentiable function of bounded variation on the real line. Let $F(-\infty) = G(-\infty)$, $F(\infty) = G(\infty)$. Suppose that $\int_{\mathbb{R}} |x|^s |d(F(x) - G(x))| < \infty$ and $|G'(x)| \leq A(1 + |x|)^s$, $(x \in \mathbb{R})$ for some $s \geq 2$, where A is a constant. Then, for any x and $T > 1$,*

$$|F(x) - G(x)| \leq C(s)(1 + |x|)^{-s}\left\{\int_{-T}^{T} \frac{\widehat{F}(t) - \widehat{G}(t)}{|t|}\,dt + \int_{-T}^{T} \frac{|\tilde{æ}_s(t)|}{|t|}\,dt + \frac{A}{T}\right\}.$$

Here

$$\tilde{æ}_s(t) = \int_{-\infty}^{\infty} e^{itx} d(x^s[F(x) - G(x)]).$$

The following lemma helps to deal with $\tilde{æ}_s(t)$.

Lemma 8.4 *Let $G(x)$ be a function of bounded variation on the real line. Suppose that $\lim_{|t| \to \infty} G(x) = 0$ and $\int_{\mathbb{R}} |x|^m |dG(x)| < \infty$ for some $m \geq 1$. Then $x^m G(x)$ is a function of bounded variation on the real line and*

$$(-it)^m \int_{-\infty}^{\infty} e^{itx} d(x^m G(x)) = m! \sum_{j=0}^{m} \frac{(-t)^j}{j!} \frac{d^j}{dt^j}\widehat{G}(t).$$

Both lemmas were specifically tailored with the normal approximation in mind, when $G(x) = \Phi(x)$ or $G(x)$ is the Edgeworth asymptotic expansion. Lemma 8.4 shows that some integration by parts is involved. When $F, G \in \mathcal{F}$ have bounded densities it is easier to apply integration by parts directly. Indeed, if a random variable ξ with distribution F and bounded density has finite second moment and $n \geqslant 3$, then by (8.1) and (1.13)

$$\left| \int_{-\infty}^{\infty} (\widehat{F^n}(t))' dt \right| \leqslant n\mathbb{E} \,|\,\xi\,| \int_{-\infty}^{\infty} |\,\widehat{F}(t)\,|^2 |\,\widehat{F}(t)\,|^{n-3} dt \leqslant C(F).$$

Example 8.1 We prove a non-uniform estimate for the difference of two normal densities $\phi_a(x) = \Phi'_a(x)$ and $\phi_b(x) = \Phi'_b(x)$. Let $0 < a < b$. Then integrating by parts the inversion formula (8.2) for $x \neq 0$, we get

$$| \phi_a(x) - \phi_b(x) | = \frac{1}{2\pi} \left| \int_{-\infty}^{\infty} e^{-itx} (e^{-at^2/2} - e^{-bt^2/2}) dt \right|$$

$$= \frac{1}{2\pi|x|} \left| \int_{-\infty}^{\infty} e^{-itx} (e^{-at^2/2} - e^{-bt^2/2})' dt \right|$$

$$\leqslant \frac{1}{2\pi|x|} \int_{-\infty}^{\infty} |\,(e^{-at^2} - e^{-bt^2})'\,| dt. \qquad (8.15)$$

In Proposition 8.1 we already proved that $|\exp\{-at^2/2\} - \exp\{-bt^2/2\}| \leqslant C(b - a)t^2 \exp\{-at^2/2\}$. Similarly, we obtain

$$\left| \left(e^{-at^2/2} - e^{-bt^2/2} \right)' \right| \leqslant |\,(-at + bt)e^{-at^2/2}\,| + b|\,t\,||\,e^{-bt^2/2} - e^{-at^2/2}\,|$$

$$\leqslant C(b - a)(|\,t\,| + b|\,t\,|^3)e^{-at^2/2}.$$

Substituting the last estimate into (8.15) we arrive at $|\,x\,||\,\phi_a(x) - \phi_b(x)\,| \leqslant (b - a)a^{-1}$, which also holds for $x = 0$. It is not difficult to prove directly, applying (8.2), that $\sup_x |\,\phi_a(x) - \phi_b(x)\,| \leqslant (b - a)a^{-3/2}$. Combining both estimates we can write the non-uniform estimate in a more standard form

$$\left(1 + \frac{|x|}{\sqrt{a}} \right) |\,\phi_a(x) - \phi_b(x)\,| \leqslant \frac{C(b - a)}{a\sqrt{a}}, \qquad x \in \mathbb{R}.$$

Of course, the second derivative will lead to non-uniform estimates containing x^2, etc.

8.7 Problems

8.1 Let F have bounded density $f(x)$ and $\widehat{F}(t)$ be absolutely integrable over $\mathbb{R}$. Prove that

$$\int_{-\infty}^{\infty} f^2(x)dx = \frac{1}{2\pi} \int_{-\infty}^{\infty} |\widehat{F}(t)|^2 dt.$$

8.2 Let $F\{-1\} = F\{1\} = 1/4$, $F\{0\} = 1/2$. Prove that approximation by the normal density is of the order $O(n^{-3/2})$.

8.3 Let $F^n \in \mathcal{F}_+$ have bounded density $f_n(x)$, finite variance σ^2 and finite fourth moment. Prove that, for $n \geqslant 4$,

$$\sup_x \left| f_n(x) - \frac{1}{\sqrt{2\pi}n\sigma} e^{-x^2/2n\sigma^2} \right| \leqslant \frac{C(F)}{n\sqrt{n}}.$$

8.4 Let $F \in \mathcal{F}_+$ have bounded density, finite variance σ^2 and finite fourth moment. Prove that, for $n \geqslant 4$,

$$|F^n - \Phi_\sigma^n|_K \leqslant \frac{C(F)}{n}.$$

8.5 Let $F \in \mathcal{F}_+$ have bounded density and finite variance σ^2. Prove that

$$\| F^n - F^{n+1} \| \leqslant \frac{C(F)}{n}.$$

8.6 Let F_i have a Laplace distribution with parameters 0 and $\lambda_i > 0$, $(i = 1, \ldots, n)$. Then

$$\widehat{F}_i(t) = \frac{\lambda_i^2}{\lambda_i^2 + t^2}, \quad \int_{\mathbb{R}} xF_i\{dx\} = 0, \quad \int_{\mathbb{R}} x^2 F_i\{dx\} = \frac{2}{\lambda_i^2}.$$

Let $0 < \lambda_1 \leqslant \lambda_2 \leqslant \cdots \leqslant \lambda_n \leqslant C, n \in \mathbb{N}$. Prove that

$$\left| \prod_{i=1}^{n} F_i - \Phi_\sigma \right|_K \leqslant \frac{C}{n}.$$

Here $\sigma^2 = \sum_{i=1}^{n}(2/\lambda_i^2)$.

8.7 In the previous problem, obtain an estimate for the total variation norm.

Bibliographical Notes

The estimate (8.4) was proved in [142]. Lemmas 8.1 and 8.2 can be found in [79] p. 29 and [5] p. 139, respectively. Both lemmas from Sect. 8.6 were proved in [105]. Observe that they were not presented in [106].

Chapter 9
The Esseen Type Estimates

In previous chapters, tools for estimating accuracy of approximation were tailored specifically to match structural properties of the measures involved. Different inversion formulas for lattice and absolutely continuous distributions were used. In this chapter, we investigate the classical universal characteristic function method for the Kolmogorov norm, which is usually associated with the names of Esseen and Le Cam. It allows us to compare distributions with different structures under the existence of moments of low order or even when no finite moment exists.

9.1 General Inversion Inequalities

In this section, we present inequalities allowing us to switch from measures to their Fourier transforms and outline benefits and drawbacks of their application. Similarly to previous chapters, we denote by $F(x) = F\{(-\infty, x]\}$ the distribution function of the distribution (measure) F. The following inversion formula describes the relation between $F(x)$ and its characteristic function $\widehat{F}(t)$: if $a < b$ are points of continuity of $F(x)$, then

$$F(b) - F(a) = \frac{1}{2\pi} \lim_{T \to \infty} \int_{-T}^{T} \frac{e^{-itb} - e^{-ita}}{-it} \, dt. \qquad (9.1)$$

Though (9.1) is of limited practical use, it provides an insight into the essential step of the characteristic function method: estimation of the difference of Fourier transforms in a neighborhood of zero.

© Springer International Publishing Switzerland 2016
V. Čekanavičius, *Approximation Methods in Probability Theory*, Universitext,
DOI 10.1007/978-3-319-34072-2_9

We begin with the general version of the inversion inequality. For nonnegative finite measure G on $\mathbb{R}$ and $h \in (0, \infty)$ set $|G|_{0-} = 0$ and·

$$|G|_{h-} = \sup_{x \in \mathbb{R}} G((x, x + h]) = \sup_{x \in \mathbb{R}} G([x, x + h)) = \sup_{x \in \mathbb{R}} G((x, x + h)).$$

Recall that, for $M \in \mathcal{M}$, we denote by $M = M^+ - M^-$ the Jordan-Hahn decomposition of M.

Lemma 9.1 *Let* $M_1, M_2 \in \mathcal{M}$ *with* $M_1\{\mathbb{R}\} = 0$, *and set* $M = M_1 + M_2$. *For* $y \in [0, \infty)$, *let*

$$\rho(y) = \min\{|M^+|_{y-}, |M^-|_{y-}\}.$$

Then, for arbitrary $\vartheta \in (0, \infty)$ *and* $r \in (0, 1)$, *we have*

$$|M|_K \leq \frac{1}{2r}\|M_1\| + \frac{1}{2\pi r}\int_{|t|<1/\vartheta}\left|\frac{\widehat{M_2}(t)}{t}\right| dt + \frac{1+r}{2r}\rho(4\,\eta(r)\vartheta),$$

where $\eta(r) \in (0, \infty)$ *is defined by the equation*

$$\frac{1+r}{2} = \frac{2}{\pi}\int_0^{\eta(r)}\frac{\sin^2(x)}{x^2}\,dx.$$

Lemma 9.1 is usually combined with the following result.

Lemma 9.2 *For* $G \in \mathcal{F}$, $W \in \mathcal{M}$ *with* $W\{\mathbb{R}\} = 0$, *and* $\vartheta \in (0, \infty)$, *we have*

$$|(WG)^+|_{\vartheta-} \leq \frac{1}{2}\|W\|\,|G|_{\vartheta-}. \tag{9.2}$$

Lemma 9.1 can be adapted to fit certain needs. For example, we can estimate the smoothing effect by some concentration function.

Lemma 9.3 *Let* $h > 0$, $W \in \mathcal{M}$, $W\{\mathbb{R}\} = 0$, $H \in \mathcal{F}$ *and* $|\widehat{H}(t)| \leq C\widehat{G}(t)$, *for* $|t| \leq 1/h$ *and some* $G \in \mathcal{F}_+$. *Then*

$$|WH|_K \leq C\int_{|t|\leq 1/h}\left|\frac{\widehat{W}(t)\widehat{H}(t)}{t}\right| dt + C\|W\|Q(H, h)$$

$$\leq C\left(\sup_{|t|\leq 1/h}\frac{|\widehat{W}(t)|}{|t|}\cdot\frac{1}{h} + \|W\|\right)Q(G, h).$$

Proof We apply Lemma 9.1 with $W_1 = 0$, $W = W_2 = WH$, and $r = 0.5$. Then by (1.20) and (9.2) we have

$$\rho(4\,\eta(r)h) \leq C|(WH)^+|_{4\eta(r)h} \leq C\|W\|Q(H, 4\eta(r)h) \leq C\|W\|Q(H, h).$$

Moreover, applying (1.23) and (1.24), we prove that

$$Q(H, h) \leq Ch \int_{|t| \leq 1/h} |\widehat{H}(t)| \, dt \leq Ch \int_{|t| \leq 1/h} \widehat{G}(t) \, dt \leq CQ(G, h)$$

and

$$\int_{|t| \leq 1/h} |\widehat{H}(t)| \, dt \leq C\frac{1}{h} h \int_{|t| \leq 1/h} \widehat{G}(t) \, dt \leq \frac{1}{h} CQ(G, h).$$

This, obviously, completes the proof of the lemma. □

Next, we formulate five other versions of Lemma 9.1 for arbitrary distributions $F, G \in \mathcal{F}$.

Lemma 9.4 *Let $F \in \mathcal{F}$, $G \in \mathcal{F}$, $h > 0$. Then*

$$|F - G|_K \leq C \int\limits_{-1/h}^{1/h} \left| \frac{\widehat{F}(t) - \widehat{G}(t)}{t} \right| dt + C \min\{Q(F, h), Q(G, h)\}. \tag{9.3}$$

Lemma 9.5 *Let ξ and η be two random variables with distributions F and G, respectively. Then for every $h \geq 0$,*

$$|F - G|_K \leq P(|\xi - \eta| > h) + \min\{Q(F, h), Q(G, h)\}.$$

Note that, in general, we do not assume $G(x)$ to be continuous. The remainder term is simpler for G with bounded density.

Lemma 9.6 *Let $F, G \in \mathcal{F}$ and let $\sup_x G'(x) \leq A$. Then, for every $T > 0$ and every $b > 1/(2\pi)$,*

$$|F - G|_K \leq b \int\limits_{-T}^{T} \left| \frac{\widehat{F}(t) - \widehat{G}(t)}{t} \right| dt + C(b)\frac{A}{T}. \tag{9.4}$$

Here $C(b)$ is a positive constant depending only on b.

We present one more version of the Esseen type inequality, which is not a direct corollary of Lemma 9.1, though has a similar structure.

Lemma 9.7 *For $F, G \in \mathcal{F}$ and $T > 0$ the following inequality holds:*

$$|F - G|_K \leq \frac{1}{2\pi} \int\limits_{-T}^{T} \frac{|\widehat{F}(t) - \widehat{G}(t)|}{|t|} \, dt + \frac{1}{T} \int\limits_{-T}^{T} |\widehat{F}(t)| \, dt + \frac{1}{T} \int\limits_{-T}^{T} |\widehat{G}(t)| \, dt. \tag{9.5}$$

Finally, we formulate one version of an Esseen type inversion lemma suitable for a precise estimation of constants.

Lemma 9.8 *Let $F, G \in \mathcal{F}$. Then, for any $T \geq T_1 > 0$,*

$$|F - G|_K \leq \frac{1}{T} \int_{-T_1}^{T} |K(t/T)| |\widehat{F}(t) - \widehat{G}(t)| dt + \frac{1}{T} \int_{T_1 < |t| \leq T} |K(t/T)| |\widehat{F}(t)| dt$$

$$+ \int_{-T_1}^{T} \left| \frac{1}{T} K\left(\frac{t}{T}\right) - \frac{i}{2\pi t} \right| |\widehat{G}(t)| dt + \frac{1}{2\pi} \int_{|t| > T_1} \frac{|\widehat{G}(t)|}{|t|} dt.$$

Here

$$K(t) = \frac{1}{2}(1 - |t|) + \frac{i}{2}\left((1 - |t|)\cos(\pi t) + \frac{\text{sign} t}{\pi}\right), \quad -1 \leq t < 1.$$

It is also known that

$$|K(t)| \leq \frac{1.0253}{2\pi |t|}, \qquad \left| K(t) - \frac{i}{2\pi t} \right| \leq \frac{1}{2}\left(1 - |t| + \frac{\pi^2 t^2}{18}\right).$$

General as they are, these lemmas do not cover the inversion formulas from previous chapters.

Example 9.1 Let $F = ((1 - p)I + pI_1)^n$, $G = \exp\{np(I_1 - I)\}$, $n \in \mathbb{N}$, $p \leq 1/2$, $np \geq 1$. Then, as noted in Example 5.3

$$|\widehat{F}(t) - \widehat{G}(t)| \leq Cnp^2 \sin^2(t/2) \exp\{-np \sin^2(t/2)\}$$

and from the Tsaregradskii inequality (4.2) it follows that

$$|F - G|_K \leq Cp. \tag{9.6}$$

Next apply (9.3) with $h = 1/\pi$. The integral on the right-hand side of (9.3), up to a constant coincides with the Tsaregradskii inequality. For the estimate of $Q(G, h)$ we apply (1.23) and (3.3) and obtain

$$|F - G|_K \leq Cp + \frac{C}{\sqrt{np}}.$$

It is obvious that, for $p \leq n^{-1/3}$, the last estimate is significantly worse than (9.6).

Typical application. To estimate $|F - G|_K$, $F, G \in \mathcal{F}$ we

- Estimate $|\widehat{F}(t) - \widehat{G}(t)|/|t|$ in a neighborhood of zero. For distributions having finite moments, (1.12) can be applied.
- Estimate the concentration function $Q(G, h)$.
- Apply Lemma 9.1.

Advantages. No assumptions on distributions. Closeness of the characteristic functions are necessary only for small t. No need to estimate $|\widehat{F}(t)|$ outside a neighborhood of zero. Usually no need to estimate $Q(F, h)$. Easy to extend for estimation of asymptotic expansions.

Drawbacks. In the case of schemes of series (triangular arrays) the accuracy of approximation is rarely of the correct order. Absolute constants in estimates are not small.

In the following sections we demonstrate how the inversion lemmas can be applied.

9.2 The Berry-Esseen Theorem

In this section we prove the Berry-Esseen theorem for iid rvs having a finite third absolute moment. Let $F \in \mathcal{F}$ and

$$\int_{-\infty}^{\infty} x F\{dx\} = 0, \quad \int_{-\infty}^{\infty} x^2 F\{dx\} = \sigma^2; \quad \int_{-\infty}^{\infty} |x|^3 F\{dx\} = \beta_3 < \infty. \tag{9.7}$$

As before, we denote by Φ_σ ($\Phi \equiv \Phi_1$) the normal distribution with zero mean and variance σ^2, i.e. with characteristic function

$$\widehat{\Phi}_\sigma(t) = \exp\left\{-\frac{\sigma^2 t^2}{2}\right\}.$$

Theorem 9.1 *Let conditions (9.7) be satisfied. Then, for any $n \in \mathbb{N}$, the following inequality holds*

$$|F^n - \Phi_\sigma^n|_K \leq C \frac{\beta_3}{\sqrt{n}\sigma^3}. \tag{9.8}$$

Proof We shall use (9.4) with $T = \sigma^2/(4\beta_3)$. Note that $\Phi_\sigma^n = \Phi_{\sqrt{n}\sigma}$ and its density is equal to

$$\phi(x) = \frac{1}{\sqrt{2\pi n}\sigma} \exp\{-x^2/(n\sigma^2)\}$$

and is bounded by $(\sqrt{2\pi n}\sigma)^{-1}$. Therefore the second summand in (9.4) is bounded by $C\beta_3 n^{-1/2}\sigma^{-3}$.

We shall expand the characteristic function $|\widehat{F}(t)|^2$ in powers of (it). Note that $|\widehat{F}(t)|^2$ corresponds to the difference of two independent random variables $(\xi - \tilde{\xi})$, both having distribution F. Thus,

$$|\widehat{F}(t)|^2 \leqslant 1 - \sigma^2 t^2 + \frac{4}{3}\beta_3|t|^3.$$

Consequently, for $|t| \leqslant \sigma^2/(4\beta_3)$,

$$|\widehat{F}(t)| \leqslant \exp\left\{1 - \frac{1}{2}\sigma^2 t^2 + \frac{2}{3}\beta_3|t|^3\right\} \leqslant \exp\left\{-\frac{1}{3}\sigma^2 t^2\right\}.$$

Let $|t| \leqslant \sigma^2/(4\beta_3)$ and $n\beta_3|t|^3 > 1$. Then

$$|\widehat{F^n}(t) - \widehat{\Phi_\sigma^n}(t)| \leqslant 2n \exp\left\{-\frac{n}{3}\sigma^2 t^2\right\} \leqslant 2n\beta_3|t|^3 \exp\left\{-\frac{n}{3}\sigma^2 t^2\right\}.$$

If $|t| \leqslant \sigma^2/(4\beta_3)$ and $n\beta_3|t|^3 \leqslant 1$, then we directly estimate

$$|\widehat{F}(t) - \widehat{\Phi_\sigma}(t)| \leqslant C(\beta_3|t|^3 + \sigma^4 t^4).$$

But by the relation between moments

$$\sigma \leqslant \beta_3^{1/3}$$

and, for $n\beta_3|t|^3 \leqslant 1$,

$$\sigma^2 t^2 \leqslant (\beta_3|t|^3)^{2/3} \leqslant 1.$$

Consequently,

$$|\widehat{F^n}(t) - \widehat{\Phi_\sigma^n}(t)| \leqslant n \exp\left\{-\frac{(n-1)}{3}\sigma^2 t^2\right\}|\widehat{F}(t) - \widehat{\Phi_\sigma}(t)| \leqslant$$

$$Cn\beta_3|t|^3 \exp\left\{-\frac{n}{3}\sigma^2 t^2\right\}.$$

We prove that, for $|t| \leqslant \sigma^2/(4\beta_3)$,

$$|\widehat{F^n}(t) - \widehat{\Phi}_\sigma^n(t)| \leqslant Cn\beta_3|t|^3 \exp\left\{-\frac{n}{3}\sigma^2 t^2\right\}.$$

To complete the proof of the theorem it suffices to use the last estimate in the integral:

$$\int_{-T}^{T} \left|\frac{\widehat{F^n}(t) - \widehat{\Phi}_\sigma^n(t)}{t}\right| dt \leqslant C \int_0^\infty n\beta_3 t^2 \exp\left\{-\frac{n}{3}\sigma^2 t^2\right\} dt \leqslant C \frac{\beta_3}{\sqrt{n}\sigma^3}.$$

□

9.3 Distributions with $1 + \delta$ Moment

In the previous section, we applied the Esseen inequality to distributions having finite third absolute moments. In this section we show that the same general approach works even for distributions with infinite variances.

Throughout this section we assume that F does not depend on n, has mean zero and $1 + \delta$ absolute moment ($0 < \delta \leqslant 1$) and satisfies Cramer's (C) condition:

$$\lim_{|t| \to \infty} \sup |\widehat{F}(t)| < 1. \tag{C}$$

First we recall some facts about the (C) condition from Chap. 1. If F satisfies the (C) condition, then for any $\varepsilon = \varepsilon(F) > 0$ and for all $|t| > \varepsilon$,

$$|\widehat{F}(t)| \leqslant e^{-C(F)}. \tag{9.9}$$

We combine this property with the fact that for any non-degenerate F, for small $\varepsilon = \varepsilon(F)$ and for $|t| \leqslant \varepsilon$,

$$|\widehat{F}(t)| \leqslant \exp\{-C(F)t^2\}. \tag{9.10}$$

First we apply Lemma 9.7 to estimate subsequent convolutions of F.

Theorem 9.2 *Let $F \in \mathcal{F}$ satisfy Cramer's (C) condition, have mean zero and $1 + \delta$ absolute moment ($0 < \delta \leqslant 1$). Then, for any $n \in \mathbb{N}$,*

$$|F^n - F^{n+1}|_K \leqslant C(F)n^{-\delta}.$$

Proof Due to (9.9), for any $T > \varepsilon$,

$$\frac{1}{T}\int_0^T |\widehat{F}(t)|^n \, dt \leq \frac{1}{T}\left(\int_0^\varepsilon |\widehat{F}(t)|^n \, dt + \int_\varepsilon^T |\widehat{F}(t)|^n \, dt\right)$$

$$\leq \frac{1}{T}\int_0^\infty e^{-C(F)nt^2} \, dt + e^{-C(F)n} \leq C(F)(T^{-1} + e^{-Cn}). \qquad (9.11)$$

Similarly,

$$\int_{-T}^T \frac{|\widehat{F}^n(t)(\widehat{F}(t) - 1)|}{|t|} \, dt \leq 2\int_0^\varepsilon \frac{|\widehat{F}^n(t)(\widehat{F}(t) - 1)|}{|t|} \, dt + \frac{4}{\varepsilon}\int_\varepsilon^T |\widehat{F}(t)|^n \, dt$$

$$\leq 2\int_0^\varepsilon \frac{|\widehat{F}^n(t)(\widehat{F}(t) - 1)|}{|t|} \, dt + C(F)Te^{-C(F)n}. \qquad (9.12)$$

In a neighborhood of zero, we use the existence of a finite $1 + \delta$ moment and mean zero. Applying (1.33) we prove

$$|\widehat{F}(t) - 1| = \left|\int_{-\infty}^\infty (e^{itx} - 1) F\{dx\}\right| = \left|\int_{-\infty}^\infty (e^{itx} - 1 - itx) F\{dx\}\right|$$

$$\leq \int_{-\infty}^\infty |e^{itx} - 1 - itx| F\{dx\} \leq \frac{1}{2}\int_{|tx|\leq 1} t^2x^2 \, F\{dx\} + 2\int_{|tx|>1} |tx| F\{dx\}$$

$$\leq 3\int_{-\infty}^\infty |tx|^{1+\delta} F\{dx\} \leq C(F)|t|^{1+\delta} \qquad (9.13)$$

and, therefore,

$$\int_0^\varepsilon \frac{|\widehat{F}^n(t)(\widehat{F}(t) - 1)|}{|t|} \, dt \leq C(F)\int_0^\infty e^{-C(F)nt^2} t^\delta \, dt \leq C(F)n^{-\delta}.$$

Substituting the last estimate into (9.12) we obtain

$$\int\limits_{-T}^{T} \frac{|\widehat{F^n}(t)(\widehat{F}(t) - 1)|}{|t|} \, dt \leqslant C(F)n^{-\delta} + C(F)Te^{-C(F)n}.$$

It remains to use (9.5) with $T = n$ and to apply (9.11). $\square$

Next, we illustrate the application of Lemma 9.3 by considering exponential smoothing.

Theorem 9.3 *Let the conditions of Theorem 9.2 be satisfied, $k \in \mathbb{N}$. Then*

$$| (F - I)^k \exp\{n(F - I)\} |_K \leqslant C(F, k)n^{-k(1+\delta)/2}. \tag{9.14}$$

Proof We apply Lemma 9.3 with $W = (F - I)^k$, $H = \exp\{n(F - I)\}$ and $h = n^{-k}$. Observe that

$$\| W \| \leqslant \| F - I \|^k \leqslant (\| F \| + \| I \|)^k = 2^k. \tag{9.15}$$

Next, we estimate the characteristic function $\widehat{H}(t)$. There exists an $\varepsilon = \varepsilon(F)$ such that for $|t| \leqslant \varepsilon$

$$| \exp\{\widehat{F}(t) - 1\} | \leqslant \exp\{-C(F)t^2\}, \tag{9.16}$$

since $\exp\{\widehat{F}(t) - 1\}$ is a non-degenerate distribution, see the remark before Theorem 9.2. Next, observe that for $|t| > \varepsilon$, $|\widehat{F}(t)| \leqslant \exp\{-C(F)\}$ and, therefore,

$$| \exp\{\widehat{F}(t) - 1\} | \leqslant \sum_{j=0}^{\infty} \frac{|\widehat{F}(t)|^j}{j!}e^{-1} = \exp\{|\widehat{F}(t)| - 1\} \leqslant \exp\{-C_1(F)\}. \tag{9.17}$$

Let $T = 1/h = n^k$. Then, taking into account (1.23), (9.17) and (9.16), we obtain

$$Q(\exp\{n(F - I)\}, h) \leqslant \frac{C}{T} \int_0^T | \exp\{n(\widehat{F}(t) - 1)\} | dt \leqslant \frac{C}{T}\left(\int_0^\varepsilon + \int_\varepsilon^T \right)$$

$$\leqslant \frac{C(F)}{T} \int_0^\infty e^{-C(F)nt^2}dt + C(F)e^{-C_1(F)n}$$

$$\leqslant \frac{C(F, k)}{n^k}. \tag{9.18}$$

Similarly to the proof of (9.12) we get

$$\int_{\varepsilon}^{T} \frac{|\widehat{F}(t) - 1|^k| \exp\{n(\widehat{F}(t) - 1)\}|}{|t|} dt \leqslant \frac{2^k}{\varepsilon} \int_{\varepsilon}^{T} e^{-C_1(F)n} dt$$

$$\leqslant C(F, k)Te^{-C_1(F)n} \leqslant \frac{C(F, k)}{n^k}. \qquad (9.19)$$

From (9.16) and (9.13) it follows that

$$\int_{0}^{\varepsilon} \frac{|\widehat{F}(t) - 1|^k| \exp\{n(\widehat{F}(t) - 1)\}|}{|t|} dt \leqslant C(F, k) \int_{0}^{\varepsilon} |t|^{k(1+\delta)-1} e^{-C(F)nt^2} dt$$

$$\leqslant C(F, k)n^{-k(1+\delta)/2}. \qquad (9.20)$$

Lemma 9.3 and estimates (9.15), (9.18), (9.19) and (9.20) complete the proof of (9.14). $\square$

9.4 Estimating Centered Distributions

The integral term of the inversion inequalities indirectly implies that the distributions involved should have at least one finite moment, since then (1.12) can be applied. This, however, is untrue. In this section, we demonstrate that the characteristic function method can be applied to an arbitrary distribution, provided it is suitably centered. Note also that, unlike the previous examples, F in this section can depend on n, that is, we consider the general scheme of series.

Theorem 9.4 *For any $n \in \mathbb{N}$ and any $F \in \mathcal{F}$ the following estimate holds*

$$\inf_{u} | F^n - I_{-nu} \exp\{n(FI_u^{\cdot} - I)\} |_K \leqslant Cn^{-1/3}. \qquad (9.21)$$

First we discuss the purpose of centering. For any $F \in \mathcal{F}$ we can find a finite interval containing the main probabilistic mass of the initial distribution. Therefore we can decompose F into a mixture of two distributions: one concentrated in the interval and the other concentrated outside the interval. Then we can shift (center) F ensuring that the first distribution has mean zero.

More precisely, let $0 \leqslant p \leqslant C < 1$, and let $F \in \mathcal{F}$. Then there exist $A, V \in \mathcal{F}$, h_-, h_+ and $u \in \mathbb{R}$ such that

$$FI_u = (1 - p)A + pV \quad FI_u\{(-\infty, -h_-]\} \geqslant p/2, \quad FI_u\{[h_+, \infty)\} \geqslant p/2, \qquad (9.22)$$

$$A\{[-h_-, h_+]\} = 1, \quad V\{(-h_-, h_+)\} = 0, \quad \int x A\{dx\} = 0. \qquad (9.23)$$

Set $h = \max(h_-, h_+)$. The distribution V is concentrated outside the open finite interval, which is not shorter than h. Therefore, if $h > 0$, then by definition

$$1 - Q(FI_u, h/2) \geqslant p/2.$$

Now, by the properties of the concentration function (1.20), (1.21), and (1.22) we get

$$Q((I_u F)^n, h) \leqslant C(np)^{-1/2}, \quad Q(\exp\{n(I_u F - I)\}, h) \leqslant C(np)^{-1/2}. \tag{9.24}$$

Distribution A is concentrated on a finite interval. Consequently, it has moments of all orders. Set

$$\sigma^2 = \int x^2 A\{dx\}.$$

The properties of A are summarized in the following lemma.

Lemma 9.9 *Let A be defined as above. Then*

(a) for all $t \in \mathbb{R}$

$$|\widehat{A}(t) - 1| \leqslant \sigma^2 t^2/2, \tag{9.25}$$

(b) if $h > 0$ then for all $|t| \leqslant 1/h$

$$\sigma^2 t^2 \leqslant 1, \quad |\exp\{\widehat{A}(t) - 1\}| \leqslant \exp\{-\sigma^2 t^2/3\}, \quad |\widehat{A}(t)| \leqslant \exp\{-\sigma^2 t^2/3\}. \tag{9.26}$$

Proof Estimate (9.25) directly follows from (1.12). For the proof of (9.26) observe that, if $|t| \leqslant 1/h$, then

$$\left| \widehat{A}(t) - 1 + \frac{\sigma^2 t^2}{2} \right| = \left| \int\limits_{|x| \leqslant h} \left(e^{itx} - 1 - itx + \frac{t^2 x^2}{2} \right) A\{dx\} \right|$$

$$\leqslant \frac{1}{6} \int\limits_{|x| \leqslant h} |tx|^3 A\{dx\} \leqslant \frac{1}{6} \int\limits_{|x| \leqslant h} |tx|^2 A\{dx\} = \frac{\sigma^2 t^2}{6}.$$

The estimate (9.26) follows from

$$|\widehat{A}(t)| \leqslant \left| \widehat{A}(t) - 1 + \frac{\sigma^2 t^2}{2} \right| + \left| 1 - \frac{\sigma^2 t^2}{2} \right| \leqslant \frac{\sigma^2 t^2}{6} + 1 - \frac{\sigma^2 t^2}{2} \leqslant \exp\left\{ -\frac{\sigma^2 t^2}{3} \right\}.$$

Similarly,

$$Re\left(1 - \widehat{A}(t)\right) = \frac{\sigma^2 t^2}{2} + Re\left(1 - \widehat{A}(t) - \frac{\sigma^2 t^2}{2}\right)$$

$$\geq \frac{\sigma^2 t^2}{2} - \left| Re\left(\widehat{A}(t) - 1 + \frac{\sigma^2 t^2}{2}\right)\right| \geq \frac{\sigma^2 t^2}{3}.$$

☐

Now we can proceed with the proof of the theorem.

Proof of Theorem 9.4 Note that (9.21) can be written as

$$\inf_u |(FI_u)^n - \exp\{n(FI_u - I)\}|_K \leq Cn^{-1/3}.$$

Let $p = n^{-1/3}$ and let us choose $A \in \mathcal{F}$ and h in the decomposition $FI_u = (1 - p)A + pV$ as defined by (9.22) and (9.23). Note that, without loss of generality, we can choose sufficiently large n. For example, we can take $n > 8$. Indeed, if $n \leq 8$, then the left-hand side of (9.21) is bounded by $2 \leq Cn^{-1/3}$.

· From the triangle inequality it follows that

$$|(FI_u)^n - \exp\{n(FI_u - I)\}|_K \leq J_1 + J_2.$$

Here

$$J_1 = |((1 - p)A + pV)^n - \exp\{n(1 - p)(A - I)\}((1 - p)I + pV)^n|_K,$$

and

$$J_2 = |\exp\{n(1 - p)(A - I)\}(((1 - p)I + pV)^n - \exp\{np(V - I)\})|_K.$$

By the properties of norms and (2.19)

$$J_2 \leq \|((1 - p)I + pV)^n - \exp\{np(V - I)\}\| \leq Cp = Cn^{-1/3}. \tag{9.27}$$

We apply (9.3) to estimate J_1. Taking into account (9.24) we then obtain

$$|J_1|_K \leq C \int_0^{1/h} \left|\frac{\widehat{J}_1(t)}{t}\right| dt + Cn^{-1/3}.$$

Observing that $V \in \mathcal{F}$ and, consequently, its characteristic function is less than or equal to 1, we estimate $\widehat{J}_1(t)$

$$
\begin{aligned}
|\widehat{J}_1(t)| &\leq \sum_{k=0}^{n} \binom{n}{k}(1-p)^k p^{n-k} |\widehat{V}(t)^{n-k}| \left| \widehat{A}(t)^k - e^{n(1-p)(\widehat{A}(t)-1)} \right| \\
&\leq \sum_{k=1}^{n} \binom{n}{k}(1-p)^k p^{n-k} \left| \widehat{A}(t)^k - e^{k(\widehat{A}(t)-1)} \right| \\
&\quad + \sum_{k=0}^{n} \binom{n}{k}(1-p)^k p^{n-k} \left| e^{k(\widehat{A}(t)-1)} - e^{n(1-p)(\widehat{A}(t)-1)} \right|.
\end{aligned}
\tag{9.28}
$$

Applying (9.25) and (9.26) we obtain

$$
\left| \widehat{A}(t)^k - \exp\{k(\widehat{A}(t)-1)\} \right| \leq Ck(\sigma^2 t^2) \exp\left\{ -\frac{k\sigma^2 t^2}{3} \right\}.
$$

Thus, if $k \geq 1$, then

$$
\int_0^{1/h} \frac{|\widehat{A}(t)^k - \exp\{k(\widehat{A}(t)-1)\}|}{t}\, dt \leq \frac{C}{k} \leq \frac{C}{k+1}.
$$

Next, by (1.47) we prove that

$$
\begin{aligned}
\sum_{k=1}^{n} \binom{n}{k}(1-p)^k p^{n-k} & \int_0^{1/h} \frac{|\widehat{A}(t)^k - \exp\{k(\widehat{A}(t)-1)\}|}{t}\, dt \\
&\leq C \sum_{k=1}^{n} \binom{n}{k}(1-p)^k p^{n-k}(k+1)^{-1} \leq \frac{C}{np} \leq Cn^{-1/3}.
\end{aligned}
\tag{9.29}
$$

Similarly,

$$
\begin{aligned}
&| \exp\{k(\widehat{A}(t)-1)\} - \exp\{n(1-p)(\widehat{A}(t)-1)\}| \\
&\qquad \leq C|k-n(1-p)|\, \sigma^2 t^2 \exp\{-\min\{k, n(1-p)\}\sigma^2 t^2/3\}.
\end{aligned}
$$

Let $k \geq 1$. Then

$$\int_0^{1/h} \frac{| \exp\{k(\widehat{A}(t) - 1)\} - \exp\{n(1-p)(\widehat{A}(t)-1)\} |}{|t|} \, dt$$

$$\leq C \frac{|k - n(1-p)|}{k+1} + C \frac{|k - n(1-p)|}{n(1-p)}. \qquad (9.30)$$

For $k = 0$, we can prove (9.30) directly. It suffices to note that

$$\sigma^2 = \int_{-\infty}^{\infty} x^2 A\{dx\} = \int_{-h}^{h} x^2 A\{dx\} \leq h^2.$$

Therefore

$$\int_0^{1/h} n(1-p)\sigma^2 t \, dt \leq Cn(1-p)\frac{\sigma^2}{h^2} \leq Cn(1-p).$$

Thus, we have established that (9.30) holds for $0 \leq k \leq n$. Let ξ be the binomial random variable with parameters n and $1 - p$. Then, taking into account (1.47) and (9.30), we obtain

$$\sum_{k=0}^{n} \binom{n}{k} (1-p)^k p^{n-k} \int_0^{1/h} \frac{\left| e^{k(\widehat{A}(t)-1)} - e^{n(1-p)(\widehat{A}(t)-1)} \right|}{t} \, dt$$

$$\leq C \sum_{k=0}^{n} \binom{n}{k} (1-p)^k p^{n-k} |k - n(1-p)| \left(\frac{1}{n(1-p)} + \frac{1}{k+1} \right)$$

$$\leq \frac{C}{n(1-p)} \mathbb{E} \, |\xi - \mathbb{E}\xi| + C \sqrt{\mathbb{E}(\xi - \mathbb{E}\xi)^2} \left(\sum_{k=0}^{n} \binom{n}{k} (1-p)^k p^{n-k} \frac{1}{(k+1)^2} \right)^{1/2}$$

$$\leq C \sqrt{\frac{p}{n}} + C \sqrt{\frac{np(1-p)}{n^2 p^2}} \leq Cn^{-1/3}. \qquad (9.31)$$

The theorem's assertion follows from (9.27), (9.28), (9.29) and (9.31). $\square$

9.5 Discontinuous Distribution Functions

For the sake of completeness we end this chapter by presenting rather cumbersome inversion formulas for distribution functions with a specific emphasis on discontinuity points. In this section we assume that $G(x)$ is a bounded variation function defined on $\mathbb{R}$ with a set of discontinuity points $A_G = \{\ldots, x_{-2}, x_{-1}, x_0, x_1, x_2, \ldots\}$. Let $F(x)$ be a distribution function with the set of discontinuity points A_F satisfying condition (A):

$$A_F \supseteq A_G \cap \{x : x_{min} \leqslant x \leqslant x_{max}\}. \tag{A}$$

Here x_{min} is the largest x such that $F(x) = 0$ and x_{max} is the smallest x such that $F(x+) = 1$. Later in this section we assume that both functions F and G are left-continuous, that is $F(x) = F\{(-\infty, x)\}$. Note that analogous results also hold when both functions are right-continuous.

Theorem 9.5 *Case 1. If $F(-\infty) = G(-\infty)$, condition (A) is satisfied and G is a purely discontinuous function, then the following estimates hold*

$$|F - G|_K \leqslant \frac{J_T + \text{æ}_F U(x)}{2U(x) - 1},$$

$$|F - G|_K \leqslant \frac{J_T + (\ell_{T_1} + R_F(T_1) + \bar{\text{æ}}_G/T_1)U(x)}{2U(x) - 1}.$$

Here

$$J_T = \frac{1}{2\pi} \int_{-T}^{T} \frac{|\widehat{F}(t) - \widehat{G}(t)|}{|t|}\, dt, \qquad \ell_{T_1} = \frac{T_1}{2\pi} \int_{-\infty}^{\infty} \frac{|\widehat{F}(t) - \widehat{G}(t)|}{|t|\sqrt{T_1^2 + t^2}}\, dt,$$

$$\text{æ}_F = \max_j (F(x_{j+1} - 0) - F(x_j + 0)), \qquad \bar{\text{æ}}_G = 1 + 1.25\|G\|,$$

$$U(x) = \int_{|u|<x} p(u)\,du, \qquad p(u) = \frac{1}{2\pi}\left(\frac{\sin(u/2)}{u/2}\right)^2,$$

T_1, T and x are arbitrary numbers satisfying

$$T_1 \geqslant 2, \qquad \frac{\ln T_1}{T_1} \leqslant \frac{\beta}{3}, \qquad T \geqslant \frac{2x}{\beta}, \qquad x > \alpha,$$

α is a unique positive root of the equation $2U(x) - 1 = 0$, $\beta = \inf_j(x_{j+1} - x_j)$ and

$$R_F(T_1) = \sup_j \left[(F(x_{j+1} - 0) - F(x_{j+1} - \varepsilon_{T_1})) + (F(x_j + \varepsilon_{T_1}) - F(x_j + 0)) \right],$$

$$\varepsilon_{T_1} = \frac{\ln T_1}{T_1}.$$

Case 2. If $F(-\infty) = G(-\infty)$, *condition (A) is satisfied, G is not a purely discontinuous function having a uniformly bounded derivative in the continuity intervals*

$$\sup_j \ \sup_{x_j < x < x_{j+1}} \ G'(x) < q < \infty, \quad (q \text{ is some constant}),$$

then

$$|F - G|_K \leq \frac{J_T + \mathrm{æ}_{F-G} U(x)}{2U(x) - 1},$$

$$|F - G|_K \leq \frac{J_T + (\ell_{T_1} + R_F(T_1) + \mathrm{æ}_{G,T_1}/T_1)U(x)}{2U(x) - 1}.$$

Here

$$\mathrm{æ}_{F-G} = \sup_j \ \sup_{(x,y) \in (x_j, x_{j+1})} \ |F(y) - G(y) - F(x) + G(x)|, \quad \mathrm{æ}_{G,T_1} = 1 + \|G\| + q \ln T_1$$

and T_1, T, x are arbitrary numbers satisfying

$$T_1 \geq 2, \quad \frac{\ln T_1}{T_1} \leq \frac{\beta}{6}, \quad T \geq \frac{4x}{\beta}, \quad x > \alpha.$$

Case 3. If $F(-\infty) = G(-\infty)$, *condition (A) is satisfied and, for all $0 < y < y'$, there exists a finite number q_G satisfying*

$$|G(y') - G(y)| \leq q_G(y' - y),$$

then, for any $x > \alpha$ and $T \geq 4x/\beta$,

$$|F - G|_K \leq \frac{J_T + (\mathrm{æ}_F + q_G x/T)U(x)}{2U(x) - 1}.$$

The proof of Theorem 9.5 can be found in [1]. Observe that all estimates in Theorem 9.5 contain explicit expressions and constants. Moreover, as noted in [1], one can take $\alpha = 1.7$.

We prove one supplementary result, which demonstrates that, in some cases, ℓ_{T_1} can be replaced by J_T.

Lemma 9.10 *Let $F, G \in \mathcal{F}$ and $H \in \mathcal{F}_+$, $0 < T < T_1$. Let, for some $a > 0$ and all $t \in \mathbb{R}$,*

$$|\widehat{F}(t) - \widehat{G}(t)| \leq a\widehat{H}(t).$$

Then, for any $\tau > 0$,

$$\ell_{T_1} \leq J_T + \frac{13a}{2\pi}Q(H, \tau)\left(\frac{T_1}{T\sqrt{T_1^2 + T^2}}\frac{1}{\tau} + 2 + \ln\frac{T_1}{T}\right).$$

Proof It is obvious that

$$\frac{T_1}{2\pi}\int_{-T}^{T}\frac{|\widehat{F}(t) - \widehat{G}(t)|}{|t|\sqrt{T^2 + t^2}}\,dt \leq J_T.$$

Therefore

$$\ell_{T_1} \leq J_T + \frac{1}{2\pi}\int_{|t|>T}\frac{T_1|\widehat{F}(t) - \widehat{G}(t)|}{|t|\sqrt{T_1^2 + t^2}}\,dt \leq J_T + \frac{a}{2\pi}\int_{-\infty}^{\infty}g(t)\widehat{H}(t)\,dt.$$

Here

$$g(t) = \begin{cases} \dfrac{T_1}{|t|\sqrt{T_1^2 + t^2}}, & |t| \geq T, \\[2ex] 0, & |t| < T. \end{cases}$$

Then

$$\sup_t g(t) = \frac{T_1}{T\sqrt{T_1^2 + T^2}}, \qquad \sup_{|s|\geq|t|} g(s) = \begin{cases} \dfrac{T_1}{|t|\sqrt{T_1^2 + t^2}}, & |t| \geq T, \\[2ex] \dfrac{T_1}{T\sqrt{T_1^2 + T^2}}, & |t| < T. \end{cases}$$

Observe that

$$\int_0^\infty \sup_{|s|\geq|t|} g(s)dt = \int_0^T + \int_T^{T_1} + \int_{T_1}^\infty$$

$$\leq \frac{T_1}{\sqrt{T_1^2 + T^2}} + \int_T^{T_1} \frac{1}{t}dt + \int_{T_1}^\infty \frac{T_1}{t^2}dt \leq 2 + \ln\frac{T_1}{T}.$$

It remains to apply Lemma 1.1. $\square$

9.6   Problems

9.1 Prove the Esseen theorem for the sum of independent non-identically distributed summands:

$$\left| \prod_{j=1}^n F_j - \Phi_\sigma \right|_K \leq C \frac{\sum_1^n \beta_3(F_i)}{(\sum_1^n \sigma^2(F_i))^3}.$$

Here $i = 1, 2, \ldots, n$ and $F_i \in \mathcal{F}$ and

$$\int_{-\infty}^\infty x F_i\{dx\} = 0, \quad \int_{-\infty}^\infty x^2 F_i\{dx\} = \sigma^2(F_i); \quad \int_{-\infty}^\infty |x|^3 F_i\{dx\} = \beta_3(F_i) < \infty$$

and Φ_σ denotes the normal distribution with variance $\sigma^2 = \sum_1^n \sigma^2(F_i)$, i.e.

$$\widehat{\Phi}_\sigma(t) = \exp\left\{ -\frac{1}{2} \sum_{i=1}^n \sigma^2(F_i)t^2 \right\}.$$

9.2 Prove that the rate $O(n^{-1/2})$ in (9.8), in general, cannot be improved. *Hint.* Consider $| F^{2n}\{(-\infty, 0)\} - \Phi(0) |$, where $F = 0.5I_{-1} + 0.5I_1$.

9.3 Let $F, G \in \mathcal{F}_s$ have absolutely integrable characteristic functions and let their densities be bounded by A. Prove that, for any $T > 0$,

$$|F - G|_K \leq \frac{1}{2\pi} \int_{-T}^T \frac{|\widehat{F}(t) - \widehat{G}(t)|}{|t|}dt + \frac{2A}{T}.$$

9.4 Suppose F does not depend on n, has mean zero and finite variance and satisfies Cramer's (C) condition. Prove that,

$$|F^n - \exp\{n(F - I)\}|_K \leq C(F)n^{-1}.$$

9.5 Let $F \in \mathcal{F}, a > 0$. Prove that

$$\inf_u | (FI_u - I) \exp\{a(FI_u - I)\} |_K \leqslant Ca^{-3/4}. \tag{9.32}$$

9.6 Prove that

$$\inf_u | (FI_u)^{n+1} - (FI_u)^n |_K \leqslant Cn^{-1/3}.$$

Bibliographical Notes

The inversion inequalities of Chapter 11 are various modifications of Esseen's inequality, see [59], and Le Cam's lemma, see [92] and Chapter 15 in [93]. Lemmas 9.1, 9.3 and 9.7 were respectively proved in [40, 42] and [19]. Note also Theorem 5.1 and 5.2 in [106] and Theorems 1.1 and 1.2 from Chapter 3 of [5]. Subsequent convolutions under the (C) condition were investigated in [36], see also [166]. Proofs of (9.9) and (9.10) can be found in [106], p. 14. Theorem 9.4 and Lemma 9.5 were proved in [92], see also [67, 80], Theorem 1.10 on p. 16 in [5] and Lemma 2 on p. 402 in [93]. Theorem 9.5 was proved in [1]. The inversion inequality from Lemma 9.8 is a special case of Prawitz's result, see [65, 109]. Recent years have witnessed a certain renaissance in establishing the best absolute constants for many classical estimates, such as the Berry-Esseen theorem. Lemma 9.8 plays a very important role in this research. It was successfully used by Shevtsova and her colleagues to obtain the best upper bound constants for the normal approximation in [86, 87, 139, 140], see also [100].

Chapter 10
Lower Estimates

In this chapter we consider the characteristic function method, when moments of higher order exist. For the Stein method, see Sect. 11.8.

10.1 Estimating Total Variation via the Fourier Transform

In general, one usually uses the following relation

$$|M|_K \leqslant \|M\|$$

and obtains lower estimates for the uniform metric. However, the following very simple relation between the total variation norm and the Fourier transform allows direct estimates:

$$|\widehat{M}(t)| \leqslant \|M\|. \tag{10.1}$$

Inequality (10.1) holds for all t. Therefore we can estimate $|\widehat{M}(t)|$ by suitably choosing t.

We shall demonstrate the above technique on one special case. We shall assume that F does not depend on n in any way, i.e. we assume that we have the scheme of sequences.

Theorem 10.1 *Let $F \in \mathcal{F}_+ \cap \mathcal{F}_Z$ have finite fourth absolute moment and $F \neq I$. Then*

$$\| F^n - \exp\{n(F-I)\} \| \geqslant C(F)n^{-1}. \tag{10.2}$$

© Springer International Publishing Switzerland 2016
V. Čekanavičius, *Approximation Methods in Probability Theory*, Universitext,
DOI 10.1007/978-3-319-34072-2_10

Proof Applying (1.12) and taking into account the symmetry of F we obtain

$$\widehat{F}(t) = 1 - \frac{\sigma^2 t^2}{2} + \theta a_4 t^4.$$

Here σ^2 and a_4 denote the variance and fourth moment of F. Thus,

$$\frac{\sigma^2 t^2}{2} - a_4 t^4 \le 1 - \widehat{F}(t) \le \frac{\sigma^2 t^2}{2} + a_4 t^4.$$

Let

$$c_0 = c_0(F) = \frac{\sigma^4}{16 a_4}, \quad t_0^2 = \frac{4 c_0}{an}.$$

Here the constant $a > 1$ will be chosen later. Then

$$\frac{c_0}{an} \le 1 - \widehat{F}(t_0) \le \frac{3 c_0}{an}.$$

Further on, for the sake of brevity, we omit the dependence on t_0. We have

$$\widehat{F}^n - \exp\{n(\widehat{F} - 1)\} = \frac{n(\widehat{F} - 1)^2}{2} + \frac{n(\widehat{F} - 1)^2}{2}(e^{(n-1)(\widehat{F}-1)} - 1)$$

$$+ ne^{(n-1)(\widehat{F}-1)}\left(\widehat{F} - \frac{1}{2}(\widehat{F} - 1)^2 - e^{\widehat{F}-1}\right)$$

$$+ [\widehat{F}^n - e^{n(\widehat{F}-1)} - ne^{(n-1)(\widehat{F}-1)}(\widehat{F} - e^{\widehat{F}-1})].$$

Taking into account (1.37) and (1.33) and omitting, for the sake of brevity, the dependence on t_0 we prove

$$|\widehat{F}^n - e^{n(\widehat{F}-1)} - ne^{(n-1)(\widehat{F}-1)}(\widehat{F} - e^{\widehat{F}-1})|$$

$$\le \sum_{m=2}^{n} \binom{m-1}{2} |\widehat{F}|^{n-m} |e^{(m-s-1)(\widehat{F}-1)}| |\widehat{F} - e^{\widehat{F}-1}|^2$$

$$\le |\widehat{F} - e^{\widehat{F}-1}|^2 \sum_{m=2}^{n} \binom{m-1}{2} \le \binom{n}{2}\frac{1}{4}|\widehat{F} - 1|^4 \le \frac{11 c_0^4}{a^4 n^2},$$

$$\left| ne^{(n-1)(\widehat{F}-1)} \left(1 + (\widehat{F}-1) - \frac{1}{2}(\widehat{F}-1)^2 - e^{\widehat{F}-1} \right) \right|$$

$$\leq n \left| 1 + (\widehat{F}-1) - \frac{1}{2}(\widehat{F}-1)^2 - e^{\widehat{F}-1} \right|$$

$$\leq n \frac{|\widehat{F}-1|^3}{6} \leq \frac{9c_0^3}{4a^3 n^2},$$

$$\left| n \frac{(\widehat{F}-1)^2}{2} \left(e^{(n-1)(\widehat{F}-1)} - 1 \right) \right| \leq n(n-1) \frac{|\widehat{F}-1|^3}{2} \leq \frac{27c_0^3}{2a^3 n}$$

$$\frac{c_0^2}{2a^2 n} \leq \frac{n}{2} |n(\widehat{F}-1)|^2.$$

Combining all estimates we see that, for $t = t_0$,

$$|\widehat{F}^n - \exp\{n(\widehat{F}-1)\}| \geq \frac{c_0^2}{2a^2 n} \left(1 - \frac{27c_0}{a} - \frac{9c_0}{2an} - \frac{22c_0^2}{a^2 n} \right).$$

To finish the proof of (10.2) we should choose $a = C(F)$ sufficiently large. For example, we can take $a = 55(1 + c_0)$. $\square$

10.2 Lower Estimates for the Total Variation

Let $M, H \in \mathcal{M}$. The general idea of the lower estimates can be expressed in the following way

$$\| MH \| \leq \| M \| \| H \|.$$

Dividing both sides by $\| H \|$ we get the lower estimate for $\| M \|$. Consequently, we need H to have good properties. Usually convolution with some 'good' measure is called smoothing. There are various possible choices for H. We choose H to be quite close to the normal distribution. To get a lower estimate for $\| MH \|$ one applies Parseval's identity and the fact that the total variation norm is invariant under scaling.

For $M \in \mathcal{M}$ and $t \in \mathbb{R}$ let

$$\widehat{\psi}_1(t) = e^{-t^2/2}, \quad \widehat{\psi}_2(t) = t e^{-t^2/2}, \tag{10.3}$$

$$V_j(a, b) = \int_{-\infty}^{\infty} \widehat{\psi}_j(t) \widehat{M}(t/b) \exp\{-ita/b\} \, dt. \tag{10.4}$$

Lemma 10.1 *Let $M \in \mathcal{M}$, $b \geq 1$, $a \in \mathbb{R}$, $s = 1, 2$. Then*

$$\| M \| \geq (2\pi)^{-1/2} | V_j(a, b) |.$$

Proof It is known that

$$\int_{-\infty}^{\infty} \widehat{\psi}_j(t) e^{ity} \, dt = \sqrt{2\pi} \psi_j(y), \quad \psi_1(y) = e^{-y^2/2}, \quad \psi_2(y) = iye^{-y^2/2}.$$

Let $M_{a,b}$ be a measure such that

$$\widehat{M}_{a,b}(t) = \int_{-\infty}^{\infty} e^{itx} M_{a,b}\{dx\} = \widehat{M}(t/b) \exp\{-ita/b\}. \tag{10.5}$$

Then, changing the order of integration, we prove

$$| V_j(a, b) | = \left| \int_{-\infty}^{\infty} \widehat{\psi}_j(t) \int_{-\infty}^{\infty} e^{itx} M_{a,b}\{dx\} dt \right| = \left| \int_{-\infty}^{\infty} \int_{-\infty}^{\infty} \widehat{\psi}_j(t) e^{itx} dt M_{a,b}\{dx\} \right|$$

$$= \sqrt{2\pi} \left| \int_{-\infty}^{\infty} \psi_j(x) M_{a,b}\{dx\} \right| \leq \sqrt{2\pi} \max_x | \psi_j(x) | \| M_{a,b} \|$$

$$\leq \sqrt{2\pi} \| M_{a,b} \| = \sqrt{2\pi} \| M \|.$$

For the last equality we used the fact that the total variation norm is invariant under shifting and scaling. $\square$

Why do we need two possible $V_j(a, b)$? The answer is determined by the method of application. Just as in the previous section, we shall expand $\widehat{M}(t)$ in powers of t. The result will be of the form:

$$| \widehat{V}_j(a, b) | \geq C_1(M, b) \left| \int_{-\infty}^{\infty} t^s \widehat{\psi}_j(t) \, dt \right| - C_2(M, b) \int_{-\infty}^{\infty} | t |^{s+1} | \widehat{\psi}_j(t) | \, dt - \ldots$$

Observe that the first integral on the right-hand side equals zero for odd functions. Therefore we should choose $\widehat{\psi}_1(t)$ if s is even and $\widehat{\psi}_2(t)$ if s is odd.

Observe that the above argument is based on the existence of a sufficient number of finite moments. This is one of the drawbacks of the method. We formulate the main facts about estimates based on Lemma 10.1.

Typical application. Lemma 10.1 can be applied to estimate $\| F^n - G^n \|$ when F and G have more than two matching moments.

- Choose a to be equal to the mean of F and b to be proportional to $\sqrt{n}\sigma$, where σ^2 is the variance of F.
- Replace $\widehat{F}^n - \widehat{G}^n$ by $n\widehat{F}^{n-1}(\widehat{F} - \widehat{G})$ and the latter by $n(\widehat{F} - \widehat{G})$.
- Expand $\widehat{F}(t/b) - \widehat{G}(t/b)$ in powers of t. Let s be the maximum power of t in the expansion.
- Choose $\widehat{\psi}_1(t)$ if s is even and $\widehat{\psi}_2(t)$ if s is odd.

Advantages. Requires only Bergström's and Taylor's expansions. As a rule, the correct order of the accuracy of approximation can be proved.

Drawbacks. In comparison to upper estimates, constants in the lower estimates are very small. As a rule, the finiteness of the third absolute moment is required.

We illustrate the application of Lemma 10.1 by considering $F \in \mathcal{F}_+$.

Theorem 10.2 *Let $n \in \mathbb{N}$, $F \in \mathcal{F}_+$ and $\int_{-\infty}^{\infty} x^4 F\{dx\} < \infty$. Then*

$$\| F^n (F - I) \| \geqslant \frac{C(F)}{n}.$$

Proof We choose $a = 0$ and $b = h\sqrt{n}\sigma$. The quantity $h > 0$ will be chosen later. We denote the variance of F by σ^2 and the finite fourth moment by a_4. We have

$$|\widehat{F}^n(t/b) - 1| \leqslant n|\widehat{F}(t/b) - 1| \leqslant \frac{n\sigma^2 t^2}{2b^2} = \frac{nt^2}{2h^2 n} = \frac{t^2}{2h^2}.$$

Similarly,

$$\widehat{F}(t/b) - 1 = \frac{-\sigma^2 t^2}{2b^2} + \theta\frac{a_4 t^4}{b^4 4!} = \frac{-t^2}{2h^2 n} + \theta\frac{a_4 t^4}{24\sigma^4 h^4 n^2}$$

and

$$|\widehat{F}(t/b) - 1| \leqslant \frac{t^2 \sigma^2}{2b^2} = \frac{t^2}{2h^2 n}.$$

Therefore

$$\widehat{F}^n(t/b)(\widehat{F}(t/b) - 1) = (\widehat{F}(t/b) - 1) + (\widehat{F}^n(t/b) - 1)(\widehat{F}(t/b) - 1)$$

$$= -\frac{t^2}{2h^2 n} + \frac{\theta a_4 t^4}{24\sigma^4 h^4 n^2} + \frac{\theta t^4}{4h^4 n}$$

and

$$
|V_1(0,b)| \geqslant \left| \int_{-\infty}^{\infty} \frac{t^2}{2h^2 n} e^{-t^2/2} dt \right| - \int_{-\infty}^{\infty} e^{-t^2/2} \left(\frac{a_4 t^4}{24\sigma^4 h^4 n^2} + \frac{t^4}{4h^4 n} \right) dt
$$

$$
\geqslant \frac{C_3}{h^2 n} - \frac{C_4 a_4}{\sigma^4 h^4 n^2} - \frac{C_5}{h^4 n} = \frac{C_3}{h^2 n} \left(1 - \frac{C_6 a_4}{\sigma^4 h^2 n} - \frac{C_7}{h^2} \right) \geqslant \frac{C(F)}{n} \qquad (10.6)
$$

if we choose $h = C(F)$ to be a sufficiently large constant. □

10.3   Lower Estimates for Densities

Lemma 10.1 can be modified to estimate densities.

Lemma 10.2 *Let $F, G \in \mathcal{F}$ have densities $f(x)$ and $g(x)$, respectively. Let $b \geqslant 1$, $a \in \mathbb{R}, j = 1, 2$ and let $\widehat{\psi}_j(t)$ and $V_j(a,b)$ be respectively defined by (10.3) and (10.4) with $M = F - G$. Then*

$$
\sup_{x \in \mathbb{R}} |f(x) - g(x)| \geqslant (2\pi b)^{-1} |V_j(a,b)|.
$$

Proof Let $M_{a,b}$ be defined by (10.5). Then $M_{a,b}$ has a density $M'_{a,b}(x) = b[f(b(x - a)) - g(b(x - a))]$ and, therefore,

$$
|V_j(a,b)| = \sqrt{2\pi} \left| \int_{-\infty}^{\infty} \psi_j(x) M_{a,b}\{dx\} \right|
$$

$$
= \sqrt{2\pi} b \left| \int_{-\infty}^{\infty} \psi_j(x)[f(b(x - a)) - g(b(x - a))] dx \right|
$$

$$
\leqslant \sqrt{2\pi} b \max_x |f(b(x - a)) - g(b(x - a))| \int_{-\infty}^{\infty} |\psi_j(x)| dx
$$

$$
\leqslant 2\pi b \max_x |f(x) - g(x)|.
$$

Here, as before, $\psi_1(y) = e^{-y^2/2}$, $\psi_2(y) = iy e^{-y^2/2}$. □

Example 10.1 Let $n \in \mathbb{N}$, $F \in \mathcal{F}_+$ have density $f(x)$ and $\int_{-\infty}^{\infty} x^4 F\{dx\} < \infty$. Then

$$
\sup_{x \in \mathbb{R}} |f_n(x) - f_{n+1}(x)| \geqslant \frac{C(F)}{n\sqrt{n}}.
$$

Here $f_n(x)$ is the density of F^n. Indeed, we choose $a = 0$ and $b = h\sqrt{n}\sigma$ and apply Lemma 8.1 and (10.6) with sufficiently large $h = C(F)$.

10.4 Lower Estimates for Probabilities

Lemma 10.1 can be modified for the local norm of $M \in \mathcal{M}_Z$.

Lemma 10.3 *Let $M \in \mathcal{M}_Z$, $b \geqslant 1$, $a \in \mathbb{R}$, $j = 1, 2$ and let $\widehat{\psi}_j(t)$ and $V_j(a,b)$ be respectively defined by (10.3) and (10.4). Then*

$$\| M \|_\infty \geqslant 0.1 | V_j(a,b) | |b|^{-1}.$$

Proof Let $M_{a,b}$ correspond to

$$\widehat{M}_{a,b}(t) = \int_{-\infty}^\infty e^{itx} M_{a,b}\{dx\} = \widehat{M}(t/b) \exp\{-ita/b\}.$$

Then changing the order of integration, just as in the proof of Lemma 10.1, we obtain

$$| V_j(a,b) | = \sqrt{2\pi} \left| \int_{-\infty}^\infty \psi_j(x) M_{a,b}\{dx\} \right| = \sqrt{2\pi} \left| \sum_{k \in \mathbb{Z}} \psi_j\left(\frac{k-a}{b}\right) M\{k\} \right|$$

$$\leqslant \sqrt{2\pi} \| M \|_\infty \sum_{k \in \mathbb{Z}} \left| \psi_j\left(\frac{k-a}{b}\right) \right|.$$

Here $\psi_1(y) = e^{-y^2/2}$, $\psi_2(y) = iye^{-y^2/2}$. The proof is completed by noting that

$$\sum_{k \in \mathbb{Z}} \psi_1\left(\frac{k-a}{b}\right) \leqslant 1 + b\sqrt{2\pi}, \quad \sum_{k \in \mathbb{Z}} \left| \psi_2\left(\frac{k-a}{b}\right) \right| \leqslant 2(e^{-1/2} + b).$$

□

We demonstrate this approach on the Poisson approximation to the binomial law.

Theorem 10.3 *Let $0 < p \leqslant 1/2$, $n \in \mathbb{N}$. Then*

$$\| ((1-p)I + pI_1)^n - \exp\{np(I_1 - I)\} \|_\infty \geqslant C \min\left\{np^2, \sqrt{\frac{p}{n}}\right\}.$$

Proof In principle, we repeat the same steps of proof as in previous sections. However we need to center distributions properly. Let $M = ((1-p)I + pI_1)^n - \exp\{np(I_1 - I)\}$.
Then by (1.40)

$$\widehat{M}(t)e^{-itnp} = (1 - p + pe^{it} - \exp\{p(e^{it} - 1)\})$$

$$\times \sum_{k=1}^n e^{-itp} e^{-it(n-1)p} (1 - p + pe^{it})^{k-1} \exp\{(n-k)p(e^{it} - 1)\}. \quad (10.7)$$

We have

$$1 - p + pe^{it} - \exp\{p(e^{it} - 1)\} = 1 + p(e^{it} - 1) - \sum_{j=0}^{\infty} \frac{p^j}{j!}(e^{it} - 1)^j$$

$$= -\frac{p^2}{2}(e^{it} - 1)^2 + R_1(t) = \frac{p^2 t^2}{2} + R_1(t) + R_2(t).$$

Here

$$|R_1(t)| \leq C_8 p^3 |e^{it} - 1|^3 \leq C_9 p^3 |t|^3, \quad |R_2(t)| \leq C_{10} p^2 |t|^3.$$

The next step is to substitute these estimates into (10.7). We also replace e^{-itp} by unity and take into account that $|e^{-itp} - 1| \leq p|t|$. Consequently,

$$\widehat{M}(t)e^{-itmp} = \frac{p^2 t^2}{2} \sum_{k=1}^{n} e^{-itp}e^{-it(n-1)p}(1 - p + pe^{it})^{k-1}e^{(n-k)p(e^{it}-1)} + \theta n(|R_1(t) + R_2(t)|)$$

$$= \frac{p^2 t^2}{2} \sum_{k=1}^{n} e^{-it(n-1)p}(1 - p + pe^{it})^{k-1}e^{(n-k)p(e^{it}-1)} + R_3(t). \tag{10.8}$$

Here

$$|R_3(t)| \leq C_{11} np^2 |t|^3.$$

Next, we replace all summands in (10.8) by unity. Note that for any centered characteristic function with two finite moments $\widehat{G}(t)$, from (1.12) it follows that

$$|\widehat{G}(t) - 1| \leq \frac{t^2}{2}\sigma^2(G),$$

where $\sigma^2(G)$ is the variance of G. Let us consider

$$G = I_{-(n-1)p}((1 - p)I + pI_1)^{k-1} \exp\{(n - k)p(I_1 - I)\}.$$

Note that G corresponds to the sum of independent binomial and Poisson random variables. Hence, it has variance equal to $(k - 1)p(1 - p) + (n - k)p \leq np$ and

$$\left| e^{-it(n-1)p}(1 - p + pe^{it})^{k-1}e^{(n-k)p(e^{it}-1)} - 1 \right| \leq \frac{npt^2}{2}.$$

Substituting the last estimate into (10.8) we obtain

$$\widehat{M}(t)e^{-itmp} = \frac{np^2 t^2}{2} + R_3(t) + R_4(t).$$

Here

$$|R_4(t)| \le \frac{np^2t^2}{2} \cdot \frac{npt^2}{2}.$$

The next step is application of Lemma 10.3:

$$|V_1(a,b)| \ge \left| \int_{-\infty}^{\infty} \frac{np^2t^2}{2b^2} e^{-t^2/2}\, dt \right| - \int_{-\infty}^{\infty} \left|R_3\left(\frac{t}{b}\right)\right| e^{-t^2/2}\, dt - \int_{-\infty}^{\infty} \left|R_4\left(\frac{t}{b}\right)\right| e^{-t^2/2}\, dt$$

$$\ge C_{12}\frac{np^2}{b^2} - C_{13}\frac{np^2}{b^3} - C_{14}\frac{n^2p^3}{b^4} = C_{12}\frac{np^2}{b^2}\left(1 - C_{15}\frac{1}{b} - C_{16}\frac{np}{b^2}\right). \qquad (10.9)$$

Let $np < 1$. Then

$$1 - C_{15}\frac{1}{b} - C_{16}\frac{np}{b^2} \ge 1 - (C_{15} + C_{16})\frac{1}{b} \ge C_{17}$$

for a sufficiently large absolute constant $b \ge 1$. Therefore, for $np < 1$, from (10.9) we obtain

$$|V_1(a,b)| \ge C_{18}np^2.$$

Let $np \ge 1$. Then taking $b = \sqrt{np}$ we obtain

$$|V_1(a,b)| \ge \frac{C_{12}p}{h^2}\left(1 - \frac{C_{15}}{h\sqrt{np}} - \frac{C_{16}}{h^2}\right) \ge \frac{C_{12}p}{h^2}\left(1 - \frac{C_{15} + C_{16}}{h}\right) \ge C_{19}p,$$

for a sufficiently large absolute constant $h > 1$. Substituting these estimates into Lemma 10.3 we get

$$\|M\|_\infty \ge Cnp^2, \quad \text{if} \quad np < 1,$$

$$\|M\|_\infty \ge C\sqrt{\frac{p}{n}}, \quad \text{if} \quad np \ge 1,$$

which is equivalent to the statement of the theorem. $\square$

10.5 Lower Estimates for the Kolmogorov Norm

Once again one needs to modify Lemma 10.1.

Lemma 10.4 *Let $M \in \mathcal{M}$, $b \geq 1$, $a \in \mathbb{R}$, $j = 1, 2$ and let $\widehat{\psi}_j(t)$ and $V_j(a, b)$ be respectively defined by (10.3) and (10.4). Then*

$$|M|_K \geq (4\pi)^{-1}|V_j(a, b)|. \tag{10.10}$$

Proof Let $\psi_1(x) = e^{-x^2/2}$, $\psi_2(x) = xe^{-x^2/2}$ and let $M_{a,b}$ be defined by (10.5). Observe that $M_{a,b}(x)$ has bounded variation and $\psi_j(x)$ is continuous. Let $N > 0$. Integrating by parts we obtain

$$\int_{-N}^{N} \psi_j(x) M_{a,b}\{dx\} = -\int_{-N}^{N} M_{a,b}(x-)(\psi_j'(x)) dx + \psi_j(N) M_{a,b}(N) - \psi_j(-N) M_{a,b}(-N),$$

see, for example [141], Theorem 11 from Chapter 2, or [82], p. 354. Letting $N \to \infty$ we arrive at the following relation

$$\left| \int_{-\infty}^{\infty} \psi_j(x) M_{a,b}\{dx\} \right| = \left| -\int_{-\infty}^{\infty} M_{a,b}(x-)(\psi_j'(x)) dx \right| \leq |M_{a,b}|_K \int_{-\infty}^{\infty} |\psi_j'(x)| dx$$

$$\leq 2\sqrt{2\pi} |M_{a,b}|_K = 2\sqrt{2\pi} |M|_K.$$

The proof is easily completed if we recall that, in the proof of Lemma 10.1, it was already shown that

$$|V_j(a, b)| = \sqrt{2\pi} \left| \int_{-\infty}^{\infty} \psi_j(x) M_{a,b}\{dx\} \right|.$$

$\square$

Remark 10.1 The difference between Lemmas 10.1 and 10.4 is that the latter has a smaller constant. On the other hand, $\|M\| \geq |M|_K$. Thus, if the absolute constant is not of much importance, one can always use Lemma 10.4.

We illustrate the application of Lemma 10.4 by considering the Poisson approximation to the binomial law.

Theorem 10.4 *Let $0 < p \leq 1/2$, $n \in \mathbb{N}$. Then*

$$|((1 - p)I + pI_1)^n - \exp\{np(I_1 - I)\}|_K \geq C \min(np^2, p). \tag{10.11}$$

Proof In the proof of Theorem 10.3, for $a = np$ and suitably chosen $b > 1$ we obtained

$$|V_1(a, b)| \geq C_{18} np^2, \quad \text{if} \quad np < 1,$$

$$|V_1(a, b)| \geq C_{19} p, \quad \text{if} \quad np \geq 1.$$

Applying Lemma 10.4, we complete the proof. $\square$

In many cases, the characteristic function method described in this chapter allows us to obtain lower bounds of the correct order. Compare, for example (10.11) and (2.19), (5.7).

10.6 Problems

10.1 Prove (10.1).

10.2 Let F have a Laplace distribution with parameters 0 and $\lambda > 0$, that is

$$\widehat{F}(t) = \frac{\lambda^2}{\lambda^2 + t^2}.$$

Applying (10.1) prove that, for any $n \in \mathbb{N}$,

$$\| F^n(F - I) \| \geq \frac{1}{2en}.$$

10.3 Let F be defined as in the previous problem and let $m(x)$ be the density of $M = F^n(F - I)^2$. Prove that there exists a $C_L(\lambda)$ such that

$$\sup_x | m(x) | \geq \frac{C_L(\lambda)}{n^2 \sqrt{n}}.$$

10.4 Let $F \in \mathcal{F}_Z$, $F\{0\} = p_0 \in (0, 1)$ and assume that F does not depend on n in any way. Prove that

$$\| (F - I) \exp\{n(F - I)\} \|_\infty \geq C(F)n^{-1}.$$

10.5 Let a non-degenerate $F \in \mathcal{F}_Z \cap \mathcal{F}_+$ have four finite moments and not depend on n in any way. Prove that

$$| F^n - \exp\{n(F - I)\} |_K \geq C(F)n^{-1}.$$

10.6 Let $F = (1 - p)I + pI_1$, $0 < p < 1/2$, $np > 1$. Prove that

$$| F^n - F^{n+1} |_K \geq C\sqrt{\frac{p}{n}}.$$

10.7 Let $0 < p_i < 1$, $i = 1, 2, \ldots, n$, $p_1 + p_2 + \cdots + p_n > 1$. Prove that, for some absolute constant C_P,

$$\left\| \prod_{i=1}^n ((1 - p_i)I + p_i I_1) - \exp\left\{ \sum_{i=1}^n p_i(I_1 - I) \right\} \right\| \geq \frac{C_P \sum_{i=1}^n p_i^2}{\sum_{i=1}^n p_i}.$$

Bibliographical Notes

The idea to use smoothing and Parseval's identity is well-known and widely used, especially for the normal approximation, see for example [106] Section 5.4 or [97]. An integer-valued smoothing measure was used in [88]. Sections 10.2, 10.3, 10.4 and 10.5 are based on lemmas from [137], see also [39].

Chapter 11
The Stein Method

Methodological aspects of the Stein method are exceptionally well discussed in the literature, and for more advanced applications the reader is advised to consult the books and papers referenced at the end of this chapter. Here we present just a basic idea of how the method works for the normal approximation and a short introduction to lattice random variables. On the other hand, we include some results that might be viewed as complementary material to the standard textbooks on Stein's method.

11.1 The Basic Idea for Normal Approximation

Let $f : \mathbb{R} \to \mathbb{R}$ be any continuously differentiable function and let η be the standard normal variable. Then

$$\mathbb{E} f'(\eta) = \mathbb{E} \, \eta f(\eta). \tag{11.1}$$

The equality (11.1) can be verified by integration by parts. As it turns out, the converse is also true, that is, if (11.1) holds for all continuous and piecewise continuously differentiable functions, then η is the normal random variable. Thus, (11.1) can be viewed as a characterization of the normal law. Therefore one expects that if a random variable S is close to the normal one, $\mathbb{E} f'(S) - \mathbb{E} Sf(S)$ will be small.

The next step is to relate (11.1) to some probabilistic metric. Let, for a bounded continuously differentiable real valued function $h, f(x)$ be a solution of the first order linear differential equation

$$f'(x) - xf(x) = h(x) - \mathbb{E} \, h(\eta). \tag{11.2}$$

© Springer International Publishing Switzerland 2016
V. Čekanavičius, *Approximation Methods in Probability Theory*, Universitext,
DOI 10.1007/978-3-319-34072-2_11

Then

$$\mathbb{E}\, h(S) - \mathbb{E}\, h(\eta) = \mathbb{E} f'(S) - \mathbb{E}\, Sf(S). \tag{11.3}$$

What properties does f possess in this case? Let, in this section, $\|f\| = \sup_x |f(x)|$. Then the solution to (11.2) satisfies the inequalities

$$\|f\| \leqslant \sqrt{2\pi}\,\|h\|, \quad \|f'\| \leqslant 2\|h\|, \quad \|f''\| \leqslant 2\|h'\|.$$

Next, we demonstrate how to estimate the right-hand side of (11.3), when $S = \sum_{i=1}^{n} \xi_i$, ξ_i are independent, $\mathbb{E}\,\xi_i = 0$ and $\sum_{i=1}^{n} \mathbb{E}\,\xi_i^2 = 1$. We will also assume that f is twice continuously differentiable. Let $S^i = S - \xi_i$. Then by Taylor's expansion

$$\mathbb{E}\, Sf(S) = \sum_{i=1}^{n} \mathbb{E}\,\xi_i f(S^i + \xi_i) = \sum_{i=1}^{n} \mathbb{E}\,\xi_i [f(S^i) + f'(S^i)\xi_i + \theta\|f''\|\,\xi_i^2/2]$$

$$= \sum_{i=1}^{n} \mathbb{E}\,\xi_i^2 \mathbb{E} f'(S^i) + \frac{\theta\|f''\|}{2} \sum_{i=1}^{n} \mathbb{E}\,|\xi_i^3|.$$

Similarly,

$$\mathbb{E} f'(S) = \sum_{i=1}^{n} \mathbb{E}\,\xi_i^2 \mathbb{E} f'(S) = \sum_{i=1}^{n} \mathbb{E}\,\xi_i^2 [\mathbb{E} f'(S^i) + \theta\|f''\|\,\mathbb{E}\,|\xi_i|]$$

$$= \sum_{i=1}^{n} \mathbb{E}\,\xi_i^2 \mathbb{E} f'(S^i) + \theta\|f''\| \sum_{i=1}^{n} \mathbb{E}\,|\xi_i|\,\mathbb{E}\,|\xi_i|^2$$

$$= \sum_{i=1}^{n} \mathbb{E}\,\xi_i^2 \mathbb{E} f'(S^i) + \theta\|f''\| \sum_{i=1}^{n} \mathbb{E}\,|\xi_i|^3.$$

Therefore

$$|\mathbb{E}\, h(S) - \mathbb{E}\, h(\eta)| \leqslant 3\|h'\| \sum_{i=1}^{n} \mathbb{E}\,|\xi_i|^3. \tag{11.4}$$

Observe that (11.4) is a special version of the CLT. Indeed, it suffices to take $\xi_i = (Y_i - \mathbb{E}\, Y_i)/\sigma$, where σ^2 is the variance of $Y_1 + \cdots + Y_n$ and we arrive at a more traditional expression of CLT.

Of course, (11.4) is a very simple example of the Stein method. In general, Taylor's expansion is hardly a solution. One alternative is to search for S^* such that $\mathbb{E}\, Sf(S) = \mathbb{E} f(S^*)$ (the so-called zero coupling) or $\mathbb{E}\, Sf(S) = a\mathbb{E} f(S^*)$ (the size bias coupling). Stein's operator $\mathbb{E} f'(S) - \mathbb{E}\, Sf(S)$ can be simplified if we construct the exchangeable pair, that is on the same probability space where S is defined we construct S^* such that (S, S^*) and (S^*, S) have the same distribution and $\mathbb{E}(S^*|S) = (1-a)S$, for some a. A lot of research has also been done on when S is the sum of weakly dependent random variables.

11.2 The Lattice Case

First we note that various choices of $f(\cdot)$ in $\mathbb{E}f(\xi) - \mathbb{E}f(\eta)$ lead to estimates for the total variation, local and Wasserstein norms. Since later in this chapter we study distributions concentrated on $\mathbb{Z}_+$ let $f : \mathbb{Z}_+ \to \mathbb{R}$. Set $\|f\|_\infty = \sup_{j \geq 0} |f(j)|$, $\Delta f(j) = f(j+1) - f(j)$, $\Delta^2 f(j) = \Delta(\Delta f(j))$. The local norm corresponds to the point indicator function $f(j) = \mathbb{I}_k(j)$. Here $\mathbb{I}_k(k) = 1$ and $\mathbb{I}_k(j) = 0, j \neq k$. For other norms we have

$$\| \mathcal{L}(\xi) - \mathcal{L}(\eta) \| = \sup_{\|f\|_\infty \leq 1} |\mathbb{E}f(\xi) - \mathbb{E}f(\eta)|,$$

$$\| \mathcal{L}(\xi) - \mathcal{L}(\eta) \|_W = \sup_{\|\Delta f\|_\infty \leq 1} |\mathbb{E}f(\xi) - \mathbb{E}f(\eta)|.$$

Note that similar relations also hold in the general case, see, for example, [16], Appendix A1.

Instead of the usual measure definition, the Stein method does not involve characteristic functions and is based instead on the properties of the special difference equation. Let η and ξ be $\mathbb{Z}_+$-valued random variables with finite means. Assume that we want to estimate $\mathbb{E}f(\xi) - \mathbb{E}f(\eta)$ for some function $f : \mathbb{Z}_+ \to \mathbb{R}$. Stein's method then is realized in three consecutive steps.

1. For any bounded function $g : \mathbb{Z}_+ \to \mathbb{R}$, a linear operator $\mathcal{A}$ such that $\mathbb{E}(\mathcal{A}g)(\eta) = 0$ is established. The operator $\mathcal{A}$ is called Stein's operator.
2. The Stein equation

$$(\mathcal{A}g)(j) = f(j) - \mathbb{E}f(\eta), \quad j \in \mathbb{Z}_+ \tag{11.5}$$

is solved with respect to $g(j)$. As a rule, solutions to the Stein equation have some useful properties.
3. The obvious identity $\mathbb{E}f(\xi) - \mathbb{E}f(\eta) = \mathbb{E}(\mathcal{A}g)(\xi)$ translates the initial problem to estimation of $\mathbb{E}(\mathcal{A}g)(\xi)$.

Below we outline the main features of the Stein method for lattice variables.

Typical application. Approximation of the sum of integer-valued random variables $S = \xi_1 + \xi_2 + \cdots + \xi_n$ by some integer-valued random variable η. The three-step approach:

(a) establishing Stein's operator $(Ag)(\eta)$;
(b) solving the Stein equation (11.5) and estimating $\| \Delta g \|_\infty$;
(c) estimating $| \mathbb{E} (Ag)(S) |$ through $\| \Delta g \|_\infty$.

Advantages. The Stein method is flexible and can be adapted for dependent random variables and stochastic processes. The method works extremely well for Poisson and other one-parametric approximations. In the estimates, the absolute constants are smaller than those obtained via the characteristic function method. The same estimate of $| \mathbb{E} (Ag)(S) |$ by $\| \Delta g \|_\infty$ can be used for the total variation, local and Wasserstein norms.

Drawbacks. The Stein equation (11.5) can be satisfactorily solved for some distributions only. The method is oriented toward nonnegative random variables and is of limited use if we want to benefit from the symmetry of distributions. In many cases, some analogue of the Barbour-Xia inequality (5.13) is needed.

We elaborate on the last statement. In view of (11.8), one seeks to estimate $\mathbb{E} (Ag)(S)$ by $\| \Delta g \|_\infty$. However, two-parametric approximations can lead to a term $\| \Delta^2 g \|_\infty$, where $\Delta^2 g(j) = \Delta(\Delta g(j))$. The obvious estimate $\| \Delta^2 g \|_\infty \leqslant 2\| \Delta g \|_\infty$ is usually too rough. On the other hand, observe that

$$\left| \sum_{j\in\mathbb{Z}} \Delta^2 g(j) P(S = j) \right| = \left| \sum_{j\in\mathbb{Z}} \Delta g(j)[P(S = j - 1) - P(S = j)] \right|$$

$$\leqslant \| \mathcal{L}(S + 1) - \mathcal{L}(S) \| \| \Delta g \|_\infty.$$

Now, as a rule, the Barbour-Xia inequality (5.13) can be applied. We repeat the formulation of (5.13) in terms of random variables. Let $S = \xi_1 + \xi_2 + \cdots + \xi_n$ be the sum of independent integer-valued random variables. Then

$$\| \mathcal{L}(S + 1) - \mathcal{L}(S) \| \leqslant 2\sqrt{\frac{2}{\pi}} \left(\sum_{j=1}^{n} \left[1 - \frac{1}{2}\| \mathcal{L}(\xi_j + 1) - \mathcal{L}(\xi_j) \| \right] \right)^{-1/2}. \quad (11.6)$$

Note that to obtain an estimate similar to (11.6) for dependent summands is usually a very serious problem.

The essential step in Stein's method is solving equation (11.5). The following lemma gives an idea of the properties of its solutions.

Lemma 11.1 *If*

$$(\mathcal{A}g)(j) = \alpha_j g(j+1) - \beta_j g(j),\tag{11.7}$$

where $\beta_0 = 0$ and $\alpha_k - \alpha_{k-1} \leqslant \beta_k - \beta_{k-1}$ ($k = 1, 2, \ldots$), then the solution to (11.5) satisfies

$$|\Delta g(j)| \leqslant 2\|f\|_\infty \min\left\{\frac{1}{\alpha_j}, \frac{1}{\beta_j}\right\}, \quad j \in \mathbb{Z}_+.\tag{11.8}$$

If $f : \mathbb{Z}_+ \to [0, 1]$, then $2\|f\|_\infty$ in (11.8) should be replaced by 1.

Regrettably Stein's equations are rarely of the form (11.7). Poisson, negative binomial and binomial laws satisfy (11.7), but not their convolutions. Roughly speaking, there are two possible solutions to this problem. One can stick to constructing approximations satisfying (11.7) or use a perturbation approach, as discussed in Sect. 11.6.

11.3 Establishing Stein's Operator

First of all, note that Stein's operator is not unique. For example, $C\mathcal{A}$, $C \neq 0$, is also a Stein operator for the same random variable. Typically the choice of the form of operator is determined by the method of proof.

We begin with the most popular approach, which is called the density approach (discrete version). Let η be a $\mathbb{Z}_+$-valued random variable with $p_k = P(\eta = k)$ and finite mean. Let $g : \mathbb{Z}_+ \to \mathbb{R}$ be any bounded function. For the sake of simplicity, assume that $p_k > 0$. Then the obvious identity

$$\sum_{j=0}^\infty p_j\left(\frac{(j+1)p_{j+1}}{p_j}g(j+1) - jg(j)\right) = 0$$

allows us to choose

$$(\mathcal{A}g)(j) = \frac{(j+1)p_{j+1}}{p_j}g(j+1) - jg(j), \quad j \in \mathbb{Z}_+.$$

The density approach works well for Poisson, binomial and negative binomial random variables. However, it is not easily applicable to their sums, to say nothing about more structurally complex distributions.

A different approach to the problem is to use the characteristic or probability generating function to establish recursions between probabilities. Note that to obtain the probability generating function from the characteristic function it suffices to replace e^{it} by z.

We will abandon the assumption that all p_k are positive and denote the probability generating function of η by $G_\eta(z)$

$$G_\eta(z) := \sum_{k=0}^{\infty} p_k z^k.$$

Then formally

$$G'_\eta(z) = \frac{d}{dz} G_\eta(z) = \sum_{k=1}^{\infty} k p_k z^{k-1} = \sum_{k=0}^{\infty} (k+1) p_{k+1} z^k. \qquad (11.9)$$

If $G'_\eta(z)$ can be expressed through $G_\eta(z)$ then by collecting coefficients of z^k the recursion follows. Such an approach is preferable if explicit formulas for probabilities are difficult to analyze.

Let us consider integer-valued CP distribution with the probability generating function

$$G(z) = \exp\left\{ \sum_{j=1}^{\infty} \lambda_j (z^j - 1) \right\} \qquad (11.10)$$

and $\sum_{j=1}^{\infty} j\lambda_j < \infty$. Then

$$G'(z) = G(z) \sum_{j=1}^{\infty} j\lambda_j z^{j-1} = \sum_{k=0}^{\infty} p_k z^k \sum_{j=1}^{\infty} j\lambda_j z^{j-1} = \sum_{k=0}^{\infty} z^k \sum_{m=0}^{k} p_m (k-m+1)\lambda_{k-m+1}.$$

Comparing the last expression to the right-hand side of (11.9) we obtain a recursive relation, for all $k \in \mathbb{Z}_+$, as

$$\sum_{m=0}^{k} p_m (k-m+1)\lambda_{k-m+1} - (k+1)p_{k+1} = 0.$$

As usual for Stein's operator we rewrite the sum

$$0 = \sum_{k=0}^{\infty} g(k+1) \left[\sum_{m=0}^{k} p_m (k-m+1)\lambda_{k-m+1} - (k+1)p_{k+1} \right]$$

$$= \sum_{m=0}^{\infty} p_m \left[\sum_{k=m}^{\infty} g(k+1)(k-m+1)\lambda_{k-m+1} - mg(m) \right]$$

$$= \sum_{m=0}^{\infty} p_m \left[\sum_{j=1}^{\infty} j\lambda_j g(j+m) - mg(m) \right].$$

Thus Stein's operator for the compound Poisson distribution is

$$(\mathcal{A}g)(j) = \sum_{l=1}^{\infty} l\lambda_l g(j+l) - jg(j)$$

$$= \sum_{l=1}^{\infty} l\lambda_l g(j+1) - jg(j) + \sum_{m=2}^{\infty} m\lambda_m \sum_{l=1}^{m-1} \Delta g(j+l), \quad j \in \mathbb{Z}_+. \quad (11.11)$$

The same argument holds for signed CP approximation, i.e. when some λ_i are negative. Then we should assume that $\sum_{j=1}^{\infty} j |\lambda_j| < \infty$.

11.4   The Big Three Discrete Approximations

In the discrete case, the Stein method works best for the Poisson, negative binomial and binomial random variables. Here is a summary of the main Stein method related facts for those variables and bounded $f : \mathbb{Z}_+ \to \mathbb{R}$:

- Let η be a Poisson random variable with parameter $\lambda > 0$

$$P(\eta = k) = \frac{\lambda^k}{k!} e^{-\lambda}, \quad k = 0, 1, 2, \ldots$$

Then

$$(\mathcal{A}g)(j) = \lambda g(j+1) - jg(j), \quad j \in \mathbb{Z}_+ \quad (11.12)$$

and the solution to (11.5) satisfies

$$\| \Delta g \|_\infty \leq \frac{2\|f\|}{\max(1, \lambda)}. \quad (11.13)$$

- Let η be the negative binomial distribution with parameters $p \in (0, 1)$, $\gamma > 0$,

$$P(\eta = j) = \frac{\Gamma(\gamma + j)}{\Gamma(\gamma) j!} p^\gamma q^j, \quad j \in \mathbb{Z}_+, \quad q = 1 - p. \quad (11.14)$$

Here $\Gamma(\cdot)$ denotes the gamma function. Then

$$(\mathcal{A}g)(j) := q(\gamma + j)g(j+1) - jg(j), \quad j \in \mathbb{Z}_+ \quad (11.15)$$

and the solution to (11.5) satisfies

$$\| \Delta g \|_\infty \le \frac{2\|f\|}{\gamma q}. \tag{11.16}$$

- Let $0 < p < 1$, $N \in \mathbb{N}$, $q = 1 - p$, and η have the binomial distribution

$$P(\eta = j) = \binom{N}{j} p^j q^{N-j}, \quad j = 0, 1, \dots, N. \tag{11.17}$$

Then

$$(\mathcal{A}g)(j) = (N - j)pg(j + 1) - jqg(j), \quad j = 0, 1, \dots N \tag{11.18}$$

and the solution to (11.5) satisfies

$$\| \Delta g \|_\infty \le \frac{2\|f\|}{Npq}. \tag{11.19}$$

The translation from estimates of Stein's operator to estimates in total variation is straightforward. First we formulate such a translation result for the negative binomial distribution.

Lemma 11.2 *Let S be an integer-valued random variable concentrated on $\mathbb{Z}_+$. Let η have the negative binomial distribution defined by (11.14). If, for any bounded function g*

$$|\mathbb{E}\,(\mathcal{A}g)(S)\,| \le \varepsilon\|\,\Delta g\,\|_\infty,$$

then

$$\|\,\mathcal{L}(S) - \mathcal{L}(\eta)\,\| \le \frac{2\varepsilon}{\gamma q}.$$

Here $(\mathcal{A}g)$ is defined by (11.15).

For the binomial approximation we should add the tail of the distribution.

Lemma 11.3 *Let S be an integer-valued random variable concentrated on $\mathbb{Z}_+$. Let η have the binomial distribution defined by (11.17). If, for any bounded function g*

$$|\mathbb{E}\,(\mathcal{A}g)(S)\,| \le \varepsilon\|\,\Delta g\,\|_\infty,$$

then

$$\|\,\mathcal{L}(S) - \mathcal{L}(\eta)\,\| \le 2\Big(\varepsilon \min\Big(1, \frac{1}{Npq}\Big) + P(S > N)\Big).$$

The Poisson approximation case is similar to Lemma 11.2 and is applied in the next section.

11.5   The Poisson Binomial Theorem

As an example we will consider the Poisson approximation to the sum of independent Bernoulli (indicator) variables. We assume that η has a Poisson distribution with the parameter λ and $\pi(k) = P(\eta = k)$. Apart from the total variation we will prove estimates for the local and Wasserstein norms. Therefore we need additional information on corresponding solutions to (11.5).

Lemma 11.4 *Let g_f be the solution to the Stein equation*

$$\lambda g(j+1) - jg(j) = f(j) - \mathbb{E}f(\eta), \quad j = 0, 1, \dots$$

Then:

(a) for bounded $f : \mathbb{Z}_+ \to \mathbb{R}$, g_f has the following properties

$$\| g_f \|_\infty \leqslant 2\|f\|_\infty \min\{1, \sqrt{2/e}\lambda^{-1/2}\}, \quad \| \Delta g_f \|_\infty \leqslant 2\|f\|_\infty \min\{1, \lambda^{-1}\};$$
$$(11.20)$$

(b) for $f : \mathbb{Z}_+ \to \mathbb{R}$, satisfying $\|f\|_1 := \sum_{k=0}^\infty |f(k)| < \infty$, g_f has the following properties:

$$\| g_f \|_1 \leqslant 2\|f\|_1 \lambda^{-1/2}, \quad \| \Delta g_f \|_1 \leqslant 2\|f\|_1 \lambda^{-1};$$
$$(11.21)$$

(c) for $f : \mathbb{Z}_+ \to \mathbb{R}$, satisfying $\| \Delta f \|_\infty \leqslant 1$, g_f has the following properties:

$$\| g_f \|_\infty \leqslant 3, \quad \| \Delta g_f \|_\infty \leqslant 3\lambda^{-1/2}.$$
$$(11.22)$$

Remark 11.1   Observe that in case (c) we do not assume f to be bounded.

All estimates are now proved according to the same scheme: for an integer-valued random variable ξ we estimate the right-hand side of

$$\left| \mathbb{E}f(\xi) - \mathbb{E}f(\eta) \right| = \left| \mathbb{E}\left\{ \lambda g(\xi+1) - \xi g(\xi) \right\} \right|.$$
$$(11.23)$$

Then we establish:

(a) the estimate in total variation by taking in (11.23) the supremum over all $\|f\|_\infty \leqslant 1$ and applying (11.20);

(b) a local estimate by taking in (11.23) the supremum over all $f(j) = \mathbb{I}_k(j)$ and applying (11.21). Here $\mathbb{I}_k(k) = 1$ and $\mathbb{I}_k(j) = 0, j \neq k$;

(c) an estimate for the Wasserstein norm by taking in (11.23) the supremum over
all $f(j)$ satisfying $\| \Delta f \|_\infty \leqslant 1$ and applying (11.22).

Let $S = \mathbb{I}_1 + \mathbb{I}_2 + \ldots + \mathbb{I}_n$. Here $P(\mathbb{I}_j = 1) = p_j = 1 - P(\mathbb{I}_j = 0)$ are Bernoulli
(indicator) variables. We also assume that all $\mathbb{I}_j$ are independent. Let $\lambda = \sum_{k=1}^{n} p_k$.

Theorem 11.1 *Let $n \in \mathbb{N}$, $p_i \in (0,1)$, $\lambda \geqslant 1$. Then*

$$\| \mathcal{L}(S) - \mathcal{L}(\eta) \| \leqslant 2\lambda^{-1} \sum_{1}^{n} p_i^2, \qquad (11.24)$$

$$\| \mathcal{L}(S) - \mathcal{L}(\eta) \|_\infty \leqslant \lambda^{-1} \tilde{\tau}^{-1/2} \sum_{i=1}^{n} p_i^2, \qquad (11.25)$$

$$\| \mathcal{L}(S) - \mathcal{L}(\eta) \|_W \leqslant 3\lambda^{-1/2} \sum_{i=1}^{n} p_i^2. \qquad (11.26)$$

Here $\tilde{\tau} := \lambda - \sum_{k=0}^{n} p_k^2 - \max_i p_i(1 - p_i)$.

Proof Let $S^i := S - \mathbb{I}_i$. Then

$$\mathbb{E}\,g(S+1) = p_i\mathbb{E}\,g(S^i + 2) + (1 - p_i)\mathbb{E}\,g(S^i + 1), \quad \mathbb{E}\,\mathbb{I}_i g(S) = p_i\mathbb{E}\,g(S^i + 1).$$

Therefore

$$\mathbb{E}\,\{\lambda g(S + 1) - Sg(S)\}$$

$$= \sum_{i=1}^{n} p_i\mathbb{E}\,g(S+1) - \sum_{i=1}^{n} \mathbb{E}\,\mathbb{I}_i g(S) = \sum_{i=1}^{n} p_i\mathbb{E}\,\{g(S+1) - g(S^i + 1)\}$$

$$= \sum_{i=1}^{n} p_i\left\{ \mathbb{E}\,p_i g(S^i + 2) + \mathbb{E}\,(1 - p_i)g(S^i + 1) - \mathbb{E}\,g(S^i + 1) \right\}$$

$$= \sum_{i=1}^{n} p_i^2(\mathbb{E}\,g(S^i + 2) - \mathbb{E}\,g(S^i + 1)) = \sum_{i=1}^{n} p_i^2\mathbb{E}\,\Delta g(S^i + 1).$$

Consequently, (11.23) yields

$$\left| \sum_{k=0}^{\infty} f(k)\,(P(S = k) - \pi(k)) \right| \leqslant \| \Delta g \|_\infty \sum_{i=1}^{n} p_i^2.$$

By (11.20) and (11.22) we respectively prove (11.24) and (11.26).

Without much additional calculation we get the local estimate. Indeed, by definition of the mean we have

$$\left| \mathbb{E} \, \Delta g(S^i + 1) \right| \leqslant \sum_{k=0}^{n} | \, \Delta g(k) \, | \, P(S^i = k - 1) \leqslant \| \, \Delta g \, \|_1 \max_{i,k} P(S^i = k).$$

Assuming that $f(j) = \mathbb{I}_k(j)$, from (11.21) we get $\| \, \Delta g \, \|_1 \leqslant 2\lambda^{-1}$. Next, by (3.2) and (3.8) we obtain

$$P(S^i = k) \leqslant \frac{1}{2\pi} \int_{-\pi}^{\pi} \exp\{-2\tilde{\tau} \sin^2(t/2)\} dt \leqslant \frac{1}{2\sqrt{\tilde{\tau}}}. \tag{11.27}$$

The last estimate follows from the properties of Bessel functions. Combining the last estimates we prove (11.25). $\square$

It is easy to check that for the binomial distribution the estimate in total variation, local estimate and estimate in the Wasserstein norm metric are of the order $O(p)$, $O(\sqrt{p/n})$ and $O(p\sqrt{np})$, respectively. Note also that (11.24) and (11.25) have explicit constants and, in this sense, are more accurate than (2.20) and (3.5).

11.6   The Perturbation Approach

Observe that Stein's operator for CP approximation (11.11) is not of the form (11.7). On the other hand, the first two summands in (11.11) are similar to Stein's operator for Poisson approximation. Therefore it is natural to expect that the solution to the corresponding Stein equation will have similar properties. However, to solve the Stein equation for CP approximation is problematic. The perturbation approach allows us to check for properties similar to (11.8) without solving the Stein equation directly. Generally speaking, we expect Stein's operator to be of the form $\mathcal{A} + U$, where $\mathcal{A}$ is Stein's operator for some known-distribution and the operator U is in some sense small (a perturbation of the initial operator).

We formulate perturbation lemmas for all three cases, when solutions to the Stein equation have nice properties. We begin with CP approximation, which can be treated as Poisson perturbation.

Let the probability generating function of G be given in (11.10). Let

$$\lambda := \sum_{j=1}^{\infty} j\lambda_j, \qquad \upsilon := \frac{\sum_{m=2}^{\infty} m(m-1) | \, \lambda_m \, |}{\lambda}. \tag{11.28}$$

Lemma 11.5 *Let S be a random variable concentrated on $\mathbb{Z}_+$ and let G be the CP approximation with the probability generating function as in (11.10). Let, for any bounded function g and Stein's operator $\mathcal{A}$ defined by (11.11),*

$$|\mathbb{E}(\mathcal{A}g)(S)| \leq \varepsilon \|\Delta g\|_\infty.$$

If $\lambda > 0$ and $\upsilon < 1/2$ then

$$\|\mathcal{L}(S) - G\| \leq \frac{2\varepsilon}{1 - 2\upsilon} \min(1, \lambda^{-1}). \tag{11.29}$$

Remark 11.2 Observe that in terms of measures we can express G in the following way

$$G = \exp\left\{\sum_{k=1}^{\infty} \lambda_k(I_k - I)\right\}, \quad \lambda_k \in \mathbb{R}.$$

Remark 11.3 If the assumptions of Lemma 11.5 are satisfied then, for any bounded f, the solution of the Stein equation $(\mathcal{A}g)(j) = f(j) - \sum_{i=0}^{\infty} f(i)G\{i\}$, $(j = 0, 1, \ldots)$ satisfies the inequalities

$$\|g\|_\infty \leq \frac{2}{1 - 2\upsilon}(1 \wedge \lambda^{-1/2})\|f\|_\infty, \quad \|G\| \leq \frac{1}{1 - 2\upsilon}, \tag{11.30}$$

$$\|Ug\|_\infty \leq \frac{2\upsilon}{1 - 2\upsilon}\|f\|_\infty, \quad \|\Delta g\|_\infty \leq \frac{2}{1 - 2\upsilon}(1 \wedge \lambda^{-1})\|f\|_\infty. \tag{11.31}$$

Here $(Ug)(j) = \sum_{m=2}^{\infty} m\lambda_m \sum_{l=1}^{m-1} \Delta g(j + l)$.

Next, we formulate a perturbation lemma for the negative binomial approximation. As in the above, we assume that S is concentrated on $\mathbb{Z}_+$.

Lemma 11.6 *Let $\gamma > 0$, M be a measure of finite variation defined on $\mathbb{Z}_+$ and let for any bounded function $g : \mathbb{Z}_+ \to \mathbb{R}$*

$$\sum_{k=0}^{\infty} [q(\gamma + k)g(k + 1) - kg(k) + (Ug)(k)]M\{k\} = 0.$$

Here the operator U is defined on the set of all bounded functions with support $\mathbb{Z}_+$, such that $\|Ug\|_\infty \leq \tilde{\varepsilon}\|\Delta g\|_\infty$, $2\tilde{\varepsilon} < \gamma q$. Let

$$|\mathbb{E}[q(\gamma + S)g(S + 1) - Sg(S) + (Ug)(S)]| \leq \varepsilon \|\Delta g\|_\infty,$$

then

$$\| \mathcal{L}(S) - \mathrm{M} \| \le \frac{2\varepsilon}{\gamma q - 2\tilde{\varepsilon}}.$$

Finally, we formulate a perturbation lemma for the binomial perturbation. Let $0 < p < 1$, $q = 1 - p$, $\tilde{N} > 1$, $\tilde{\gamma} = \tilde{\varepsilon}/(\lfloor \tilde{N} \rfloor pq)$. Here, as usual, $\lfloor \cdot \rfloor$ denotes the integer part of the indicated argument.

Lemma 11.7 *Let g be any bounded function* $g : \mathbb{Z}_+ \to \mathbb{R}$, $g(j) = 0$ *if* $j \notin \{0, \ldots \lfloor \tilde{N} \rfloor\}$. *Let* M *be a measure of finite variation defined on* $\mathbb{Z}_+$ *and let*

$$\sum_{k=0}^{\infty} [(\tilde{N} - k)pg(k + 1) - kg(k) + (Ug)(k)]\mathrm{M}\{k\} = 0.$$

Here the operator U is defined on the set of all bounded functions with support $\mathbb{Z}_+$, *such that* $\| Ug \|_\infty \le \tilde{\varepsilon} \| \Delta g \|_\infty$. *Let*

$$| \mathbb{E} [(\tilde{N} - S)pg(S + 1) - Sg(S) + (Ug)(S)] | \le \varepsilon \| \Delta g \|_\infty$$

and $2\tilde{\gamma} < 1$. *Then*

$$\| \mathcal{L}(S) - \mathrm{M} \| \le \frac{2}{1 - 2\tilde{\gamma}} \left(\varepsilon \min \left(1, \frac{1}{\lfloor \tilde{N} \rfloor pq} \right) + \sum_{j=\lfloor \tilde{N} \rfloor + 1}^{\infty} |\mathrm{M}\{j\}| + P(S > \lfloor \tilde{N} \rfloor) \right).$$

We demonstrate how to use the perturbation technique by considering a two-parametric signed CP approximation to the Poisson binomial distribution. As in the previous section, let $S = \mathbb{I}_1 + \mathbb{I}_2 + \ldots + \mathbb{I}_n$, where $\mathbb{I}_j$ are independent Bernoulli variables $P(\mathbb{I}_j = 1) = p_j = 1 - P(\mathbb{I}_j = 0)$. Let $S^i = S - \mathbb{I}_i$, $\lambda = \sum_{k=1}^{n} p_k$.

The signed CP approximation D is chosen to match two factorial moments of S. Its moment generating function is equal to

$$\widehat{D}(z) = \exp\left\{ \lambda(z - 1) - \frac{1}{2} \sum_{i=1}^{n} p_i^2 (z - 1)^2 \right\} = \exp\{\lambda_1 (z - 1) + \lambda_2 (z^2 - 1)\}.$$

Here

$$\lambda_1 = \sum_{i=1}^{n} p_i (1 + p_i), \quad \lambda_2 = -\frac{1}{2} \sum_{i=1}^{n} p_i^2.$$

Let

$$\tilde{\tau} = \sum_1^n p_k(1-p_k) - \max_i p_i(1-p_i), \qquad \upsilon = \frac{\sum_1^n p_i^2}{\lambda}.$$

Theorem 11.2 *Let* $\max_{1 \leqslant i \leqslant n} p_i < 1/2$. *Then*

$$\| \mathcal{L}(S) - D \| \leqslant \frac{2}{(1-2\upsilon)\lambda\sqrt{\tilde{\tau}}} \sum_{i=1}^n p_i^3. \tag{11.32}$$

Proof Taking into account that $p_i < 1/2$, we observe that the main assumption of Lemma 11.5 is satisfied:

$$\upsilon = \frac{\sum_1^n p_i^2}{\lambda} < \frac{1}{2}.$$

Next, observe that by (11.11)

$$(\mathcal{A}g)(j) = \lambda_1 g(j+1) + 2\lambda_2 g(j+2) - jg(j) = \lambda g(j+1) - jg(j) + 2\lambda_2 \Delta g(j+1).$$

Consequently,

$$\mathbb{E}(\mathcal{A}g)(S) = \sum_{i=1}^n p_i \Big\{ \mathbb{E}g(S+1) - \mathbb{E}g(S^i+1) - p_i \mathbb{E}\Delta g(S+1) \Big\}$$

$$= \sum_{i=1}^n p_i \Big\{ p_i \mathbb{E}g(S^i+2) + (1-p_i)\mathbb{E}g(S^i+1) - \mathbb{E}g(S^i+1) - p_i \mathbb{E}\Delta g(S+1) \Big\}$$

$$= \sum_{i=1}^n p_i^2 \Big\{ \mathbb{E}\Delta g(S^i+1) - \mathbb{E}\Delta g(S+1) \Big\}$$

$$= \sum_{i=1}^n p_i^2 \Big\{ \mathbb{E}\Delta g(S^i+1) - p_i \mathbb{E}\Delta g(S^i+2) - (1-p_i)\mathbb{E}\Delta g(S^i+1) \Big\}$$

$$= -\sum_{i=1}^n p_i^3 \mathbb{E}\Delta^2 g(S^i+1).$$

We do not have appropriate estimate for the second difference of g. The problem is solved indirectly. Taking $g(j) = 0$, for $j \leqslant 0$, we obtain

$$\Big| \mathbb{E}\Delta^2 g(S^j+1) \Big| = \Big| \mathbb{E}\Big(g(S^j+2) - 2g(S^j+1) + g(S^j) \Big) \Big|$$

$$= \Big| \sum_{k=-2}^{\infty} P\{S^j = k\}(g(k+2) - g(k+1)) - \sum_{k=-2}^{n} P(S^j = k)(g(k+1) - g(k)) \Big|$$

$$= \left| \sum_{k=-2}^{n} (g(k+2) - g(k+1)) \Big(P(S^j = k) - P(S^j = k+1) \Big) \right|$$

$$\leq \| \Delta g \|_\infty \sum_{k=-1}^{n} \left| P(S^j = k) - P(S^j = k+1) \right|. \tag{11.33}$$

Observe that

$$\sum_{k=-1}^{n} \left| P(S^j = k) - P(S^j = k+1) \right| = \| \mathcal{L}(S^j + 1) - \mathcal{L}(S^j) \|.$$

Therefore it is possible to apply (11.6). However, a better estimate exists if we use the unimodality of S. Taking into account (11.27), we get

$$\sum_{k=-1}^{n} \left| P(S^j = k) - P(S^j = k+1) \right| \leq 2 \max_k P(S^j = k) \leq \frac{1}{\sqrt{\tau}}.$$

Combining all estimates we arrive at

$$|\mathbb{E} (\mathcal{A}g)(S)| \leq \frac{\sum_{i=1}^{n} p_i^3}{\sqrt{\tau}} \| \Delta g \|_\infty.$$

It remains to apply (11.29). $\square$

Note that (11.32) is always at least of the order of accuracy $O(n^{-1/2})$.

11.7   Estimating the First Pseudomoment

In this section, we present one adaptation of the Stein method, which allows us to replace the bounded function $f(j)$ by the unbounded $f(j)(j - \lambda)$. As an approximation we consider the CP (signed) measure from the previous section, that is, with the probability generating function

$$G(z) = \exp \left\{ \sum_{j=1}^{\infty} \lambda_j(z^j - 1) \right\}, \qquad \sum_{j=1}^{\infty} j |\lambda_j| < \infty$$

and Stein's operator defined as in Lemma 11.5 with λ defined by (11.28).

We consider a random variable S concentrated on $\mathbb{Z}_+$ with two finite moments and $\mathbb{E} S = \lambda$. Note that the last assumption means that $\lambda > 0$.

Lemma 11.8 *Let $\upsilon < 1/2$ and, for any bounded $g : \mathbb{Z}_+ \to \mathbb{R}$, assume that*

$$|E(\mathcal{A}g)(S)| \leqslant \varepsilon_0 \|g\|_\infty. \qquad (11.34)$$

Then, for any bounded $f : \mathbb{Z}_+ \to \mathbb{R}$,

$$\left| \sum_{j=0}^{\infty} f(j)(j-\lambda)(P(S=j) - G\{j\}) \right| \leqslant 14(1-2\upsilon)^{-2}\varepsilon_0\|f\|_\infty.$$

Remark 11.4 We can always choose $f(j)$ to be such that the sum in Lemma 11.8 becomes

$$\sum_{j=0}^{\infty} |j - \lambda| \, | P(S=j) - G\{j\} |.$$

This expression coincides with the definition of the first absolute pseudomoment of $\mathcal{L}(\xi) - \pi$.

Proof For the sake of brevity let $P(j) := P(S = j)$, $G\{f\} = \sum_{j=1}^{\infty} f(j)G\{j\}$. We introduce two auxiliary functions

$$h_1(j) = (j - \lambda)g(j), \quad \varphi_1(j) = (j - \lambda)(\mathcal{A}g)(j) - (\mathcal{A}h_1)(j).$$

Multiplying the Stein equation

$$(\mathcal{A}g)(j) = f(j) - G\{f\}$$

by $(j - \lambda)P(j)$ and, taking into account that $\mathbb{E}\,\xi = \sum_{k=1}^{\infty} jP(j) = \lambda$, we obtain

$$\sum_{j=0}^{\infty} (\mathcal{A}g)(j)(j-\lambda)P(j) = \sum_{j=0}^{\infty} f(j)(j-\lambda)P(j) - G\{f\}\sum_{j=0}^{\infty}(j-\lambda)P(j),$$

$$= \sum_{j=0}^{\infty} f(j)(j-\lambda)P(j).$$

Similarly,

$$\sum_{j=0}^{\infty} (\mathcal{A}g)(j)(j-\lambda)G\{j\} = \sum_{j=0}^{\infty} f(j)(j-\lambda)G\{j\}.$$

The difference of the last two equations gives us the following result

$$\sum_{j=0}^{\infty} f(j)(j-\lambda)(P(j) - G\{j\}) = \sum_{j=0}^{\infty} (Ag)(j)(j-\lambda)(P(j) - G\{j\})$$

$$= \sum_{j=0}^{\infty} (Ah_1)(j)(P(j) - G\{j\}) + \sum_{j=0}^{\infty} [(j-\lambda)(Ag)(j) - (Ah_1)(j)](P(j) - G\{j\})$$

$$= \sum_{j=0}^{\infty} (Ah_1)(j)(P(j) - G\{j\}) + \sum_{j=0}^{\infty} \varphi_1(j)(P(j) - G\{j\}). \tag{11.35}$$

We shall prove that $h_1(j)$ is bounded if g is the solution to the Stein equation $(Ag)(j) = f(j) - G\{f\}$. Indeed, we can rewrite the Stein equation

$$\lambda g(j+1) - j g(j) + (Ug)(j) = f(j) - G\{f\}$$

in the form

$$\lambda \Delta g(j) - (\lambda - j)g(j) + (Ug)(j) = f(j) - G\{f\}.$$

Therefore

$$|h_1(j)| = |\lambda - j||g(j)| \leqslant |f(j) - G\{f\}| + |(Ug)(j)| + \lambda |\Delta g(j)| < \infty.$$

We estimate each summand separately. Observe that by (11.31)

$$|f(j) - G\{f\}| \leqslant \|f\|_\infty \left(1 + \sum_{j=0}^{\infty} |G\{j\}|\right) \leqslant \|f\|_\infty \left(1 + \frac{1}{1 - 2\upsilon}\right).$$

Moreover, taking into account (11.30), we get

$$|(Ug)(j)| \leqslant \frac{2\upsilon}{1 - 2\upsilon}\|f\|_\infty, \quad \lambda |\Delta g(j)| \leqslant \frac{2}{1 - 2\upsilon}\|f\|_\infty.$$

Combining the last estimates we obtain

$$\|h_1(j)\|_\infty \leqslant \|f\|_\infty \left(1 + \frac{1}{1 - 2\upsilon} + \frac{2\upsilon}{1 - 2\upsilon} + \frac{2}{1 - 2\upsilon}\right) = \frac{4}{1 - 2\upsilon}\|f\|_\infty.$$

Consequently,

$$\sum_{j=0}^{\infty} (Ah_1)(j)G\{j\} = 0$$

and by assumption (11.34)

$$|\mathbb{E}\,(\mathcal{A}h_1)(S)\,| \leq \frac{4\varepsilon_0}{1-2\upsilon}\,\|f\|_\infty.$$

Substituting the two expressions into (11.35) we obtain

$$\left|\sum_{j=0}^\infty f(j)(j-\lambda)(P(j)-G\{j\})\right| \leq \frac{4\varepsilon_0}{1-2\upsilon}\,\|f\|_\infty + \left|\sum_{j=0}^\infty \varphi_1(j)(P(j)-G\{j\})\right|.$$

$$(11.36)$$

Next, we estimate $\|\varphi_1\|_\infty$ from above. From (11.11) it follows that

$$(\mathcal{A}g)(j)(j-\lambda) = \sum_{l=1}^\infty l\lambda_l g(j+l)(j-\lambda) - j(j-\lambda)g(j),$$

$$(\mathcal{A}h_1)(j) = \sum_{l=1}^\infty l\lambda_l(j+l-\lambda)g(j+l) - j(j-\lambda)g(j).$$

Therefore

$$|\varphi_1(j)| = |(j-\lambda)(\mathcal{A}g)(j) - (\mathcal{A}h_1)(j)| = \left|\sum_{l=1}^\infty l^2\lambda_l g(j+l)\right|$$

$$\leq \|g\|_\infty \sum_{l=1}^\infty l^2|\lambda_l| \leq \frac{2}{(1-2\upsilon)\sqrt{\lambda}}\|f\|_\infty \sum_{l=1}^\infty l^2|\lambda_l|. \quad (11.37)$$

Here we used (11.30) for the last estimate.

We estimate $\sum_{l=1}^\infty l^2|\lambda|$ through λ. First, note that $\lambda_1 > 0$. Indeed, $l^2 \leq 2l(l-1)$, $l > 1$ and $\upsilon < 1/2$ is equivalent to

$$2\sum_{l=2} l(l-1)|\lambda_l| \leq \lambda.$$

Therefore

$$2\sum_{l=2}^\infty l(l-1)|\lambda_l| \leq \lambda = \lambda_1 + \sum_{l=2}^\infty l\lambda_l \leq \lambda_1 + \sum_{l=2}^\infty l(l-1)|\lambda_l|$$

or

$$0 \leq \sum_{l=2}^\infty l(l-1)|\lambda_l| \leq \lambda_1.$$

Next, we show that $2\lambda_1 \leqslant 3\lambda$. Indeed, if $\sum_{l=2}^{\infty} l\lambda_l \geqslant 0$, then $\lambda_1 \leqslant \lambda$. Now let $\sum_{l=2}^{\infty} l\lambda_l < 0$. Then

$$2\left| \sum_{l=2}^{\infty} l\lambda_l \right| \leqslant 2 \sum_{l=2}^{\infty} l|\lambda_l| \leqslant 2 \sum_{l=2}^{\infty} l(l-1)|\lambda_l| \leqslant \lambda = \lambda_1 - \left| \sum_{l=2}^{\infty} l\lambda_l \right|.$$

Consequently,

$$\left| \sum_{l=2}^{\infty} l\lambda_l \right| \leqslant \lambda_1/3$$

and

$$\lambda_1 = \lambda + \left| \sum_{l=2}^{\infty} l\lambda_l \right| \leqslant \lambda + \lambda_1/3.$$

Thus, in any case, $0 < \lambda_1 \leqslant 3\lambda/2$. Finally,

$$\sum_{l=1}^{\infty} l^2 |\lambda_l| \leqslant \frac{3\lambda}{2} + 2 \sum_{l=2}^{\infty} l(l-1)|\lambda_l| \leqslant \frac{3\lambda}{2} + \lambda = \frac{5\lambda}{2}.$$

Substituting the last estimate into (11.37) we obtain

$$\| \varphi_1 \|_\infty \leqslant \frac{5\sqrt{\lambda}}{(1-2\upsilon)} \| f \|_\infty.$$

Next, observe that

$$\left| \sum_{j=0}^{\infty} \varphi_1(j)(P(j) - G\{j\}) \right| \leqslant \| \varphi_1 \|_\infty \sum_{j=0}^{\infty} |P(j) - G\{j\}|.$$

From (11.30) it follows that

$$\sum_{j=0}^{\infty} |P(j) - G\{j\}| \leqslant |\mathbb{E}(\mathcal{A}g)(S)| \leqslant \varepsilon_0 \| g \|_\infty \leqslant \varepsilon_0 \frac{2}{(1-2\upsilon)\sqrt{\lambda}}.$$

Therefore

$$\left| \sum_{j=0}^{\infty} \varphi_1(j)(P(j) - G\{j\}) \right| \leqslant \frac{10\varepsilon_0}{(1-2\upsilon)^2} \| f \|_\infty.$$

Substituting the last estimate into (11.36) we get

$$\left| \sum_{j=0}^{\infty} f(j)(j-\lambda)(P(j) - G\{j\}) \right| \leq \|f\|_{\infty} \frac{\varepsilon_0(14-8\upsilon)}{(1-2\upsilon)^2} \leq \frac{14\varepsilon_0}{(1-2\upsilon)^2} \|f\|_{\infty}.$$

□

Example 11.1 We consider a signed Poisson approximation to the Poisson binomial distribution. Let S, S^i, υ and D be the same as in Theorem 11.2, $\max_i p_i < 1/2$ and $\lambda = \sum_{i=1}^{n} p_i \geq 8$. In the proof of Theorem 11.2 we already established that

$$\left| \mathbb{E}\,(\mathcal{A}g)(S) \right| = \left| \sum_{i=1}^{n} p_i^3 \mathbb{E}\,\Delta^2 g(S^i + 1) \right|.$$

Extending (11.33) one step further we prove that

$$\left| \mathbb{E}\,\Delta^2 g(S^j + 1) \right| = \left| \sum_{k=-2}^{n} (g(k+2) - g(k+1))(P(S^j = k) - P(S^j = k+1)) \right|$$

$$= \left| \sum_{k=-2}^{\infty} g(k+2)(P(S^j = k) - 2P(S^j = k+1) + P(S^j = k+2)) \right|$$

$$\leq \|g\|_{\infty} \sum_{k=-2}^{\infty} |P(S^j = k) - 2P(S^j = k+1) + P(S^j = k+2)|$$

$$= \|g\|_{\infty} \| \mathcal{L}(S^j)(I_1 - I)^2 \| = \|g\|_{\infty} \left\| \prod_{k \neq j}((1-p_k)I + p_k I_1)(I_1 - I)^2 \right\|$$

$$\leq \|g\|_{\infty} \left\| \prod_{k \in S_1}(I + p_k(I_1 - I))(I - I_1) \right\| \cdot \left\| \prod_{k \in S_2}(I + p_k(I_1 - I))(I - I_1) \right\|.$$

Here the sets of indices S_1 and S_2 are chosen so that

$$\sum_{k \in S_{1,2}} p_k \geq \frac{\lambda}{2} - 2\max_j p_j = \frac{\lambda}{4} + \frac{\lambda}{4} - 2 \geq \frac{\lambda}{4}.$$

The last inequality is due to assumption $\lambda \geq 8$. Observe that

$$\| ((1-p_k)I + p_k I_1) \| = 1 + |1 - 2p_k| = 2 - 2p_k.$$

Applying (5.13) we, therefore, obtain

$$\left| \mathbb{E}\,\Delta^2 g(S^j + 1) \right| \leq \frac{32\|g\|_{\infty}}{\pi \lambda}$$

and, consequently,

$$\left| \mathbb{E}\,(\mathcal{A}g)(S) \right| \le \frac{32\|\,g\,\|_\infty}{\pi} \frac{\sum_{i=1}^n p_i^3}{\lambda}.$$

An application of Lemma 11.8 results in the following estimate

$$\left| \sum_{j=0}^\infty f(j)(j-\lambda)(P(S=j)-D\{j\}) \right| \le \frac{448}{\pi(1-2\upsilon)^2} \frac{\sum_{i=1}^n p_i^3}{\lambda}\|f\|_\infty.$$

11.8   Lower Bounds for Poisson Approximation

Lower bounds via the Stein method are proved by choosing an explicit function with useful properties. We illustrate this approach considering a Poisson approximation to the sum of Bernoulli variables. Note that an upper bound for this approximation was established in Theorem 11.1.

Let us consider the Poisson binomial distribution, that is, let $S = \mathbb{I}_1 + \mathbb{I}_2 + \ldots + \mathbb{I}_n$, where $P(\mathbb{I}_j = 1) = p_j = 1 - P(\mathbb{I}_j = 0)$ are Bernoulli (indicator) variables. We assume that all $\mathbb{I}_j$ are independent. Let $\lambda = \sum_{k=1}^n p_k$ and let η be a Poisson random variable with parameter λ.

Theorem 11.3 *Let $n \in \mathbb{N}$, $p_i \in (0,1)$, $\lambda \ge 1$. Then*

$$\| \mathcal{L}(S) - \mathcal{L}(\eta) \| \ge \frac{1}{16}\lambda^{-1} \sum_1^n p_i^2.$$

Proof Let $S^i := S - \mathbb{I}_i$. In the proof of Theorem 11.1 we proved that for any bounded function $g(\cdot)$

$$\mathbb{E}\{\lambda g(S+1) - Sg(S)\} = \sum_{i=1}^n p_i^2 \mathbb{E}\,\Delta g(S^i + 1).$$

Recalling that $\mathbb{E}\{\lambda g(\eta+1) - \eta g(\eta)\} = 0$ we obtain

$$\left| \mathbb{E}\{\lambda g(S+1) - Sg(S)\} \right| = \left| \mathbb{E}\{\lambda g(S+1) - Sg(S)\} - \mathbb{E}\{\lambda g(\eta+1) - \eta g(\eta)\} \right|$$

$$= \left| \sum_{j=0}^\infty (\lambda g(j+1) - jg(j))(P(S=j) - P(\eta=j)) \right|$$

$$\le \sup_{j \in \mathbb{Z}_+} |\lambda g(j+1) - jg(j)| \,\| \mathcal{L}(S) - \mathcal{L}(\eta) \|.$$

Therefore

$$\sup_{j\in\mathbb{Z}_+} |\lambda g(j+1) - jg(j)| \|\mathcal{L}(S) - \mathcal{L}(\eta)\| \geq \sum_{j=1}^{n} p_j^2 - \sum_{j=1}^{n} p_j^2(1 - \mathbb{E}\,\Delta g(S^i + 1)).$$

$$(11.38)$$

The next step is to choose the function $g(j)$ so that

$$\sup_{j\in\mathbb{Z}_+} |\lambda g(j+1) - jg(j)| \leq C\lambda, \qquad \mathbb{E}\,\Delta g(S^i + 1) \approx 1.$$

Let

$$g(j) = (j - \lambda) \exp\{-(j - \lambda)^2/(b\lambda)\}.$$

The constant $b > 0$ will be chosen later. We define an auxiliary function $\varphi(x) = x\exp\{-x^2/(b\lambda)\}$. Standard calculus shows that

$$\varphi'(x) = \exp\{-x^2/(b\lambda)\}(1 - 2x^2/(b\lambda)), \quad -2e^{-3/2} \leq \varphi'(x) \leq 1.$$

Therefore by Lagrange's theorem

$$-2e^{-3/2} \leq g(j+1) - g(j) = \varphi(j + 1 - \lambda) - \varphi(j - \lambda) \leq 1.$$

Observe that, if $\lambda g(j+1) - jg(j) \geq 0$, then

$$\lambda g(j+1) - jg(j) = \lambda(g(j+1) - g(j)) - (j - \lambda)^2 \exp\{-(j - \lambda)^2/(b\lambda)\}$$
$$\leq \lambda(g(j+1) - g(j)) \leq \lambda.$$

If $\lambda g(j+1) - jg(j) \leq 0$, then

$$-(\lambda g(j+1) - jg(j)) = -\lambda(g(j+1) - g(j)) + (j - \lambda)^2 \exp\{-(j - \lambda)^2/(b\lambda)\}$$
$$\leq 2\lambda e^{-3/2} + \sup_{x\geq 0} x\exp\{-x/(b\lambda)\} \leq 2\lambda e^{-3/2} + b\lambda e^{-1}.$$

Combining the last two estimates we obtain

$$\sup_{j\in\mathbb{Z}_+} |\lambda g(j+1) - jg(j)| \leq \lambda \max(1, 2e^{-3/2} + be^{-1}).$$ $$(11.39)$$

Next, we estimate the right-hand side of (11.38). It is easy to check that

$$1 - e^{-y}(1 - 2y) \leq 3y, \quad y \geq 0.$$

Therefore

$$1 - \Delta g(S^i + 1) = \int_{S^i + 1 - \lambda}^{S^i + 2 - \lambda} (1 - \varphi'(u)) du \leq \int_{S^i + 1 - \lambda}^{S^i + 2 - \lambda} \frac{3u^2}{b\lambda} du$$

$$= (b\lambda)^{-1}[3(S^i - \lambda)^2 + 9(S^i - \lambda) + 7].$$

Observe that $\mathbb{E} \, S^i - \lambda = -p_i$ and

$$\mathbb{E} \, (S^i - \lambda)^2 = \mathbb{E} \, (S^i - \mathbb{E} \, S^i - p_i)^2 = Var S^i + p_i^2 = \sum_{j \neq i}^{n} p_j(1 - p_j) + p_i^2 \leq \lambda + p_i^2.$$

Consequently,

$$1 - \mathbb{E} \, \Delta g(S^i + 1) \leq (b\lambda)^{-1}(3\lambda + 3p_i^2 - 9p_i + 7) \leq (b\lambda)^{-1}(3\lambda + 7).$$

Substituting the last estimate and (11.39) into (11.39) we obtain

$$\lambda \max(1, 2e^{-3/2} + be^{-1}) \| \mathcal{L}(S) - \mathcal{L}(\eta) \| \geq [1 - (b\lambda)^{-1}(3\lambda + 7)] \sum_{j=1}^{n} p_j^2. \quad (11.40)$$

We assumed that $\lambda \geq 1$. Therefore from (11.40) it follows that

$$\| \mathcal{L}(S) - \mathcal{L}(\eta) \| \geq \lambda^{-1} \sum_{j=1}^{n} p_j^2 \frac{1 - 10/b}{0.446261 + 0.36788b}.$$

It remains to choose $b = 20$. $\square$

Remark 11.5 In Theorem 11.3 we assumed that $\lambda \geq 1$. In principle, the same approach can be applied for the case $\lambda < 1$, resulting in a lower bound of $C \sum_{i=1}^{n} p_i^2$.

11.9 Problems

11.1 Prove (11.18) by using recursive properties of the binomial probabilities.

11.2 Prove (11.15) via the probability generating function approach.

11.3 Let η have a geometric distribution, $P(\eta = j) = q^j p, j = 0, 1, 2, \ldots$. Prove that η has the following two different Stein operators:

$$(\mathcal{A}g)(j) = q(j+1)g(j+1) - jg(j), \qquad (\mathcal{A}g)(j) = qg(j+1) - g(j).$$

11.4 Let $S = \mathbb{I}_1 + \mathbb{I}_2 + \ldots + \mathbb{I}_n$, where $P(\mathbb{I}_j = 1) = 1 - P(\mathbb{I}_j = 0) = p_j \leqslant 1/6$ and all indicators are independent, and let π_3 have the moment generating function

$$\widehat{\pi}_3(z) = \exp\left\{\sum_{i=1}^{n}\left(p_i(z-1) - \frac{p_i^2}{2}(z-1)^2 + \frac{p_i^3}{3}(z-1)^3\right)\right\}.$$

Let $\lambda = \sum_{i=1}^{n} p_i \geqslant 1$. Prove that

$$\|\mathcal{L}(S) - \pi_3\| \leqslant C\lambda^{-2}\sum_{i=1}^{n}p_i^4.$$

11.5 Investigate the approximation of S from the previous exercise by the binomial distribution with parameters n and $p = \sum_{i=1}^{n} p_i/n$, treating it as a CP perturbation.

11.6 Let S be defined as in the previous problem and let η have the negative binomial distribution with Stein's operator as in (11.15) and parameters satisfying

$$\gamma = n, \quad \frac{nq}{p} = \sum_{i=1}^{n}p_i.$$

Prove that

$$\|\mathcal{L}(S) - \mathcal{L}(\eta)\| \leqslant \frac{2\sum_{i=1}^{n}p_i^2}{\sum_{i=1}^{n}} + \frac{2\sum_{i=1}^{n}}{n}.$$

11.7 Let S be the sum of independent Bernoulli variables as in the previous problem, $\lambda = \sum_{k=1}^{n} p_k < 1$ and let η be a Poisson random variable with parameter λ. Prove that

$$\|\mathcal{L}(S) - \mathcal{L}(\eta)\| \geqslant C\sum_{i=1}^{n}p_i^2.$$

Bibliographical Notes

The Stein method was introduced for the normal approximation in [143]. Chen's adaptation of the Stein method was introduced in [47]. Note that the Stein method for lattice variables is also called the Stein-Chen method. Introductory texts on Stein's method can be found in [7, 127, 144]. See also Chapter 2 in [128] and [117]. Overviews of the method are presented in [8, 9, 48, 50]. Generalizations of the density approach are given in [94, 95].

The classical applications of Poisson approximation are given in [16]. Solutions to the discrete Stein equation are investigated in [29]. The solution to the Stein equation for a CP distribution was considered in [13]. The perturbation technique was introduced in [14] and developed in [6, 15], see also [158]. Estimates for pseudomoments were investigated in [38], the Wasserstein norm in

[160]. The normal approximation is discussed in [51]. For zero biasing, see [66]. Exchangeable pairs are discussed in [145]. For some new results, when the random variables are dependent, see [52, 53, 131]. It must be noted that the Stein method has also been very fruitfully united with Malliavin calculus, see [101].

Estimate (11.8) follows from the more general Theorem 2.10 in [29]. Estimate (11.27) was proved in [11], Lemma 1. In Lemma 11.4, we presented slightly more rough constants than can be obtained by solving the Stein equation, see [7], p. 66. The idea to use the probability generating functions to establish probabilistic recursions was taken from [156]. Theorem 11.3 is a special case of Theorem 2 from [10].

Chapter 12
The Triangle Function Method

In this chapter, we discuss the triangle function method, which is related to the Esseen type inversion inequality, but contains much more elaborate estimates for the concentration function. The advantages and drawbacks of the triangle function method can be summarized in the following way.

> **Advantages.** The method can be applied in very general situations. No moment assumptions. In many cases, there is no alternative method which can ensure the same order of accuracy.
>
> **Drawbacks.** Difficult to use. Absolute constants in estimates are not explicit. Can be applied only to measures with compound structure, such as $|\varphi(F)|_K$. For example, the method cannot be applied to $|F - G|_K$, where F is discrete and G is continuous.

Many 'tools' of this chapter are given without proofs. All omitted proofs can be found in [5].

12.1 The Main Lemmas

The triangle function method was developed for the estimation of compound distributions. In this section we assume that $W \in \mathcal{M}$ and $W\{\mathbb{R}\} = 0$. For $h > 0$ let us introduce the following pseudo-metric:

$$|W|_h = \sup_z |W\{[z, z + h]\}|. \tag{12.1}$$

© Springer International Publishing Switzerland 2016
V. Čekanavičius, *Approximation Methods in Probability Theory*, Universitext,
DOI 10.1007/978-3-319-34072-2_12

Observe that

$$|W|_K \leq \sup_{h \geq 0} |W|_h \leq 2|W|_K.$$

If $F \in \mathcal{F}$, then $|F|_h = Q(F, h)$. Let $\mathbb{I}\{[z, z+h]\}$ be the indicator function, i.e.

$$\mathbb{I}\{[z, z+h]\}(x) = \begin{cases} 1, & \text{if } x \in [z, z+h], \\ 0, & \text{if } x \notin [z, z+h]. \end{cases}$$

Then

$$|W|_h = \sup_z \left| \int_{-\infty}^{\infty} \mathbb{I}\{[z, z+h]\}(x)\, W\{dx\} \right|.$$

The triangle function method replaces of the indicator function by a function with better properties. Let us introduce the set of triangle functions

$$\{ f_{z,h,\tau}(x) : z \in \mathbb{R}, \, 0 < \tau \leq h \}.$$

Here $f_{z,h,\tau}(x)$ increases linearly from 0 to 1 on $[z, z+\tau]$, decreases linearly to 0 on $[z+\tau, z+h]$ and is equal to zero for $x \notin [z, z+h]$.

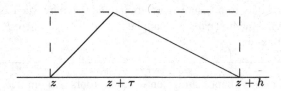

$$z \qquad\qquad z+\tau \qquad\qquad\qquad z+h$$

Let

$$|W|_{h,\tau} = \sup_{z \in \mathbb{R}} \left| \int_{-\infty}^{\infty} f_{z,h,\tau}(x)\, W\{dx\} \right|. \tag{12.2}$$

The pseudo-metric $|\cdot|_{h,\tau}$ has the following useful properties. Let $W, V \in \mathcal{M}$, $W\{\mathbb{R}\} = 0$, $a > 0$. Then

$$|WV|_{h,\tau} \leq \|V\| \, |W|_{h,\tau}, \qquad |aW|_{h,\tau} = a|W|_{h,\tau}, \tag{12.3}$$

$$|W|_{h,\tau} \leq |W|_{h,\omega} + |W|_{\omega,\tau}. \tag{12.4}$$

It is obvious that $f_{z,h,\tau}(x)$ is not a very good replacement for $\mathbb{I}\{[z, z + \tau]\}$. On the other hand, we can replace the indicator function by the sum of triangle functions with weights and apply (12.4). The general idea of the triangle function method is the following:

If we can estimate $|W|_{h,\tau}$ for sufficiently large sets of h and τ, then we can obtain an estimates for $|W|_h$ and for $|W|_K$.

More precisely, this idea is formulated in Lemma 12.1 below.

As a rule, the triangle function method is used when one can obtain estimate of the form

$$|\widehat{W}(t)| \leqslant C\varepsilon\widehat{D}(t), \tag{12.5}$$

where D is symmetric infinitely divisible distribution. Usually D is a CP distribution, that is, $D = \exp\{\lambda(F - I)\}$, for some symmetric $F \in \mathcal{F}_s$ and $\lambda > 0$. Set

$$\gamma_h = Q(D, h).$$

Typically, the triangle function method allows estimates of $|W|_{h,\tau}$ in terms of γ_h. Then, as follows from Lemma 12.1 below, it is possible to get the estimate for $|\cdot|_h$ and, consequently, for $|\cdot|_K$.

We recall that, for $F \in \mathcal{F}$, $F^{(-)}$ denotes the distribution, for any Borel set X, satisfying $F^{(-)}\{X\} = F\{-X\}$. Note that $\widehat{F^{(-)}}(t) = \widehat{F}(-t)$.

Lemma 12.1 *Let $a > 0$, $b > 0$ be some absolute constants, $W \in \mathcal{M}$, $W\{\mathbb{R}\} = 0$, $\gamma_0 > 0$, and let, for any h and τ such that $0 < \tau \leqslant h/2$ and $\gamma_\tau \geqslant \gamma_h/8$,*

$$\max\{|W|_{h,\tau}, |W^{(-)}|_{h,\tau}\} \leqslant C_1\varepsilon\gamma_h^a(|\ln\gamma_h| + 1)^b.$$

Then, for any $h > 0$,

$$|W|_h \leqslant C_2\varepsilon\gamma_h^a(|\ln\gamma_h| + 1)^b.$$

Remark 12.1 Taking into account that $\gamma_h \lesssim 1$, we can see that, from (12.1), it follows that $|W|_K \leqslant C(a, b)\varepsilon$.

Transition from the estimates, valid for all $0 \leqslant \tau \leqslant h/2$, to the estimates, valid for all $0 \leqslant \tau \leqslant h/2$ and $\gamma_\tau \geqslant \gamma_h/8$, is far from trivial. The following lemma, though somewhat cumbersome, gives quite a good idea of the structure of estimates allowing such a transition.

Lemma 12.2 *Let $W \in \mathcal{M}$, $W\{\mathbb{R}\} = 0$ and let H_1 be a finite nonnegative measure, for any $\tau > 0$, allowing the decomposition*

$$H_1 = H_{\tau 2} + H_{\tau 3} + H_{\tau 4}.$$

Here the nonnegative measures $H_{\tau 2}$, $H_{\tau 3}$ and $H_{\tau 4}$ satisfy relations

$$H_{\tau 3}\{\mathbb{R}\} \leq C_3(|\ln \gamma_\tau| + 1)^3,$$

$$H_{\tau 2}\{[-\tau, \tau]\} = H_{\tau 3}\{[-\tau, \tau]\} = H_{\tau 4}\{\mathbb{R} \setminus [-\tau, \tau]\} = 0.$$

Let, for any $0 \leq \tau \leq h/2$,

$$\max\{|W|_{h,\tau}, |W^{(-)}|_{h,\tau}\} \leq C_4 \varepsilon \left\{ \gamma_h^\beta (|\ln \gamma_h| + 1)^\delta (H_1\{\{x : |x| > \tau\}\})^m + \gamma_h^\alpha \right\}, \tag{12.6}$$

$$\max\{|W|_{h,\tau}, |W^{(-)}|_{h,\tau}\} \leq C_5 \varepsilon \left\{ \frac{(|\ln \gamma_\tau| + 1)^s}{\sqrt{H_{\tau 2}\{\{x : |x| > h/4\}\}}} + \gamma_h^\alpha \right\}, \tag{12.7}$$

where $0 < \beta, m, \alpha \leq C_6$ and $0 \leq \delta, s \leq C_7$ and $\varepsilon > 0$. Then, for any τ and h such that $0 < \tau \leq h/2$ and $\gamma_\tau \geq \gamma_h/8$,

$$|W|_{h,\tau} \leq C_8 \varepsilon \gamma_h^a (|\ln \gamma_h| + 1)^b. \tag{12.8}$$

Here

$$a = \min\left\{\alpha, \frac{\beta}{2m + 1}\right\}, \quad b = \max\left(\frac{2ms + \delta}{2m + 1}, 3m + \delta\right).$$

Moreover, for all $h > 0$,

$$|W|_h \leq C_9 \varepsilon \gamma_h^a (|\ln \gamma_h| + 1)^b. \tag{12.9}$$

Proof For $u > 0$ let

$$v_u(y) = H_{u2}\{\{x : |x| > y\}\}; \quad \kappa_u = \gamma_u^{-2\beta/(2m+1)} (|\ln \gamma_u| + 1)^{2(s-\delta)/(2m+1)}.$$

Let $0 < \tau \leq h/2$ and $\gamma_h \leq 8\gamma_\tau$. Then, recalling that by definition $0 \leq \gamma_t, \gamma_h \leq 1$, we get

$$|\ln \gamma_\tau| \leq |\ln \gamma_h| + \ln 8.$$

For the proof of (12.8) we consider three separate cases:

(1) $v_\tau(h/4) \geqslant \kappa_h$;
(2) $v_\tau(\tau) \leqslant \kappa_h$;
(3) $v_\tau(h/4) < \kappa_h < v_\tau(\tau)$.

In case (1), we apply (12.7). In case (2), we have

$$H_1\{\{x : |x| > \tau\}\} = H_{\tau2}\{\{x : |x| > \tau\}\} + H_{\tau3}\{\{x : |x| > \tau\}\}$$
$$\leqslant v_\tau(\tau) + C(|\ln \gamma_h| + 1)^3$$
$$\leqslant C\left(\gamma_h^{-2\beta/(2m+1)}(|\ln \gamma_h| + 1)^{2(s-\delta)/(2m+1)} + (|\ln \gamma_h| + 1)^3\right).$$

Consequently,

$$(H_1\{\{x : |x| > \tau\}\})^m \leqslant C\left\{\gamma_h^{-2\beta m/(2m+1)}(|\ln \gamma_h| + 1)^{2(s-\delta)m/(2m+1)} + (|\ln \gamma_h| + 1)^{3m}\right\}.$$

Now the estimate (12.8) follows from (12.6). In case (3), we note that $v_\tau(y)$ is non-increasing and we can find y such that

$$\tau < y < h/4, \quad v_\tau(2y) \leqslant \kappa_h \leqslant v_\tau(y/2).$$

Setting $\omega = 2y$, we can rewrite the last inequalities as

$$2\tau < \omega < h/2, \quad v_\tau(\omega) \leqslant \kappa_h \leqslant v_\tau(\omega/4).$$

From (12.4) it follows that

$$|W|_{h,\tau} \leqslant |W|_{h,\omega} + |W|_{\omega,\tau}.$$

For the estimate of $|W|_{\omega,\tau}$ it suffices to apply (12.7), replacing h by ω and taking into account that $v_\tau(\omega/4) \geqslant \kappa_h$ and $\gamma_\omega \leqslant \gamma_h$. For the estimate of $|W|_{h,\omega}$ we apply (12.6). Note that

$$H_1\{\{x : |x| > \omega\}\} \leqslant v_\tau(\omega) + H_{\tau3}\{\{x : |x| > \tau\}\} \leqslant \kappa_h + C(|\ln \gamma_h| + 1)^3.$$

Combining all estimates we get (12.8). The estimate (12.9) follows from Lemma 12.1. $\square$

Note that the idea of the proof can be applied in many similar situations which involve two estimates outside the finite interval $[-\tau, \tau]$ through some measure $H_{\tau2}$. Considering (12.6) and (12.7), we see that two different estimates for distributions concentrated outside the finite interval $[-\tau, \tau]$ are needed. This is not a coincidence. The triangle function method is close to the one used in Sect. 9.4. The main idea is to decompose the measure under estimation into two components: one measure concentrated on some finite interval and the other one concentrated outside this interval. The second measure must be estimated as in Lemma 12.2, because the standard estimate through the concentration function as in (9.3) is too rough and, in general, cannot ensure the an accuracy of order better than $n^{-1/2}$. On the other hand, for the measure concentrated on the finite interval we can use the standard approach, estimating it via a Fourier transform in a neighborhood of zero. For this we need some properties of the Fourier transforms for $f_{z,h,\tau}(x)$.

12.2   Auxiliary Tools

The technical aspects of the triangle function method can be summarized in the following statements:

1. For measures concentrated on $[-\tau, \tau]$ Parseval's identity and properties of $\widehat{f}_{z,h,\tau}(t)$ are used.
2. For estimation of the measure concentrated outside $[-\tau, \tau]$ the triangle function is replaced by some special function which has very similar properties for $x \in K_m(\mathbf{u})$. Here $\mathbf{u} = (u_1, u_2, \ldots, u_N) \in \mathbb{R}^\ell$ and

$$
K_m(\mathbf{u}) = \left\{ \sum_{i=1}^{\ell} j_i u_i : j_i \in \{-m, -m+1, \ldots, m\}; \ i = 1, \ldots, \ell \right\}. \tag{12.10}
$$

3. As a rule, the measure concentrated outside $[-\tau, \tau]$ can be decomposed as a sum of two measures: one concentrated in a neighborhood of $K_m(\mathbf{u})$ with $\mathbf{u} \in \mathbb{R}^\ell$ and the other having small total variation norm and concentrated outside this neighborhood.

Next, we elaborate on each statement, giving exact mathematical formulations. First we note that the triangle function $f_{z,h,\tau}(x)$ has a Fourier transform with quite good properties in a neighborhood of zero.

Lemma 12.3 *Let* $z \in \mathbb{R}$, $0 < \tau \leqslant h/2$. *Then*

(a) for all $t \in \mathbb{R}$

$$
|\widehat{f}_{z,h,\tau}(t)| \leqslant \min\left\{ \frac{2}{|t|}, \frac{h}{2} \right\}; \tag{12.11}
$$

(b) for $|t| > 1/\tau$

$$
|\widehat{f}_{z,h,\tau}(t)| \leqslant \frac{8\tau}{1 + (t\tau)^2}. \tag{12.12}
$$

Lemma 12.3 and Parseval's identity (1.18) allows us to get the estimate for W concentrated on $[-\tau, \tau]$:

$$\left| \int_{-\infty}^{\infty} f_{z,h,\tau}(x) \, W\{dx\} \right| \leq \frac{1}{2\pi} \int_{-\infty}^{\infty} |\widehat{f}_{z,h,\tau}(t)| \, |\widehat{W}(t)| \, dt \leq C \int_{-\infty}^{\infty} \frac{|\widehat{W}(t)|}{|t|} \, dt. \tag{12.13}$$

However, if W is concentrated outside the interval $[-\tau, \tau]$, the estimate (12.13) is too rough. The following general lemma shows that $f_{z,h,\tau}$ can be replaced by a special function ω which has good properties in a neighborhood of $K_m(\mathbf{u})$. Further a closed τ-neighborhood of the set X is denoted by $[X]_\tau$.

Lemma 12.4 *Let ℓ and m be positive integers, $\mathbf{u} = (u_1, u_2, \ldots, u_\ell) \in \mathbb{R}^\ell$, $z \in \mathbb{R}$ and $0 < \tau \leq h/2$. Then there exists a continuous function $\omega(x)$, $x \in \mathbb{R}$, with the following properties:*

$$0 \leq \omega(x) \leq f_{z,h,\tau}(x), \quad \text{for all } x \in \mathbb{R}, \tag{12.14}$$

$$\omega(x) = f_{z,h,\tau}(x,) \quad \text{for } x \in [K_m(\mathbf{u})]_{m\tau}, \tag{12.15}$$

$$\sup_{t \in \mathbb{R}} |\widehat{\omega}(t)| \leq h/2, \tag{12.16}$$

$$\int_{-\infty}^{\infty} \sup_{|s| \geq |t|} |\widehat{\omega}(t)| \, dt \leq C\ell^2 \ln(\ell m + 1); \tag{12.17}$$

and, for any $F \in \mathcal{F}_+$,

$$\int_{-\infty}^{\infty} |\widehat{\omega}(t)| \, |\widehat{F}(t)| \, dt \leq CQ(F, h)\ell^2 \ln(\ell m + 1). \tag{12.18}$$

Lemma 12.4 can be combined with the obvious

$$\int_{-\infty}^{\infty} f_{z,h,\tau}(x) \, W\{dx\} = \int_{-\infty}^{\infty} \omega(x) \, W\{dx\} + \int_{\mathbb{R}^*} (f_{z,h,\tau}(x) - \omega(x)) \, W\{dx\}. \tag{12.19}$$

Here $\mathbb{R}^* = \mathbb{R} \setminus [K_m(\mathbf{u})]_{m\tau}$. The first integral can be estimated by the Parseval identity and Lemma 12.4. Meanwhile, the estimate of the second integral depends on the choice of $\mathbf{u}$ and m, which is discussed below. The function $\omega(x)$ is used to obtain estimates of the type (12.7).

In principle, for any $F \in \mathcal{F}$, it is always possible to choose $K_1(\mathbf{u})$ in such a way that almost all probabilistic mass of F is concentrated in a neighborhood of $K_1(\mathbf{u})$. We formulate this statement as the following lemma.

Lemma 12.5 *Let $F_i \in \mathcal{F}$, $(i = 1, 2, \ldots, n)$, $\tau > 0$, $a \in (0, 1]$. Then there exists a vector $\mathbf{u} \in \mathbb{R}^{\ell}$ such that*

$$\ell \leq C(|\ln \gamma_{\tau,a}| + 1),$$

$$\sum_{i=1}^{n} F_i\{\mathbb{R} \setminus [K_1(\mathbf{u})]_{\tau}\} \leq Ca^{-1} \left(|\ln \gamma_{\tau,a}| + 1\right)^3.$$

Here

$$\gamma_{\tau,a} = Q\left(\exp\left\{a \sum_{i=1}^{n} (\tilde{F}_i - I)\right\}, \tau\right),$$

and $\tilde{F}_i \in \mathcal{F}$ has the characteristic function $\operatorname{Re} \widehat{F}_i(t)$.

We formulate an analogue of Lemma 12.5 for symmetric F.

Lemma 12.6 *Let $F \in \mathcal{F}$ be a symmetric distribution, $\gamma_{\tau} = Q(\exp\{\lambda(F - I)\}, \tau)$, $\tau > 0$, $\lambda > 0$. Then there exists a vector $\mathbf{u} \in \mathbb{R}^{\ell}$ such that*

$$\ell \leq C(|\ln \gamma_{\tau}| + 1), \tag{12.20}$$

$$F\{\mathbb{R} \setminus [K_1(\mathbf{u})]_{\tau}\} \leq \frac{C}{\lambda} \left(|\ln \gamma_{\tau}| + 1\right)^3. \tag{12.21}$$

Similar estimates can be obtained from Lemma 12.5 for other distributions or their mixtures. We need one additional lemma to estimate the characteristic function.

Lemma 12.7 *Let $H = (1 - p)F + pG$, $F, G \in \mathcal{F}$, $0 \leq p \leq 1$, F is a symmetric distribution and, for all t, $\widehat{F}(t) \geq -\alpha > -1$. Then, for any $t \in \mathbb{R}$,*

$$|\widehat{H}(t)| \leq \exp\left\{\frac{(1 - \alpha)(1 - p)}{1 + p + \alpha(1 - p)} (\operatorname{Re} \widehat{H}(t) - 1)\right\}.$$

Corollary 12.1 *Let $\alpha \in [0, 1)$ and let F be a symmetric distribution satisfying $\widehat{F}(t) \geq -\alpha$ for all $t \in \mathbb{R}$. Then, for any $t \in \mathbb{R}$,*

$$|\widehat{F}(t)| \leq \exp\left\{\frac{1 - \alpha}{1 + \alpha} (\widehat{F}(t) - 1)\right\}. \tag{12.22}$$

12.3   First Example

As the first example of application of the triangular function method we consider exponential smoothing for symmetric distributions.

Theorem 12.1 *Let $F \in \mathcal{F}_s$, $\lambda > 1$. Then*

$$| (F - I) \exp\{\lambda(F - I)\} |_K \leqslant \frac{C}{\lambda}. \tag{12.23}$$

Remark 12.2 We recall that, for any $G \in \mathcal{F}$,

$$\| (G - I) \exp\{\lambda(G - I)\} \| \leqslant \frac{C}{\sqrt{\lambda}},$$

see (2.10). Thus, we see that the symmetry of the distribution radically improves the order of accuracy, albeit for a weaker norm. Note also that suitable centering gives an intermediate result, see (9.32).

Proof We prove (12.23) step by step.

Step 0 *Preliminary investigation.* First we check what estimate can be obtained for the Fourier transform, i.e. (12.5). It can be easily proved that

$$| (\widehat{F}(t) - 1) \exp\{\lambda(\widehat{F}(t) - 1)\}| = (1 - \widehat{F}(t)) \exp\left\{-\frac{\lambda}{2}(1 - \widehat{F}(t))\right\} \exp\left\{\frac{\lambda}{2}(\widehat{F}(t) - 1)\right\}$$

$$\leqslant \frac{C}{\lambda} \exp\left\{\frac{\lambda}{2}(\widehat{F}(t) - 1)\right\}.$$

Therefore the Fourier transform is of the type (12.5) with

$$\widehat{D}(t) = \exp\left\{\frac{\lambda}{2}(\widehat{F}(t) - 1)\right\}. \tag{12.24}$$

We apply the triangle function method. Therefore we shall prove a more general result than (12.23), i.e. we prove that if $h > 0$, then

$$| (F - I) \exp\{\lambda(F - I)\} |_h \leqslant \frac{C}{\lambda} \gamma_h^{1/3} (| \ln \gamma_h | + 1)^3. \tag{12.25}$$

Here

$$\gamma_h = Q(D, h) = Q\left(\exp\left\{\frac{\lambda}{2}(F - I)\right\}, h \right).$$

As was noted above, taking the supremum over all $h > 0$ we get (12.23).

Step 1 *Decomposition of F.* Similarly to Sect. 9.4 we decompose F into a sum of distributions concentrated on a finite interval and outside that interval. According to Lemma 12.6, for any $\tau > 0$, there exists a vector $\mathbf{u} \in \mathbb{R}^{\ell}$ such that

$$\ell \leqslant C(|\ln \gamma_\tau| + 1), \tag{12.26}$$

$$F\{\mathbb{R} \setminus [K_1(\mathbf{u})]_\tau\} \leqslant \frac{C}{\lambda} \left(|\ln \gamma_\tau| + 1\right)^3. \tag{12.27}$$

Setting

$$q = F\{[-\tau, \tau]\}, \quad s = F\{[K_1(\mathbf{u})]_\tau \setminus [-\tau, \tau]\}, \quad d = F\{\mathbb{R} \setminus [K_1(U)]_\tau\},$$

we decompose F as a mixture of distributions $A, W, \Psi \in \mathcal{F}$ concentrated on the sets

$$[-\tau, \tau], \qquad [K_1(\mathbf{u})]_\tau \setminus [-\tau, \tau], \qquad \mathbb{R} \setminus [K_1(\mathbf{u})]_\tau,$$

respectively, i.e.

$$F = qA + sW + d\Psi = qA + rV,$$

$$F - I = q(A - I) + s(W - I) + d(\Psi - I) = q(A - I) + r(V - I),$$

$$A\{[-\tau, \tau]\} = W\{[K_1(\mathbf{u})]_\tau \setminus [-\tau, \tau]\} = \Psi\{\mathbb{R} \setminus [K_1(\mathbf{u})]_\tau\} = 1.$$

From (12.27) it follows that

$$d \leqslant \frac{C}{\lambda} \left(|\ln \gamma_\tau| + 1\right)^3. \tag{12.28}$$

Step 2 *Estimating measures containing* $(A-I)$. Our goal is to get the estimate for $|\cdot|_{h,\tau}$ and to use Lemma 12.2. We begin with the part of the measure containing the difference $(A - I)$. It is easy to see that

$$| (F - I) \exp\{\lambda(F - I)\} |_{h,\tau}$$
$$\leqslant | q(A - I) \exp\{\lambda(F - I)\} |_{h,\tau} + | r(V - I) \exp\{\lambda(F - I)\} |_{h,\tau}. \tag{12.29}$$

Let us estimate the first component of (12.29).

Lemma 12.8 *For all* $0 < \tau \leqslant h/2$

$$| q(A - I) \exp\{\lambda(F - I)\} |_{h,\tau} \leqslant \frac{C}{\lambda} \gamma_h.$$

Proof For the sake of brevity, set

$$g(t) = q \, |\widehat{f}_{z,h,\tau}(t)| \, |\widehat{A}(t) - 1| \exp\left\{\frac{\lambda}{2}(\widehat{F}(t) - 1)\right\}.$$

Applying Parseval's identity, we obtain

$$|q(A - I)\exp\{\lambda(F - I)\}|_{h,\tau} \leqslant$$

$$\frac{q}{2\pi} \int_{-\infty}^{\infty} |\widehat{f}_{z,h,\tau}(t)| \, |\widehat{A}(t) - 1| \exp\{\lambda(\widehat{F}(t) - 1)\} \, dt = \frac{1}{2\pi} \int_{-\infty}^{\infty} g(t)\widehat{D}(t) \, dt, \quad (12.30)$$

see (12.13). Here $\widehat{D}(t)$ is defined by (12.24). Estimating $g(t)$ we consider two cases: (a) when t is near zero; (b) when t is far from zero. (a) Let $|t| \leqslant 1/\tau$, then $\widehat{f}_{z,h,\tau}(t)$ is not small (see Lemma 12.3):

$$|\widehat{f}_{z,h,\tau}(t)| \leqslant \frac{2}{|t|}.$$

On the other hand, A is concentrated on $[-\tau, \tau]$, and from (9.26) it follows that

$$\frac{\sigma^2 t^2}{3} \leqslant 1 - \operatorname{Re}\widehat{A}(t) = 1 - \widehat{A}(t) \leqslant \frac{\sigma^2 t^2}{2}.$$

Here

$$\sigma^2 = \int x^2 A\{dx\}.$$

Therefore, for all $|t| \leqslant 1/\tau$,

$$g(t) \leqslant \frac{2q}{|t|} |\widehat{A}(t) - 1| \exp\left\{\frac{q\lambda}{2}(\widehat{A}(t) - 1)\right\} \exp\left\{\frac{p\lambda}{2}(\widehat{V}(t) - 1)\right\}$$

$$\leqslant Cq\sigma^2|t| \exp\left\{-\frac{q\lambda\sigma^2 t^2}{6}\right\} \leqslant \frac{C}{\lambda}\sqrt{\lambda q\sigma^2} \exp\left\{-\frac{q\lambda\sigma^2 t^2}{12}\right\} =: \frac{C}{\lambda} g_1(t). \tag{12.31}$$

(b) Now let $|t| > 1/\tau$. Then by (12.12)

$$|\widehat{f}_{z,h,\tau}(t)| \leqslant \frac{8\tau}{1 + (t\tau)^2}$$

and

$$g(t) \leq C \frac{\tau}{1 + (t\tau)^2} q(1 - \widehat{A}(t)) \exp\left\{\frac{\lambda}{2} q(\widehat{A}(t) - 1)\right\} \leq \frac{C}{\lambda} \frac{\tau}{1 + (t\tau)^2} =: \frac{C}{\lambda} g_2(t).$$

$$(12.32)$$

Combining (12.31) and (12.32) we see that

$$g(t) \leq \frac{C}{\lambda} \begin{cases} g_1(t), & \text{if } |t| \leq 1/\tau, \\ g_2(t), & \text{if } |t| > 1/\tau. \end{cases}$$

Of course, we can roughly estimate the integral in (12.30) by $C\lambda^{-1}$. However, we want to preserve γ_τ. Therefore we shall apply Lemma 1.1. Note that $g_1(t)$ and $g_2(t)$ are even and decreasing as $|t| \to \infty$. Therefore

$$\sup_{s:|s| \geq |t|} |g(s)| \leq \frac{C}{\lambda} \left(g_1(t) + g_2(t)\right)$$

and

$$\int_{-\infty}^{\infty} \sup_{s:|s| \geq |t|} |g(s)| \, dt \leq \frac{C}{\lambda} \int_{-\infty}^{\infty} \left(g_1(t) + g_2(t)\right) dt \leq \frac{C}{\lambda}.$$

Moreover, by (12.11)

$$\sup_t g(t) \leq \sup_t |\widehat{f}_{z,h,\tau}(t)| \, q(1 - \widehat{A}(t)) \exp\left\{\frac{q\lambda}{2}(\widehat{A}(t) - 1)\right\}$$

$$\leq \frac{C}{\lambda} \sup_t |\widehat{f}_{z,h,\tau}(t)| \leq \frac{Ch}{\lambda}.$$

Now applying Lemma 1.1 we get

$$\int_{-\infty}^{\infty} g(t)\widehat{D}(t) \, dt \leq \frac{C}{\lambda} Q(D, h) = \frac{C}{\lambda} \gamma_h,$$

which in view of (12.30) completes the proof of the lemma. $\square$.

Remark 12.3 There are other choices of γ_h. For example, it is possible to prove Lemma 12.8 with $\tilde{\gamma}_h = Q(\exp\{\lambda(F - I)\}, h)$. However, then the estimate is $C\lambda^{-1}\tilde{\gamma}_h^{1/2}$.

Step 3 *Directly estimating the measure concentrated outside $[-\tau, \tau]$.* Next, we begin the estimation of the second summand in (12.29). Taking into account Lemma 12.2, two estimates are needed. We start with the easier part.

Lemma 12.9 *For all $0 < \tau \leqslant h/2$*

$$| r(V - I) \exp\{\lambda(F - I)\} |_{h,\tau} \leqslant Cr\gamma_h. \tag{12.33}$$

Proof By (12.3) we have

$$
\begin{aligned}
| r(V - I) \exp\{\lambda(F - I)\} |_{h,\tau} &\leqslant Cr\| V - I \| \, | \exp\{\lambda(F - I)\} |_{h,\tau} \\
&\leqslant Cr| \exp\{\lambda(F - I)\} |_{h,\tau} \leqslant Cr\| \exp\{\lambda(F - I)/2\} \| \, | D |_{h,\tau} \\
&\leqslant Cr| D |_{h,\tau}.
\end{aligned}
$$

We recall that $0 \leqslant f_{z,h,\tau}(x) \leqslant 1$. Therefore

$$| D |_{h,\tau} \leqslant \sup_z \int_z^{z+h} f_{z,h,\tau}(x) \, D\{dx\} \leqslant \sup_z \int_z^{z+h} D\{dx\} = \sup_z D\{[z, z+h]\} = Q(D, h) = \gamma_h.$$

The last estimate completes the proof of (12.33). $\square$

Step 4 *Second estimate for the measure concentrated outside* $[-\tau, \tau]$. Here comes the tricky part. Set

$$v_\tau(y) = sW\{\{x : |x| > y\}\}.$$

Lemma 12.10 *For all $0 < \tau \leqslant h/2$*

$$| r(V - I) \exp\{\lambda(F - I)\} |_{h,\tau} \leqslant \frac{C}{\lambda}(| \ln \gamma_\tau | + 1)^3 (\lambda v_\tau(h/4))^{-1/2}.$$

Proof Taking into account (12.3) and the fact that the norm of any CP distribution equals 1, we obtain

$$
\begin{aligned}
&| r(V - I) \exp\{\lambda(F - I)\} |_{h,\tau} \\
&= | r(V - I) \exp\{s\lambda(W - I) + \lambda d(\Psi - I) + \lambda q(A - I)\} |_{h,\tau} \\
&\leqslant | r(V - I) \exp\{s\lambda(W - I)\} |_{h,\tau} \, \| \exp\{\lambda d(\Psi - I) + \lambda q(A - I)\} \| \\
&= | r(V - I) \exp\{s\lambda(W - I)\} |_{h,\tau} \\
&= | (s(W - I) + d(\Psi - I)) \exp\{s\lambda(W - I)\} |_{h,\tau} \\
&\leqslant | s(W - I) \exp\{s\lambda(W - I)\} |_{h,\tau} + Cd| \exp\{s\lambda(W - I)\} |_{h,\tau}. \tag{12.34}
\end{aligned}
$$

We begin with the estimate of the second summand. Taking into account that $0 \leqslant f_{z,h,\tau}(x) \leqslant 1$ and applying (1.20) we get

$$| \exp\{s\lambda(W - I)\} |_{h,\tau} \leqslant Q(\exp\{s\lambda(W - I)\}, h) \leqslant 4Q(\exp\{s\lambda(W - I)\}, h/4).$$

By the property of concentration functions (1.22) we prove

$$Q(\exp\{s\lambda(W - I)\}, h/4) \le \frac{C}{\sqrt{s\lambda W\{\{x : |x| > h/4\}\}}} = \frac{C}{\sqrt{\lambda v_\tau(h/4)}}. \qquad (12.35)$$

We decomposed F in such a way that d was small. Combining the last estimate with (12.28) we, therefore, obtain

$$d| \exp\{s\lambda(W - I)\}|_{h,\tau} \le \frac{C}{\lambda} (| \ln \gamma_\tau | + 1)^3 (\lambda v_\tau(h/4))^{-1/2}. \qquad (12.36)$$

It remains to estimate the first summand on the right-hand side of (12.34). First, let us assume that $s\lambda < 2$. Then we can take into account that $0 \le f_{z,h,\tau}(x) \le 1$ and (12.3) and prove that

$$| s(W - I) \exp\{s\lambda(W - I)\}|_{h,\tau} \le \| W - I \| s Q(\exp\{s\lambda(W - I)\}, h)$$

$$\le CsQ(\exp\{s\lambda(W - I)\}, h/4) \le Cs(s\lambda)^{-1} Q(\exp\{s\lambda(W - I)\}, h/4)$$

$$\le C\lambda^{-1} Q(\exp\{s\lambda(W - I)\}, h/4).$$

By (1.21):

$$Q(\exp\{s\lambda(W - I)\}, h/4) \le \frac{C}{\sqrt{s\lambda W\{\{x : |x| > h/4\}\}}} = \frac{C}{\sqrt{\lambda v_\tau(h/4)}}. \qquad (12.37)$$

Consequently, for $\lambda s < 2$,

$$| s(W - I) \exp\{s\lambda(W - I)\}|_{h,\tau} \le \frac{C}{\lambda} (\lambda v_\tau(h/4))^{-1/2}. \qquad (12.38)$$

Now let us assume that $\lambda s \ge 2$. We shall employ the fact that W is concentrated on $[K_1(\mathbf{u})]_\tau \setminus [-\tau, \tau]$ and, consequently, we can apply (12.19) with suitable $\omega(x)$. We choose ω to be defined as in Lemma 12.4 with $\mathbf{u}$ and ℓ as above and with $m = 4\lfloor s\lambda \rfloor + 1$. Then, just as in (12.19), we get

$$| s(W - I) \exp\{s\lambda(W - I)\}|_{h,\tau} \le \int\limits_{-\infty}^{\infty} \omega(x) s(W - I) \exp\{s\lambda(W - I)\}\{dx\}$$

$$+ \int\limits_{\mathbb{R}^*} (f_{z,h,\tau}(x) - \omega(x)) s(W - I) \exp\{s\lambda(W - I)\}\{dx\} = J_1 + J_2. \qquad (12.39)$$

Here $\mathbb{R}^* = \mathbb{R} \setminus [K_m(\mathbf{u})]_{m\tau}$. Let us estimate J_1. We apply Parseval's identity and the properties of $\omega(x)$ (12.14), (12.15), (12.16), (12.17), and (12.18). We prove

$$|J_1| \leqslant C \int_{-\infty}^{\infty} |\widehat{\omega}(t)| |s(1 - \widehat{W}(t)) \exp\{s\lambda(\widehat{W}(t) - 1)\} \, dt$$

$$\leqslant \frac{C}{\lambda} \int_{-\infty}^{\infty} |\widehat{\omega}(t)| \exp\left\{\frac{s\lambda}{2}(\widehat{W}(t) - 1)\right\} \, dt$$

$$\leqslant \frac{C}{\lambda} Q\left(\exp\left\{\frac{s\lambda}{2}(W - I)\right\}, h\right) \ell^2 \ln(\ell m + 1).$$

Observe that $Q(\cdot, h)$ has already been estimated in (12.35). Moreover, ℓ satisfies the estimates (12.20). Therefore

$$|J_1| \leqslant \frac{C}{\lambda} (\lambda v_\tau(h/4))^{-1/2} (|\ln \gamma_\tau| + 1)^2 (|\ln \gamma_\tau| + 1 + \ln m).$$

To estimate $\ln m$ observe that, by (1.21) and due to the support of W:

$$\gamma_\tau = Q(\exp\{(\lambda q(A - I) + \lambda s(W - I) + d(\Psi - I))/2\}, \tau)$$

$$\leqslant Q(\exp\{\lambda s(W - I)/2\}, \tau) \leqslant \frac{C}{\sqrt{\lambda s W\{\{x : |x| > \tau\}\}}} = \frac{C}{\sqrt{s\lambda}}.$$

Noting that $\gamma_\tau \leqslant 1$ we, consequently, prove

$$\gamma_\tau \leqslant C(\lambda s)^{-1/2}, \quad \lambda s \leqslant C\gamma_\tau^{-1/2}, \quad m \leqslant 4\lambda s + 1 \leqslant C\gamma_\tau^{-1/2}.$$

Therefore

$$\ln m \leqslant C(|\ln \gamma_\tau| + 1) \qquad (12.40)$$

and

$$|J_1| \leqslant \frac{C}{\lambda} (\lambda v_\tau(h/4))^{-1/2} (|\ln \gamma_\tau| + 1)^3. \qquad (12.41)$$

Let us return to (12.39) and estimate J_2. We use the following facts:

1. W is concentrated on the set $[K_1(\mathbf{u})]_\tau$.
2. W^k is concentrated on the set $[K_k(\mathbf{u})]_{k\tau}$.
3. $f_{z,h,\tau}(x) - \omega(x) = 0$, for $x \in [K_m(\mathbf{u})]_{m\tau}$.
4. $0 \leqslant f_{z,h,\tau}(x) - \omega(x) \leqslant 1$.

We also make use of the exponential structure of the measure

$$\exp\{s\lambda(W-I)\} = \sum_{k\leqslant 4\lambda s} \frac{(s\lambda)^k W^k}{k!} e^{-s\lambda} + \sum_{k>4\lambda s} \frac{(s\lambda)^k W^k}{k!} e^{-s\lambda}.$$

We have

$$J_2 = \sum_{k\geqslant 0} \frac{(s\lambda)^k}{k!} e^{-s\lambda} s \int_{\mathbb{R}^*} (f_{z,h,\tau}(x) - \omega(x))\,(W-I)W^k\{dx\}$$

$$= \sum_{k\geqslant m-1} \frac{(s\lambda)^k}{k!} e^{-s\lambda} s \int_{\mathbb{R}^*} (f_{z,h,\tau}(x) - \omega(x))\,(W-I)W^k\{dx\}.$$

Therefore

$$|J_2| \leqslant \sum_{k>3\lambda s} \frac{(s\lambda)^k}{k!} e^{-s\lambda} s \| (W-I)W^k \| \leqslant 2 \sum_{k>3\lambda s} \frac{(s\lambda)^k}{k!} e^{-s\lambda} s$$

$$\leqslant 2s e^{-3\lambda s} \sum_{k>3\lambda s} \frac{(es\lambda)^k}{k!} e^{-s\lambda} \leqslant 2s \exp\{(e-4)\lambda s\}$$

$$\leqslant 2s \exp\{-\lambda s\} \leqslant \frac{C}{\lambda} \exp\{-\lambda s/2\}$$

$$\leqslant \frac{C}{\lambda} \frac{1}{\sqrt{s\lambda}} \leqslant \frac{C}{\lambda} (\lambda v_\tau(h/4))^{-1/2}. \qquad (12.42)$$

Combining estimates (12.36), (12.38), (12.41) and (12.42) we complete the proof of Lemma 12.10. □

Step 5 *Combining all estimates.* We collect the estimates of Lemmas 12.8, 12.9, and 12.10 in the following lemma.

Lemma 12.11 *For all* $0 < \tau \leqslant h/2$

$$| (F-I)\exp\{\lambda(F-I)\} |_{h,\tau} \leqslant \frac{C}{\lambda} (\gamma_h + \lambda r \gamma_h)$$

and

$$| (F-I)\exp\{\lambda(F-I)\} |_{h,\tau} \leqslant \frac{C}{\lambda} (\gamma_h + (|\ln \gamma_\tau| + 1)^3 (\lambda v_\tau(h/4))^{-1/2}).$$

It remains to check that Lemma 12.11 has the same structure as Lemma 12.2. We have

$$W = (F-I)\exp\{\lambda(F-I)\}, \quad H_1 = \lambda F, \quad H_{\tau 1} = \lambda q A, \quad H_{\tau 2} = \lambda s W, \quad H_{\tau 3} = \lambda d \Psi.$$

Therefore

$$H_{\tau 2}\{\{x : |x| > h/4\}\} = \lambda v_\tau (h/4)$$

and

$$\lambda r = \lambda F\{\{x : |x| > \tau\}\} = H_1\{\{x : |x| > \tau\}\}.$$

Moreover, F is symmetric and $F = F^{(-)}$. Therefore, by Lemma 12.2 we prove (12.25).

12.4 Second Example

In this section, we consider a technically more advanced example of application of the triangle function method.

Theorem 12.2 *Let $F \in \mathcal{F}_+$, $n, N \in \mathbb{N}$ and $N \leq C$. Then*

$$|F^n(F - I)^N|_K \leq Cn^{-N}. \tag{12.43}$$

Proof In comparison to the first example, two new moments appear: first, we have the N-th convolution of $(F - I)$; second, the main smoothing distribution is not compound Poisson. The triangle function method is adapted to fit these changes. Again we present the proof step by step.

Step 0 *Preliminary investigation.* First we check what estimate can be obtained for the Fourier transform. Taking into account that $\widehat{F}(t) \leq \exp\{\widehat{F}(t) - 1\}$ we get

$$|\widehat{F}(t)^n(\widehat{F}(t)-1)^N| \leq |(\widehat{F}(t)-1)^N \exp\{n(\widehat{F}(t)-1)\}| \leq Cn^{-N} \exp\left\{\frac{n}{2}(\widehat{F}(t)-1)\right\}.$$

Thus, the Fourier transform is of the type (12.5) with

$$\widehat{D}(t) = \exp\left\{\frac{n}{2}(\widehat{F}(t) - 1)\right\}.$$

Typically for the triangle function method we will prove a more general result than (12.43), i.e. we prove that for any $h > 0$

$$|F^n(F - I)^N|_h \leq Cn^{-N}\gamma_h^{1/(2N+1)}(|\ln \gamma_h| + 1)^{6N(N+1)/(2N+1)}. \tag{12.44}$$

Here

$$\gamma_h = Q(D, h) = Q\left(\exp\left\{ \frac{n}{2}(F - I) \right\}, h \right).$$

Taking the supremum over all $h > 0$ we get (12.43).

Step 1 *Decomposition of F.* We decompose F exactly as in the previous example. For convenience we repeat that decomposition. For any $\tau > 0$, there exists a vector $\mathbf{u} \in \mathbb{R}^\ell$ such that

$$\ell \leqslant C(|\ln \gamma_\tau| + 1), \tag{12.45}$$

$$F\{\mathbb{R} \setminus [K_1(\mathbf{u})]_\tau\} \leqslant \frac{C}{n} \left(|\ln \gamma_\tau| + 1 \right)^3. \tag{12.46}$$

Moreover,

$$F = qA + sW + d\Psi = qA + rV,$$

$$A\{[-\tau, \tau]\} = W\{[K_1(\mathbf{u})]_\tau \setminus [-\tau, \tau]\} = \Psi\{\mathbb{R} \setminus [K_1(\mathbf{u})]_\tau\} = 1. \tag{12.47}$$

Here

$$q = F\{[-\tau, \tau]\}, \quad s = F\{[K_1(\mathbf{u})]_\tau \setminus [-\tau, \tau]\}, \quad d = F\{\mathbb{R} \setminus [K_1(U)]_\tau\}.$$

Though we do not know what properties Ψ has, its weight is small:

$$d \leqslant \frac{C}{n} \left(|\ln \gamma_\tau| + 1 \right)^3. \tag{12.48}$$

Step 2 *Estimating measures containing $(A - I)$.* The general idea is to consider measures containing the factor $(A - I)$ and applying Parseval's identity. For the present example, this will require a more elaborate approach. In the previous case, we could use the property of the compound Poisson distribution, allowing us to easily separate two components

$$\exp\{F - I\} = \exp\{q(A - I)\} \exp\{r(V - I)\}.$$

However, for F^n we need a different approach. We replace F^n by $(qA + rI)^n(qI + rV)^n$. Observe that

$$(qA + rI)(qI + rV) = (I + q(A - I))(I + r(V - I)) = I + q(A - I) + r(V - I)$$

and, therefore,

$$F = qA + rV = I + q(A - I) + r(V - I), \quad F - (qA + rI)(qI + rV) = -q(A - I)r(V - I).$$

Let

$$\Delta_1 := (F - I)^N \{F^n - (qA + rI)^n (qI + rV)^n\}.$$

Note that we are investigating symmetric distributions, i.e. having real characteristic functions. Therefore

$$q\widehat{A}(t) + r \geqslant q\widehat{A}(t) + r\widehat{V}(t) = \widehat{F}(t) \geqslant 0.$$

Consequently,

$$q\widehat{A}(t) + r = 1 + q(\widehat{A}(t) - 1) \leqslant \exp\{q(\widehat{A}(t) - 1)\}$$

and, similarly,

$$q + r\widehat{V}(t) \leqslant \exp\{r(\widehat{V}(t) - 1)\}.$$

Lemma 12.12 *For all* $0 < \tau \leqslant h/2$

$$|\Delta_1|_{h,\tau} \leqslant Cn^{-N} \gamma_h.$$

Proof We begin with the estimate of $\widehat{\Delta}_1(t)$. Note that

$$|(q\widehat{A}(t) + r)(q + r\widehat{V}(t))| \leqslant \exp\{q(\widehat{A}(t) - 1) + r(\widehat{V}(t) - 1)\} = \exp\{\widehat{F}(t) - 1\}.$$

Let $g(t) := n^{-N} q |\widehat{A}(t) - 1|$. Taking into account (1.45) we obtain

$$\begin{aligned}
|\widehat{\Delta}_1(t)| &\leqslant |\widehat{F}(t) - 1|^N n \exp\{(n-1)(\widehat{F}(t) - 1)\} q |\widehat{A}(t) - 1| |r| |\widehat{V}(t) - 1| \\
&\leqslant C |\widehat{F}(t) - 1|^N \exp\{n(\widehat{F}(t) - 1)/4\} \\
&\quad \times q |\widehat{A}(t) - 1| \widehat{D}(t) \\
&\quad \times nr |\widehat{V}(t) - 1| \exp\{nr(\widehat{V}(t) - 1)/4\} \\
&\leqslant Cn^{-N} q |\widehat{A}(t) - 1| \widehat{D}(t) = Cg(t)\widehat{D}(t).
\end{aligned}$$

Applying Parseval's identity, we get

$$|\Delta_1|_{h,\tau} \leqslant C \int_{-\infty}^{\infty} g(t)\widehat{D}(t) \, dt.$$

The following steps are identical to those of Lemma 12.8. Estimating $g(t)$ we distinguish between two cases: (a) when t is near zero; (b) when t is far from zero, and obtain

$$g(t) \leqslant Cn^{-N} \begin{cases} g_1(t), & \text{if } |t| \leqslant 1/\tau, \\ g_2(t), & \text{if } |t| > 1/\tau. \end{cases}$$

Here

$$g_1(t) = \sqrt{nq\sigma^2} \exp\left\{-\frac{qn\sigma^2 t^2}{12}\right\}, \quad g_2(t) = C\frac{\tau}{1 + (t\tau)^2}$$

and

$$\sigma^2 = \int x^2 A\{dx\}.$$

Note that $g_1(t)$ and $g_2(t)$ are even and decreasing as $|t| \to \infty$. Therefore

$$\sup_{s:|s|\geqslant|t|} |g(s)| \leqslant Cn^{-N} \left(g_1(t) + g_2(t)\right)$$

and

$$\int_{-\infty}^{\infty} \sup_{s:|s|\geqslant|t|} |g(s)| \, dt \leqslant Cn^{-N} \int_{-\infty}^{\infty} \left(g_1(t) + g_2(t)\right) dt \leqslant Cn^{-N}.$$

Moreover, by (12.11)

$$\sup_t g(t) \leqslant Cn^{-N} \sup_t |\widehat{f}_{z,h,\tau}(t)| \leqslant Cn^{-N}h.$$

Now applying Lemma 1.1 we get

$$\int_{-\infty}^{\infty} g(t)\widehat{D}(t) \, dt \leqslant Cn^{-N}\gamma_h.$$

□

Next, we decompose F^N:

$$(F - I)^N = r^N(V - I)^N + \sum_{j=1}^{N} \binom{N}{j} q^j(A - I)^j r^{N-j}(V - I)^{N-j}.$$

Let

$$\Delta_2 := (qA + rI)^n (qI + rV)^n \{(F - I)^N - r^N (V - I)^N\}$$

$$= (qA + rI)^n (qI + rV)^n \sum_{j=1}^{N} \binom{N}{j} q^j (A - I)^j r^{N-j} (V - I)^{N-j}.$$

Lemma 12.13 *For all* $0 < \tau \leqslant h/2$

$$|\Delta_2|_{h,\tau} \leqslant Cn^{-N} \gamma_h.$$

Proof The proof of Lemma 12.13 is similar to the proof of Lemma 12.12. We observe that

$$|\widehat{\Delta}_2(t)| \leqslant q|\widehat{A}(t) - 1| \sum_{j=1}^{N} \binom{N}{j} q^{j-1} |\widehat{A}(t) - 1|^{j-1} r^{N-j} |\widehat{V}(t) - 1|^{N-j} \widehat{D}(t)^2$$

$$\leqslant Cn^{-N+1} q|\widehat{A}(t) - 1| \widehat{D}(t)^{3/2} \leqslant Cg(t) \widehat{D}(t).$$

The remaining proof is identical to the previous one and is omitted. □

Step 3 *Estimating the measure concentrated outside* $[-\tau, \tau]$. The first step in estimating measures concentrated outside finite interval is straightforward. Set

$$\Delta_3 = (qA + rI)^n (qI + rV)^n r^N (V - I)^N.$$

Lemma 12.14 *For all* $0 < \tau \leqslant h/2$

$$|\Delta_3|_{h,\tau} \leqslant Cr^N \gamma_h. \tag{12.49}$$

Proof By (12.3) and $0 \leqslant f_{z,h,\tau}(x) \leqslant 1$ we obtain

$$|\Delta_3|_{h,\tau} \leqslant Cr^N |(qA + rI)^n (qI + rV)^n|_{h,\tau} \leqslant Cr^N Q((qA + rI)^n (qI + rV)^n, h).$$

For the proof that Q can be bounded from above by γ_h we apply the properties of concentration functions (1.23) and (1.24):

$$Q((qA + rI)^n (qI + rV)^n, h) \leqslant Ch \int_{|t| < 1/h} (q\widehat{A}(t) + r)^n (q + r\widehat{V}(t))^n \, dt$$

$$\leqslant Ch \int_{|t| < 1/h} \exp\{n(q(\widehat{A}(t) - 1) + r(\widehat{V}(t) - 1))\} \, dt$$

$$= Ch \int_{|t|<1/h} \exp\{n(\widehat{F}(t) - 1)\} \, dt$$

$$\leqslant Ch \int_{|t|<1/h} \exp\{n(\widehat{F}(t) - 1)/2\} \, dt \leqslant C\gamma_h.$$

The last estimate completes the proof of (12.49). □

Step 4 *Second estimate for the measure concentrated outside* $[-\tau, \tau]$. In principle, all ideas are the same as in the first example. Essentially we should

1. Get rid of compound measures containing A. We have already separated such measures as convolution factors and, usually, it suffices to replace them by their norms.
2. Get rid of Ψ. We expand compound measures in powers of Ψ and replace them by $\|\Psi\| = 1$.
3. Retain only the measure W and to make use of its support $[K_1(\mathbf{u})]_\tau$.

Just as in the proof of the first example set

$$v_\tau(y) = sW\{\{x : |x| > y\}\}.$$

Lemma 12.15 *For all* $0 < \tau \leqslant h/2$

$$|\Delta_3|_{h,\tau} \leqslant Cn^{-N}(|\ln \gamma_\tau| + 1)^{3+3N}(nv_\tau(h/4))^{-1/2}.$$

Proof The first essential step is to check that the proof is needed for small d only. Indeed, let $d \geqslant 1/3$. Then by (12.49)

$$|\Delta_3|_{h,\tau} \leqslant Cd^N \gamma_h \leqslant C\gamma_h \leqslant 3^{-N} Cd^N \gamma_h.$$

But, by (12.48) $d \leqslant C(|\ln \gamma_\tau| + 1)^3/n$ and by (12.37)

$$\gamma_h \leqslant CQ(\exp\{ns(W - I)/2\}, h/4) \leqslant C(nv_\tau(h/4))^{-1/2}.$$

Thus, it remains to prove the lemma for $d < 1/3$. We recall that $\|G\| = 1$, for any $G \in \mathcal{F}$. Therefore $\|(qA + rI)^n\| = 1$, $\|\Psi - I\| \leqslant 2$ and by (12.3) we have

$$|\Delta_3|_{h,\tau} \leqslant C|r^N(V - I)^N(qI + rV)^n|_{h,\tau}$$

$$= C|(s(W - I) + d(\Psi - I))^N(qI + rV)^n|_{h,\tau}$$

$$\leqslant C \sum_{k=0}^{N} \binom{N}{k} d^{N-k} |s^k(W - I)^k(qI + rV)^n|_{h,\tau}.$$

We introduce an auxiliary distribution $U \in \mathcal{F}$ such that

$$qI + rV = qI + sW + d\Psi = (1 - d)U + d\Psi, \quad U = \frac{q}{1-d}I + \frac{s}{1-d}W.$$

Once again applying (12.3) we obtain

$$|s^k(W-I)^k(qI+rV)^n|_{h,\tau} \leq \sum_{j=0}^{n} \binom{n}{j} d^{n-j}(1-d)^j |s^k(W-I)^k U^j|_{h,\tau}.$$

Combining the last estimates we see that

$$|\Delta_3|_{h,\tau} \leq \sum_{k=0}^{N} \binom{N}{k} d^{N-k} \sum_{j=0}^{n} \binom{n}{j} d^{n-j}(1-d)^j |s^k(W-I)^k U^j|_{h,\tau}. \tag{12.50}$$

For the sake of brevity set

$$\Delta_4 := s^k(W-I)^k U^j.$$

Let $j > 0$. We choose ω to be defined as in Lemma 12.4 with $\mathbf{u}$ and ℓ satisfying (12.45) and (12.46) and $m = \lfloor 6sn \rfloor + N + 1$. Then, similarly to (12.19), we prove

$$|\Delta_4|_{h,\tau} \leq \int_{-\infty}^{\infty} \omega(x) \, \Delta_4\{dx\} + \int_{\mathbb{R}^*} (f_{z,h,\tau}(x) - \omega(x)) \, \Delta_4\{dx\} =: I_1 + I_2. \tag{12.51}$$

Here $\mathbb{R}^* = \mathbb{R} \setminus [K_m(\mathbf{u})]_{m\tau}$. Let us estimate I_1. Applying Parseval's identity and the properties of $\omega(x)$ (12.14), (12.15), (12.16), (12.17), and (12.18) we obtain

$$|I_1| \leq C \int_{-\infty}^{\infty} |\widehat{\omega}(t)| |s^k(1 - \widehat{W}(t))^k| \widehat{U}(t)|^j \, dt.$$

Next, we make use of the fact that $d < 1/3$. Then

$$q + s\widehat{W}(t) = a + s\widehat{W}(t) + d - d \geq q\widehat{A}(t) + s\widehat{W}(t) + d\widehat{\Psi}(t) - d \geq \widehat{F}(t) - 1/3 \geq -1/3.$$

Therefore

$$\widehat{U}(t) \geq -\frac{1}{3(1-d)} \geq -\frac{1}{2}$$

and, by (12.22),

$$|\widehat{U}(t)| \leqslant \exp\left\{\frac{s}{9}(\widehat{W}(t) - 1)\right\}.$$

Consequently,

$$s^k(1 - \widehat{W}(t))^k|\widehat{U}(t)|^j \leqslant s^k(1 - \widehat{W}(t))^k \exp\left\{\frac{js}{36}(\widehat{W}(t) - 1)\right\} \exp\left\{\frac{js}{12}(\widehat{W}(t) - 1)\right\}$$

$$\leqslant C(k)j^{-k}\exp\left\{\frac{js}{12}(\widehat{W}(t) - 1)\right\}.$$

Taking into account the last estimates and (12.18) we prove

$$|I_1| \leqslant C(k)j^{-k}\int_{-\infty}^{\infty}|\widehat{\omega}(t)|\exp\left\{\frac{s}{12}(\widehat{W}(t) - 1)\right\}dt$$

$$\leqslant C(k)j^{-k}Q(\exp\{sj(W - I)/12\}, h)\ell^2\ln(\ell m + 1).$$

Estimating m similarly to (12.40) and taking into account (12.45) and noting that $k \leqslant N \leqslant C$ we arrive at

$$|I_1| \leqslant Cj^{-k}(|\ln\gamma_\tau| + 1)^3 Q(\exp\{sj(W - I)/12\}, h). \tag{12.52}$$

Next, we estimate I_2. Let $ns \geqslant 2$. We have

$$|I_2| \leqslant \sum_{l=0}^{j}\binom{j}{l}\left(\frac{s}{1 - d}\right)^l\left(\frac{q}{1 - d}\right)^{j-l}\int_{\mathbb{R}^*}(f_{z,h,\tau}(x) - \omega(x))|s^k(W - I)^k W^l|\{dx\}.$$

Due to the choice of m, we get

$$|I_2| \leqslant s^k\sum_{l \geqslant 6ns}\binom{j}{l}\left(\frac{s}{1 - d}\right)^l\left(\frac{q}{1 - d}\right)^{j-l}. \tag{12.53}$$

For the estimate of the sum in (12.53) we employ its probabilistic interpretation and Chebyshev's inequality. Let $S_j = \xi_1 + \cdots + \xi_j$, where ξ_j are independent Bernoulli variables taking values 0 with probabilities $q/(1 - d)$. It is not difficult to see that the right-hand side of (12.53) is equal to $s^k P(S_j \geqslant 6ns)$ and

$$s^k P(S_j \geqslant 6ns) \leqslant s^k e^{-6ns}\mathbb{E}\exp\{S_j\} = s^k e^{-6ns}\left(\frac{q}{1 - d} + \frac{se}{1 - d}\right)^j$$

$$\leqslant Cs^k\exp\{-6ns + 3sj\} \leqslant Cn^{-k}(ns)^{-1/2} \leqslant Cn^{-k}(n\nu_\tau(h/4))^{-1/2}.$$

Therefore

$$|I_2| \leqslant Cn^{-k}(ns)^{-1/2} \leqslant Cn^{-k}(nv_\tau(h/4))^{-1/2}. \tag{12.54}$$

Next, we observe that the same estimate also holds for $ns < 2$. Indeed, if $ns < 2$, then

$$|I_2| \leqslant 2s^k 2^k \int_{\mathbb{R}^*} 1\, U\{dx\} \leqslant 2^{k+1} s^k = 2^{k+1} s^k \frac{(ns)^{k+1/2}}{(ns)^{k+1/2}}$$

$$\leqslant Cn^{-k}(ns)^{-1/2} \leqslant Cn^{-k}(nv_\tau(h/4))^{-1/2}.$$

Substituting (12.52) and (12.54) into (12.51) and (12.50) we arrive at

$$|\Delta_3|_{h,\tau} \leqslant I_3 + I_4. \tag{12.55}$$

Here

$$I_3 = C \sum_{k=0}^{N} \binom{N}{k} d^{N-k} \sum_{j=0}^{n} \binom{n}{j} d^{n-j}(1-d)^j n^{-k}(nv_\tau(h/4))^{-1/2}$$

$$= C(nv_\tau(h/4))^{-1/2} \sum_{k=0}^{N} \binom{N}{k} d^{N-k} n^{-k},$$

$$I_4 = C \sum_{k=0}^{N} \binom{N}{k} d^{N-k} \Big\{ (|\ln \gamma_\tau| + 1)^3 \sum_{j=1}^{n} \binom{n}{j} d^{n-j}(1-d)^j j^{-k}$$

$$\times Q(\exp\{sj(W-I)/12\}, h) + d^n \Big\}.$$

It is not difficult to estimate I_3. Indeed by (12.48)

$$I_3 \leqslant C(nv_\tau(h/4))^{-1/2}(|\ln \gamma_\tau| + 1)^{3N} n^{-N} \sum_{k=0}^{N} \binom{N}{k}$$

$$\leqslant Cn^{-N}(|\ln \gamma_\tau| + 1)^{3N}(nv_\tau(h/4))^{-1/2}. \tag{12.56}$$

The estimate of I_4 is more complicated. We know that d is small. Therefore d^n is also small. However, we need $nv_\tau(h/4)$ and, consequently, a more elaborate approach follows. Applying (1.23) we obtain

$$\sum_{j=1}^{n} \binom{n}{j} d^{n-j}(1-d)^j j^{-k} Q(\exp\{sj(W-I)/12\}, h) + d^n$$

$$\leq C \sum_{j=1}^{n} \binom{n}{j} d^{n-j}(1-d)^j j^{-k} h \int_{-1/h}^{1/h} \exp\{sj(\widehat{W}(t)-1)/12\}\, dt + d^n$$

$$\leq Ch \int_{-1/h}^{1/h} \left\{ \sum_{j=1}^{n} \binom{n}{j} d^{n-j}(1-d)^j j^{-k} \exp\{sj(\widehat{W}(t)-1)/12\} + d^n \right\} dt.$$

Hölder's inequality gives the estimate

$$\left(\sum_{j=1}^{n} \binom{n}{j} d^{n-j}(1-d)^j j^{-k} \exp\{sj(\widehat{W}(t)-1)/12\} + d^n \right)^2$$

$$\leq \left\{ \sum_{j=1}^{n} \binom{n}{j} d^{n-j}(1-d)^j j^{-2k} + d^n \right\}$$

$$\times \left\{ \sum_{j=1}^{n} \binom{n}{j} d^{n-j}(1-d)^j \exp\{sj(\widehat{W}(t)-1)/6\} + d^n \right\} =: \widetilde{S} \times S.$$

Observe that

$$\frac{1}{j^{2k}} \leq \frac{C(k)}{(j+1)(j+2)\dots(j+2k)}$$

and

$$\binom{n}{j} \frac{1}{(j+1)(j+2)\dots(j+2k)} = \binom{n+2k}{j+2k} \frac{1}{(n+1)(n+2)\dots(n+2k)}.$$

We recall that $d < 1/3$. Therefore, for $k \leq N \leq C$,

$$\widetilde{S} \leq C(n(1-d))^{-2k} \leq Cn^{-2k}.$$

For the estimate of S observe that

$$S = \left((1-d)\exp\{s(\widehat{W}(t)-1)/6\} + d\right)^n \leq \exp\{n(1-d)(\exp\{s(\widehat{W}(t)-1)/6\} - 1)\}.$$

Applying the inequality

$$e^{-x} - 1 \leqslant -x + \frac{x^2}{2} = -x\left(1 - \frac{x}{2}\right), \quad x > 0,$$

we obtain

$$\exp\{s(\widehat{W}(t) - 1)/6\} - 1 \leqslant \frac{s(\widehat{W}(t) - 1)}{6}\left(1 - \frac{s(1 - \widehat{W}(t))}{12}\right)$$

$$\leqslant \frac{5s}{36}(\widehat{W}(t) - 1).$$

Therefore

$$S \leqslant \exp\{10sn(\widehat{W}(t) - 1)/118\}.$$

Combining all estimates and consequently applying (1.24), (12.48) and (12.37) we get

$$I_4 \leqslant C(|\ln \gamma_\tau| + 1)^3 \sum_{k=0}^{N} d^{N-k} n^{-k} Ch \int_{-1/h}^{1/h} \exp\{5sn(\widehat{W}(t) - 1)/118\}\,dt$$

$$\leqslant C(|\ln \gamma_\tau| + 1)^3 Q(\exp\{5sn(W - I)/118\}, h) \sum_{k=0}^{N} d^{N-k} n^{-k}$$

$$\leqslant Cn^{-N}(|\ln \gamma_\tau| + 1)^{3+3N} Q(\exp\{5sn(W - I)/118\}, h)$$

$$\leqslant Cn^{-N}(|\ln \gamma_\tau| + 1)^{3+3N} (nv_\tau(h/4))^{-1/2}.$$

From the last estimate, (12.56) and (12.55) we get the statement of the lemma. □

Step 5 *Combining all estimates.* Now we can perform the final step. Let us combine the estimates of Lemmas 12.12, 12.13, 12.14, and 12.15.

Lemma 12.16 *For all* $0 < \tau \leqslant h/2$

$$|F^n(F - I)^N|_{h,\tau} \leqslant Cn^{-N}(\gamma_h + (nr)^N \gamma_h)$$

and

$$|F^n(F - I)^N|_{h,\tau} \leqslant Cn^{-N}(\gamma_h + (|\ln \gamma_\tau| + 1)^{3+3N}(nv_\tau(h/4))^{-1/2}).$$

Estimate (12.44) and, consequently, the theorem's assertion now follows from Lemma 12.2. We have

$$W = (F - I)^N F^n, \quad H_1 = nF, \quad H_{\tau 1} = nqA, \quad H_{\tau 2} = nsW, \quad H_{\tau 3} = nd\Psi.$$

Therefore

$$H_{\tau 2}\{\{x : |x| > h/4\}\} = n\nu_\tau(h/4)$$

and

$$nr = nF\{\{x : |x| > \tau\}\} = H_1\{\{x : |x| > \tau\}\}.$$

Moreover, F is symmetric and $F = F^{(-)}$. $\square$

12.5   Problems

12.1   Prove Lemma 12.13.

12.2   Let $G \in \mathcal{F}_s, \lambda > 0, N \in \mathbb{N}$ and let $N \leq C$. Prove that

$$|(G - I)^N \exp\{\lambda(G - I)\}|_K \leq C\lambda^{-N}.$$

Bibliographical Notes

The triangle function method was introduced by Arak in [2] and extensively used in [5]. Lemmas 12.1 and 12.2 are slightly different versions of Lemmas 3.1 (p. 67) and 3.2 (p. 68) from [5]. Example 1 was proved in [110]. Example 2 is a special case of a result proved in [30]. Other examples of application of the triangle function method can be found in [33, 34]. Zaĭtsev in [163] extended the triangular function method to the multivariate case, see also [164, 165].

Chapter 13
Heinrich's Method for m-Dependent Variables

In this chapter, we present Heinrich's adaptation of the characteristic function method for weakly dependent random variables. Though the method can be applied for various dependencies (see the bibliographical notes at the end of the chapter) we consider m-dependent random variables only. We recall that the sequence of random variables $\{X_k\}_{k\geq1}$ is called m-dependent if, for $1 < s < t < \infty, t-s > m$, the sigma algebras generated by $X_1, \ldots, X_s$ and $X_t, X_{t+1} \ldots$ are independent. Moreover, it is clear that, by grouping consecutive summands, we can reduce the sum of m-dependent variables to the sum of 1-dependent variables. Therefore later we concentrate on 1-dependent variables, that is, on the case when X_j depends on X_{j+1} and X_{j-1} only.

13.1 Heinrich's Lemma

One of the main advantages of the characteristic function method is based on the fact that the characteristic function of the sum of independent random variables is equal to the product of the characteristic functions of the separate summands. Heinrich's method allows a similar representation for the characteristic function of the sum of weakly dependent random variables. More precisely, in a neighborhood of zero, the characteristic function of the sum can be expressed as a product of functions similar to the characteristic functions of the summands. The idea is hardly a new one, however, Heinrich presented it in a form convenient for applications.

Special analogues of centered mixed moments are needed. Let $\{Z_k\}_{k\geq1}$ be a sequence of arbitrary real or complex-valued random variables. We assume that

© Springer International Publishing Switzerland 2016
V. Čekanavičius, *Approximation Methods in Probability Theory*, Universitext,
DOI 10.1007/978-3-319-34072-2_13

$\widehat{\mathbb{E}}(Z_1) = \mathbb{E} Z_1$ and, for $k \geqslant 2$, define $\widehat{\mathbb{E}}(Z_1, Z_2, \cdots Z_k)$ by

$$\widehat{\mathbb{E}}(Z_1, Z_2, \cdots, Z_k) = \mathbb{E} Z_1 Z_2 \cdots Z_k - \sum_{j=1}^{k-1} \widehat{\mathbb{E}}(Z_1, \cdots, Z_j) \mathbb{E} Z_{j+1} \cdots Z_k.$$

Observe that

$$\widehat{\mathbb{E}}(Z_1, Z_2) = \mathbb{E} Z_1 Z_2 - \mathbb{E} Z_1 \mathbb{E} Z_2,$$

$$\widehat{\mathbb{E}}(Z_1, Z_2, Z_3) = \mathbb{E} Z_1 Z_2 Z_3 - \mathbb{E} Z_1 \mathbb{E} Z_2 Z_3 - \mathbb{E} Z_1 Z_2 \mathbb{E} Z_3 + \mathbb{E} Z_1 \mathbb{E} Z_2 \mathbb{E} Z_3.$$

Properties of the new characteristics for 1-dependent variables are given in the following lemma.

Lemma 13.1 *Let $Z_1, Z_2, \ldots, Z_k$ be 1-dependent complex or real variables.*

(a) If $\mathbb{E}|Z_j|^k < \infty, j = 1, 2, \ldots, k$, then, for any complex numbers $a_1, a_2, \ldots, a_k$,

$$\widehat{\mathbb{E}}(Z_1 + a_1, Z_2 + a_2, \ldots, Z_k + a_k) = \widehat{\mathbb{E}}(Z_1, Z_2, \ldots, Z_k).$$

(b) If $\mathbb{E}|Z_j|^2 < \infty, j = 1, 2, \ldots, k$, then

$$\left| \widehat{\mathbb{E}}(Z_1, Z_2, \ldots, Z_k) \right| \leqslant 2^{k-1} \prod_{j=1}^{k} \left(\mathbb{E}|Z_j|^2 \right)^{1/2}. \tag{13.1}$$

The next lemma is the main tool of Heinrich's method.

Lemma 13.2 (Heinrich's lemma) *Let $X_1, X_2, \ldots, X_k$ be 1-dependent real variables, $S_n = X_1 + X_2 + \cdots + X_n$ and let t be such that*

$$\max_{1 \leqslant k \leqslant n} \left(\mathbb{E}|e^{itX_k} - 1|^2 \right)^{1/2} \leqslant \frac{1}{6}. \tag{13.2}$$

Then the following representation holds

$$\mathbb{E} e^{itS_n} = h_1(t) h_2(t) \cdots h_n(t).$$

Here $h_1(t) = \mathbb{E} e^{itX_1}$ and, for $k = 2, \ldots, n$,

$$h_k(t) = 1 + \mathbb{E}(e^{itX_k} - 1) + \sum_{j=1}^{k-1} \frac{\widehat{\mathbb{E}}((e^{itX_j} - 1), (e^{itX_{j+1}} - 1), \ldots (e^{itX_k} - 1))}{h_j(t) h_{j+1}(t) \ldots h_{k-1}(t)}. \tag{13.3}$$

Moreover, for $j = 1, 2, \ldots, n$,

$$|h_j(t) - \mathbb{E}\, e^{itX_j}| \leq 6 \max_{1 \leq k \leq n} \mathbb{E}|e^{itX_k} - 1|^2. \qquad (13.4)$$

Proof By the symmetry property

$$\widehat{\mathbb{E}}\,(e^{itX_1}, e^{itX_2}, \ldots, e^{itX_k}) = \widehat{\mathbb{E}}\,(e^{itX_k}, \ldots, e^{itX_1}).$$

Therefore, setting $\mathbb{E}\, e^{itS_0} := 1$, we obtain

$$\mathbb{E}\, e^{itS_k} = \widehat{\mathbb{E}}\,(e^{itX_1}, \ldots, e^{itX_k}) + \sum_{j=1}^{k-1} \mathbb{E}\, e^{itS_j} \widehat{\mathbb{E}}\,(e^{itX_{j+1}}, \ldots, e^{itX_k})$$

$$= \mathbb{E}\, e^{itX_k} \mathbb{E}\, e^{itS_{k-1}} + \sum_{j=1}^{k-1} \mathbb{E}\, e^{itS_{j-1}} \widehat{\mathbb{E}}\,(e^{itX_j}, \ldots, e^{itX_k})$$

$$= \mathbb{E}\, e^{itX_k} \mathbb{E}\, e^{itS_{k-1}} + \sum_{j=1}^{k-1} \mathbb{E}\, e^{itS_{j-1}} \widehat{\mathbb{E}}\,((e^{itX_j} - 1), \ldots, (e^{itX_k} - 1)). \quad (13.5)$$

For the last equality we applied the first part of Lemma 13.1. Formally the expansion (13.3) follows from (13.5) if we divide it by $\mathbb{E}\, e^{itS_{k-1}}$ and use the notation $h_j(t) = \mathbb{E}\, e^{itS_j}/\mathbb{E}\, e^{itS_{j-1}}$. However, first we must prove that the division is allowed, i.e. that $|\mathbb{E}\, e^{itS_{k-1}}|, |h_1(t)|, \ldots |h_{k-2}(t)| > 0$. Observe that for this, it suffices to prove (13.4). Indeed, let

$$w_H(t) := \max_{1 \leq k \leq n} \left(\mathbb{E}|e^{itX_k} - 1|^2\right)^{1/2}. \qquad (13.6)$$

Then from (13.4) it follows that

$$6(w_H(t))^2 \geq |h_j(t) - \mathbb{E}\, e^{itX_j}| = |(h_j(t) - 1) - \mathbb{E}(e^{itX_j} - 1)| \geq \left| |h_j(t) - 1| - \mathbb{E}|e^{itX_j} - 1| \right|$$

and by (1.6)

$$|h_j(t) - 1| \leq \mathbb{E}|e^{itX_j} - 1| + 6(w_H(t))^2 \leq w_H(t) + 6(w_H(t))^2 \leq \frac{1}{3}, \qquad (13.7)$$

$$|h_j(t)| \geq 1 - |h_j(t) - 1| \geq \frac{2}{3}, \qquad \frac{1}{|h_j(t)|} \leq \frac{3}{2}.$$

For the proof of (13.4) we apply induction. Obviously, $|h_1(t) - \mathbb{E}\,e^{itX_1}| = 0$. Next, let us assume that (13.4) holds for $i < m < n$. Then $|h_i(t)| > 0$, $(i = 1, \ldots, m-1)$ and (13.3) holds for $k = 1, 2, \ldots, m$. We will prove that (13.4) then holds for $i = m$ and, consequently, (13.3) holds for $k = m+1$. Observe that, due to (13.1) and (13.2)

$$|\widehat{\mathbb{E}}\,(e^{itX_j} - 1, \ldots, e^{itX_m} - 1)| \leqslant 2^{m-j} \prod_{i=j}^{m} (\mathbb{E}\,|\,e^{itX_i} - 1\,|^2)^{1/2}$$

$$\leqslant 2^{m-j}(w_H(t))^{m-j+1} \leqslant 6(w_H(t))^2 3^{m-j}. \quad (13.8)$$

Therefore

$$|h_m(t) - \mathbb{E}\,e^{itX_m}| \leqslant 6(w_H(t))^2 \sum_{j=1}^{m-1} \frac{1}{3^{m-j}|\,h_j(t)\,|\cdots|\,h_{m-1}(t)\,|}$$

$$\leqslant 6(w_H(t))^2 \sum_{j=1}^{m-1} \frac{1}{3^{m-j}} \left(\frac{3}{2}\right)^{m-j} \leqslant 6(w_H(t))^2 \sum_{j=1}^{\infty} \frac{1}{2^j} = 6(w_H(t))^2.$$

Thus, (13.4) holds for $i = m$. Consequently, $|h_m| \geqslant 2/3 > 0$ and $|\mathbb{E}\,e^{itS_m}| > 0$. Dividing (13.5) by $\mathbb{E}\,e^{itS_m}$ we prove that (13.3) holds for $k = m+1$. $\square$

Expression (13.3) can be combined with the following result.

Lemma 13.3 *Let (13.2) be satisfied. Then*

$$\left| \ln \mathbb{E}\,e^{itS_n} - \sum_{k=1}^{n} \mathbb{E}\,(e^{itX_k} - 1) - \sum_{k=2}^{n} \widehat{\mathbb{E}}\,(e^{itX_{k-1}} - 1, e^{itX_k} - 1) \right|$$

$$\leqslant 90 \max_{1 \leqslant k \leqslant n} \left(\mathbb{E}\,|\,e^{itX_k} - 1\,|^2 \right)^{1/2} \sum_{j=1}^{n} |\,\mathbb{E}\,e^{itX_k} - 1\,|. \quad (13.9)$$

The proof of (13.9) requires expansion of $\ln h_k(t)$ in powers of $(h_k(t) - 1)$ and can be found in [70], see Corollary 3.1. If some additional information about the structure of F_n is available, it is usually possible to obtain more accurate estimates than in (13.9).

Typical application. Let F be a distribution of $S_n = X_1 + X_2 + \cdots + X_n$, where $X_1, \ldots, X_n$ are 1-dependent random variables. Let us assume that we want to estimate the closeness of F to some convolution $\prod_{j=1}^{n} G_j$, $G_j \in \mathcal{M}$, $(j = 1, 2, \ldots, n)$. Then one can

- use representation $\widehat{F}(t) = \prod_{j=1}^{n} h_j(t)$ and inequality

$$\left| \prod_{j=1}^{n} h_j(t) - \prod_{j=1}^{n} \widehat{G}_j(t) \right| \leqslant \sum_{j=1}^{n} | h_j(t) - \widehat{G}_j(t) | \prod_{k=1}^{j-1} | h_k(t) | \prod_{k=j+1}^{n} | \widehat{G}_k(t) |.$$

- Apply (13.3) for the estimation of $| h_j(t) - \widehat{G}_j(t) |$.
- Apply (13.3) or (13.9) for the estimation of $\prod_{k=1}^{j-1} | h_k(t) |$.
- Apply Lemma 9.6 for the estimate in the Kolmogorov norm.
- If $F \in \mathcal{F}_Z$, $G_j \in \mathcal{M}_Z$ and (13.2) is satisfied for all $|t| \leqslant \pi$, then (13.3) can be used for estimates of derivatives and the material of Chap. 5 can be used.

Advantages. Estimation for $\widehat{F}(t)$ is replaced by estimation of separate $h_j(t)$. For estimates in the Kolmogorov norm only estimates for small t are important. For larger t it suffices to have small estimates for approximating measures $\widehat{G}_j(t)$.

Drawbacks. Expansion of $h_j(t)$ in powers of t or $e^{it} - 1$ is a recursive procedure and requires a thorough estimation at each step. The absolute constants in estimates are not small.

13.2 Poisson Approximation

To demonstrate the main steps of Heinrich's method we begin with a simple Poisson approximation in the Kolmogorov norm. Let $X_1, X_2, \ldots, X_n$ be identically distributed 1-dependent random variables concentrated on nonnegative integers. Let F_n denote the distribution of the sum $X_1 + X_2 + \cdots + X_n$. As before we denote by ν_k the k-th factorial moment of X_1, that is, $\nu_k = \mathbb{E} X(X-1) \cdots (X-k+1)$. We assume that, for $n \to \infty$,

$$\nu_1 = o(1), \quad \nu_2 = o(\nu_1), \quad \mathbb{E} X_1 X_2 = o(\nu_1), \quad |X_1| \leqslant C_0, \quad n\nu_1 \to \infty. \tag{13.10}$$

Further on we will assume $C_0 \geqslant 1$. It is easy to check that (13.10) is stronger than Franken's condition (3.13).

Theorem 13.1 *Let assumptions (13.10) be satisfied. Then*

$$| F_n - \exp\{nv_1(I_1 - I)\} |_K = O\left(\frac{v_2 + \mathbb{E} X_1 X_2 + v_1^2}{v_1} \right). \tag{13.11}$$

Proof It is easy to check that (13.2) is satisfied for all t if n is sufficiently large. Indeed, let $w_H(t)$ be defined as in (13.6). Observe that

$$| e^{itX_1} - 1 | \leqslant X_1 | e^{it} - 1 | = 2X_1 | \sin(t/2) | \tag{13.12}$$

and

$$w_H(t) \leqslant | e^{it} - 1 | \sqrt{\mathbb{E} X_1^2} \leqslant 2\sqrt{C_0} \sqrt{v_1} | \sin(t/2) | \leqslant 2\sqrt{C_0} \sqrt{v_1}.$$

For sufficiently large n, the last estimate is less than $1/6$. From (13.3) it then follows that

$$| h_j(t) - 1 - v_1(e^{it} - 1) | \leqslant | \mathbb{E} e^{itX_j} - 1 - v_1(e^{it} - 1) | + \frac{| \widehat{\mathbb{E}} (e^{itX_{k-1}} - 1, e^{itX_k} - 1) |}{| h_{k-1}(t) |}$$

$$+ \frac{| \widehat{\mathbb{E}} (e^{itX_{k-2}} - 1, e^{itX_{k-1}} - 1, e^{itX_k} - 1) |}{| h_{k-2}(t) || h_{k-1}(t) |}$$

$$+ \sum_{j=1}^{k-3} \frac{| \widehat{\mathbb{E}} (e^{itX_j} - 1, e^{itX_{j+1}} - 1, \ldots, e^{itX_k} - 1) |}{| h_j(t) h_{j+1}(t) \ldots h_{k-1}(t) |}. \tag{13.13}$$

We expect the estimate to be at least of the order $O(v_1^2)$. Consequently, some members of the sum in (13.3) will be estimated separately. Applying (1.16) to the first summand we obtain

$$| \mathbb{E} e^{itX_j} - 1 - v_1(e^{it} - 1) | \leqslant Cv_2 \sin^2(t/2).$$

Next, observe that by (13.8)

$$| \widehat{\mathbb{E}} (e^{itX_j} - 1, \ldots, e^{itX_k} - 1) | \leqslant 2^{k-j} (w_H(t))^{k-j+1}$$

$$\leqslant 2^{2k-2j+1} (C_0 v_1)^{(k-j+1)/2} | \sin(t/2) |^{k-j+1}.$$

Due to (13.7), $1/| h_j(t) | \geqslant 3/2$ and, therefore, the last sum in (13.13) is less than

$$\sum_{j=1}^{k-3} \left(\frac{3}{2} \right)^{k-j} 2^{2k-2j+1} (C_0 v_1)^{(k-j+1)/2} | \sin(t/2) |^{k-j+1} \leqslant Cv_1^2 \sin^2(t/2),$$

for all sufficiently large n. From (13.12) it follows that

$$\frac{|\widehat{\mathbb{E}}\left(e^{itX_1} - 1, e^{itX_2} - 1\right)|}{|h_1(t)|} \leq \frac{3}{2}\left(\mathbb{E}X_1X_2|e^{it} - 1|^2 + \mathbb{E}X_1\mathbb{E}X_2|e^{it} - 1|^2\right)$$

$$= 6(\mathbb{E}X_1X_2 + v_1^2)\sin^2(t/2).$$

Recall that $|e^{itX_j} - 1| \leq |e^{itX_j}| + 1 = 2$. Then

$$\frac{|\widehat{\mathbb{E}}\left(e^{itX_1} - 1, e^{itX_2} - 1, e^{itX_3} - 1\right)|}{|h_1(t)h_2(t)|}$$

$$\leq \frac{3^2}{2}\left(\mathbb{E}X_1X_2 + \mathbb{E}X_1\mathbb{E}X_2 + \mathbb{E}X_1X_2 + \mathbb{E}X_1\mathbb{E}X_2\right)|e^{it} - 1|^2$$

$$= C(\mathbb{E}X_1X_2 + v_1^2)\sin^2(t/2).$$

Note that all random variables are identically distributed. Combining all estimates we obtain

$$|h_j(t) - 1 - v_1(e^{it} - 1)| \leq C(\mathbb{E}X_1X_2 + v_2 + v_1^2)\sin^2(t/2). \tag{13.14}$$

Next, we will estimate $\ln \widehat{F}_n(t)$. Observe that

$$\mathbb{E}|e^{itX_1} - 1|^2 \leq \mathbb{E}X_1^2|e^{it} - 1|^2 \leq C_0 v_1|e^{it} - 1|^2.$$

From (13.7) it follows that $|h_j(t) - 1| \leq 1/3$ and

$$|h_j(t) - 1| \leq \mathbb{E}|e^{itX_j} - 1| + 6w_H(t)^2 \leq Cv_1| \sin(t/2)|.$$

Therefore

$$|\ln \widehat{h}_j(t) - (h_j(t) - 1)| \leq |h_j(t) - 1|^2 \sum_{j=2}^{\infty} \frac{1}{3^{j-2}} = Cv_1^2 \sin^2(t/2).$$

Combining the last estimate with (13.14) we obtain

$$|\ln \widehat{F}_n(t) - nv_1(e^{it} - 1)| = \left|\sum_{j=1}^{n}(\ln h_j(t) - v_1(e^{it} - 1))\right|$$

$$\leq \sum_{j=1}^{n}|h_j - 1 - v_1(e^{it} - 1)| + Cnv_1^2 \sin^2(t/2)$$

$$\leq Cn(\mathbb{E}X_1X_2 + v_2 + v_1^2)\sin^2(t/2). \tag{13.15}$$

From (13.15) it follows that for all sufficiently large n

$$|\widehat{F}_n(t)| \leqslant |\exp\{nv_1(e^{it}-1)\}|\exp\{|\ln\widehat{F}_n(t)-nv_1(e^{it}-1)|\}$$
$$\leqslant \exp\{-2n\sin^2(t/2)(v_1 - C(\mathbb{E}X_1X_2 + v_2 + v_1^2))\}$$
$$= \exp\{-2n\sin^2(t/2)v_1(1-o(1))\} \leqslant \exp\{-Cnv_1\sin^2(t/2)\}.$$

Observe that

$$|\exp\{nv_1(e^{it}-1)\}| = \exp\{-2nv_1\sin^2(t/2)\}.$$

Substituting the last two estimates and (13.15) into (3.3) we obtain, for all sufficiently large n,

$$|\widehat{F}_n(t) - \exp\{nv_1(e^{it}-1)\}| \leqslant \exp\{-Cnv_1\sin^2(t/2)\}|\ln\widehat{F}_n(t) - nv_1(e^{it}-1)|$$
$$\leqslant Cn(\mathbb{E}X_1X_2 + v_2 + v_1^2)\sin^2(t/2)\exp\{-Cnv_1\sin^2(t/2)\}.$$

It remains to apply (4.1) and (1.31). □

13.3   Two-Way Runs

In this section, we demonstrate how Heinrich's method can be applied for estimates in total variation. We consider one of the most popular cases of 1-dependent Bernoulli variables, the so-called two-way runs. Let ξ_j, $j = 0,1,2,\ldots,n$ be independent identically distributed indicator variables, $P(\xi_1 = 1) = p$, $P(\xi_1 = 0) = 1 - p$. Let $\eta_j = \xi_j\xi_{j-1}$, $S = \eta_1 + \eta_2 + \cdots + \eta_n$. It is obvious that η_j are 1-dependent Bernoulli variables, $P(\eta_j = 1) = p^2 = 1 - P(\eta_j = 0)$. Let, for $k = 2,3,\ldots,n$,

$$G_k = \exp\left\{p^2(I_1-I)-\frac{p^3(2-3p)(I_1-I)^2}{2}\right\}, \quad G_1 = \exp\left\{p^2(I_1-I)-\frac{p^4(I_1-I)^2}{2}\right\}.$$

We will prove the following result.

Theorem 13.2 *Let* $0 < p \leqslant 1/12$, $n \geqslant 3$. *Then*

$$\left\|\mathcal{L}(S) - \prod_{k=1}^n G_k\right\| \leqslant C\min\left(np^4, \frac{p}{\sqrt{n}}\right).$$

For the proof we need auxiliary results. Let, in this section, $z := e^{it} - 1$. To make all expressions simpler, we will also write h_j, $\widehat{G}_j$ instead of $h_j(t)$ and $\widehat{G}_j(t)$. First note that if $p \leqslant 1/12$, then (13.2) is satisfied for all t. Indeed,

$$\mathbb{E} \, |\, e^{itX_k} - 1\,|^2 = |z|^2 p^2 \leqslant 4p^2 \leqslant \frac{1}{36}.$$

Next, we show that explicit dependency allows us to simplify all formulas.

Lemma 13.4 *Let* $0 < p \leqslant 1/12$, $k = 1, 2, \ldots$. *Then*

$$\widehat{\mathbb{E}} \, (e^{it\eta_1} - 1, e^{it\eta_2} - 1, \ldots, e^{it\eta_k} - 1) = z^k p^{k+1} (1 - p)^{k-1}.$$

Proof We apply induction. The proof obviously holds for $k = 1$, since

$$\widehat{\mathbb{E}} \, (e^{it\eta_1} - 1) = \mathbb{E} \, (e^{it\eta_1}) = zp^2.$$

Let us assume that the lemma's statement holds for $k = 1, 2, \ldots, m$. Then

$$\widehat{\mathbb{E}} \, (e^{it\eta_1} - 1, e^{it\eta_2} - 1, \ldots, e^{it\eta_{m+1}} - 1)$$
$$= \mathbb{E} \, (e^{it\eta_1} - 1, e^{it\eta_2} - 1, \ldots, e^{it\eta_{m+1}} - 1)$$
$$- \sum_{j=1}^{m} \widehat{\mathbb{E}} \, (e^{it\eta_1} - 1, \ldots, e^{it\eta_j} - 1) \mathbb{E} \, (e^{it\eta_{j+1}} - 1) \cdots (e^{it\eta_{m+1}} - 1)$$
$$= z^{m+1} p^{m+2} - \sum_{j=1}^{m} z^j p^{j+1} (1 - p)^{j-1} p^{m-j+2} z^{m-j+1}$$
$$= z^{m+1} p^{m+2} \Big(1 - p \sum_{j=1}^{m} (1 - p)^{j-1} \Big) = z^{m+1} p^{m+2} (1 - p)^m.$$

Thus, the lemma's statement is also correct for $k = m + 1$. $\square$

Substituting the lemma's statement into (13.3) we arrive at

$$h_k = 1 + p^2 z + \sum_{j=1}^{k-1} \frac{z^{k-j+1} p^{k-j+2} (1 - p)^{k-j}}{h_j h_{j+1} \cdots h_{k-1}}. \tag{13.16}$$

Here $k = 2, 3, \ldots, n$.

Lemma 13.5 *Let* $0 < p \leqslant 1/12$, $n \geqslant 3$. *Then, for all* $k = 2, \ldots, n$,

$$|h_k - 1| \leqslant 3p^2, \qquad |h_k - 1 - p^2 z| \leqslant \frac{16}{13} |z|^2 p^3 (1 - p), \tag{13.17}$$

$$|h_k - 1| \leqslant \frac{16}{13} p^2 |z|, \quad |h_k - 1 - p^2 z - p^3 (1 - p) z^2| \leqslant C p^4 |z|^3. \tag{13.18}$$

Proof We prove (13.17) by induction. For $k = 2$ the proof can be checked directly. Indeed, $h_1 = 1 + p^2 z$ and

$$|h_2 - 1 - p^2 z| = \frac{p^3(1-p)|z|^2}{|1 + p^2 z|} \leq \frac{p^3(1-p)|z|^2}{1 - 2p^2} \leq \frac{16}{13}p^2(1-p)|z|^2.$$

Assume that the second estimate of (13.17) holds for $j = 1, 2, \ldots, m - 1$. Then, for the same values of j, the first estimates of (13.17) and (13.18) hold. Indeed,

$$|h_j - 1| \leq p^2|z|^2 + \frac{16}{13}p^3|z|^2 \leq p^2|z|\left(1 + \frac{16}{13}\cdot\frac{2}{12}\right) \leq \frac{16}{13}p^2|z| \leq 3p^2.$$

Consequently,

$$\frac{1}{|h_j|} \leq \frac{1}{1 - |h_j - 1|} \leq \frac{48}{47}, \quad j = 1, 2, \ldots, m - 1. \tag{13.19}$$

Taking into account the trivial estimate $|z| \leq 2$ and (13.16), we then obtain

$$|h_m - 1 - p^2 z| \leq \sum_{j=1}^{m-1} |z|^{m-j+1} p^{m-j+2}(1-p)^{m-j}\left(\frac{48}{47}\right)^{m-j}$$

$$\leq |z|^2 p^3(1-p)\frac{48}{47}\sum_{j=0}^{\infty}\left(2p(1-p)\frac{48}{47}\right)^j$$

$$\leq |z|^2 p^3(1-p)\frac{48}{47}\sum_{j=0}^{\infty}\left(\frac{8}{47}\right)^j \leq \frac{16}{13}|z|^2 p^3(1-p).$$

Thus, we have proved that the second estimate of (13.17) (and, consequently, the first estimates and (13.19)) hold for all $k = 2, 3, \ldots, n$. The second estimate in (13.18) is proved by exactly the same argument. First we prove that

$$\left|h_k - 1 - p^2 z - \frac{z^2 p^3(1-p)}{h_{k-1}}\right| \leq \sum_{j=1}^{k-2} |z|^{k-j+1} p^{k-j+2}(1-p)^{k-j}\left(\frac{48}{47}\right)^{k-j}$$

$$\leq C|z|^3 p^4 \sum_{j=0}^{\infty}\left(2p(1-p)\frac{48}{47}\right)^j \leq C|z|^3 p^4.$$

Next, we observe that

$$\left| z^2 p^3 (1-p) - \frac{z^2 p^3 (1-p)}{h_{k-1}} \right| = \frac{|z|^2 p^3 (1-p)|h_{k-1}-1|}{h_{k-1}}$$

$$\leq \frac{3}{2}|z|^2 p^3 (1-p)\frac{16}{13}p^2 |z| \leq C |z|^3 p^4.$$

The last two estimates and the triangle inequality complete the proof of the lemma. □

Similar estimates hold for the derivatives of h_k.

Lemma 13.6 Let $0 < p \leq 1/12$, $n \geq 3$. Then, for all $k = 2,\dots,n$,

$$|h_k'| \leq 2p^2, \quad |h_k' - p^2 z'| \leq 3|z|p^3, \quad |h_k' - p^2 z' - p^3(1-p)(z^2)'| \leq 6p^4 |z|^2.$$

Proof The first two estimates follow from the last one. Indeed,

$$|h_k' - p^2 z'| \leq 2p^3 |z| + 6p^4 |z|^2 \leq p^3 |z|(2+12p) \leq 2p^3 |z|,$$

$$|h_k'| \leq 3p^3 |z| + p^2 \leq 6p^3 + p^2 \leq 2p^2.$$

Therefore it suffices to prove the last estimate. We use induction. It is easy to check that it holds for $k = 2$. Next, assume that it holds for $j = 3,\dots,k-1$. Then

$$\left| h_k' - p^2 z' - \left(\frac{z^2 p^3 (1-p)}{h_{k-1}} \right)' \right|$$

$$\leq \left| \sum_{j=1}^{k-2} \frac{(k-j+1)z^{k-j}z'p^{k-j+2}(1-p)^{k-j}}{h_j \cdots h_{k-1}} - \sum_{j=1}^{k-2} \frac{z^{k-j}p^{k-j+2}(1-p)^{k-j}}{h_j \cdots h_{k-1}} \sum_{m=j}^{k-1} \frac{h_m'}{h_m} \right|$$

$$\leq p^4 |z|^2 \left(\frac{48}{47} \right)^2 \sum_{j=1}^{k-2}(k-j+1)\left(2p\frac{48}{47} \right)^{k-j-2}$$

$$+ p^4 |z|^2 \cdot 2p^2 \left(\frac{48}{47} \right)^3 \sum_{j=1}^{k-2}(k-j)\left(2p\frac{48}{47} \right)^{k-j-2}$$

$$\leq p^4 |z|^2 \left(\frac{48}{47} \right)^2 \left[\sum_{j=0}^{\infty}(j+3)\left(\frac{8}{47} \right)^j + \frac{2}{141}\sum_{j=0}^{\infty}(j+2)\left(\frac{8}{47} \right)^j \right]. \qquad (13.20)$$

Observe that, for $0 < x < 1$,

$$\sum_{j=0}^{\infty} x^j = \frac{1}{1-x}, \quad \sum_{j=0}^{\infty} jx^j = x\left(\sum_{j=0}^{\infty} x^j\right)' = x\left(\frac{1}{1-x}\right)' = \frac{x}{(1-x)^2}.$$

Applying these equalities to (13.20) we obtain

$$\left| h_k' - p^2 z' - p^3(1-p)\left(\frac{z^2}{h_{k-1}}\right)' \right| \le 5p^4|z|^2. \tag{13.21}$$

Next, observe that

$$p^3(1-p)\left| \left(\frac{z^2}{h_{k-1}}\right)' - (z^2)' \right| \le p^3 \frac{|(z^2)'| \, ||h_{k-1}-1|}{|h_{k-1}|} + p^3 \frac{|z^2| \, ||h_{k-1}'|}{|h_{k-1}|^2}$$

$$\le 2p^3|z|\frac{48}{47} \cdot \frac{16}{13}p^2|z| + 2p^5|z|^2\left(\frac{48}{47}\right)^2 \le p^4|z|^2.$$

Combining the last estimate with (13.21) we complete the lemma's proof. $\square$

Lemma 13.7 *Let* $p \le 1/12$, $n \ge 3$. *Then, for all* $t \in \mathbb{R}$ *and* $k = 1, 2, 3, \ldots, n$, *the following estimates hold*

$$|h_k - \widehat{G}_k| \le Cp^4|z|^3, \quad |(e^{-itp^2}h_k)'| \le 2p^2|z|,$$

$$|h_k' - \widehat{G}_k'| \le Cp^4|z|^2, \quad \left| \left(e^{-itp^2}h_k - e^{-itp^2}\widehat{G}_k\right)' \right| \le Cp^4|z|^2.$$

Proof For $k = 1$ all estimates can be verified directly. Let $2 \le k \le n$. Then the analogue of (1.35) for Fourier transforms gives us

$$\widehat{G}_k = 1 + p^2 z + p^3(1-p)z^2 + \theta Cp^5|z|^3, \quad \widehat{G}_k' = p^2 z' + p^3(1-p)(z^2)' + \theta Cp^5|z|^2.$$

Therefore the first and the third lemma's estimates follow from Lemmas 13.5 and 13.6. Observe that

$$|(e^{-itp^2}h_k)'| \le |h_k' - p^2 z'| + p^2|e^{it} - h_k| \le 3p^2|z| + p|z| + p^2|1 - h_k| \le 2p^2|z|.$$

Similarly,

$$\left| \left(e^{-itp^2}h_k - e^{-itp^2}\widehat{G}_k\right)' \right| = \left| e^{-itp^2}(h_k' - \widehat{G}_k') - ip^2 e^{-itp^2}(h_h - \widehat{G}_k) \right|$$

$$\le |h_k' - \widehat{G}_k'| + p^2|h_k - \widehat{G}_k| \le Cp^4|z|^2.$$

$\square$

Next, we show that the characteristic functions are exponentially small.

Lemma 13.8 *Let $p \leqslant 1/12$, $n \geqslant 3$. For all t, $k = 1, 2, 3, \ldots, n$, we then have*

$$\max\{|h_k|, |\widehat{G}_k|\} \leqslant \exp\left\{-\frac{3}{2}p^2 \sin^2 \frac{t}{2}\right\}. \tag{13.22}$$

Proof It is easy to check that

$$|1 + p^2 z|^2 = (1 - p^2 + p^2 \cos t)^2 + p^4 \sin^2 t = 1 - 4p^2(1 - p^2) \sin^2(t/2).$$

Consequently, $|h_1| = |1 + p^2 z| \leqslant 1 - 2p^2(1 - p^2) \sin^2(t/2)$. Therefore, for $k = 2, \ldots, n$, applying the second estimate from (13.17), we obtain

$$|h_k| \leqslant |1 + p^2 z| + |h_k - 1 - p^2 z| \leqslant 1 - 2p^2\left(1 - p^2 - \frac{32p(1-p)}{13}\right)\sin^2 \frac{t}{2} \leqslant 1 - \frac{3}{2}p^2 \sin^2 \frac{t}{2}.$$

Now the estimate for h_k trivially follows. For $k = 2, \ldots, n$

$$|\widehat{G}_k| \leqslant \exp\{-2p \sin^2(t/2) + 2p^3(2 - 3p) \sin^2(t/2)\}$$

$$\leqslant \exp\{-2p^2(1 - 2p) \sin^2(t/2)\} \leqslant \exp\left\{-\frac{5}{3}p^2 \sin^2(t/2)\right\}.$$

The estimate for $|\widehat{G}_1|$ is obtained in the same way. $\square$

Now we are in a position to estimate the difference of products of the characteristic functions.

Lemma 13.9 *Let $p \leqslant 1/12$, $n \geqslant 3$, $|t| \leqslant \pi$. Then*

$$\left|\prod_{k=1}^{n}(h_k e^{-itp^2}) - \prod_{k=1}^{n}(\widehat{G}_k e^{-itp^2})\right| \leqslant C \min\left(np^4, \frac{p}{\sqrt{n}}\right) \exp\{-np^2 \sin^2(t/2)\},$$

$$\left|\left(\prod_{k=1}^{n}(h_k e^{-itp^2}) - \prod_{k=1}^{n}(\widehat{G}_k e^{-itp^2})\right)'\right| \leqslant C \min(np^4, p^2) \exp\{-np^2 \sin^2(t/2)\}.$$

Proof For the sake of brevity let

$$\tilde{h}_k = e^{-itp^2} h_k, \quad \tilde{g}_k = e^{-itp^2}\widehat{G}_k, \quad k = 1, 2, \ldots, n.$$

Then by (1.48) and Lemmas 13.7 and 13.8

$$\left| \prod_{k=1}^{n} \tilde{h}_k - \prod_{k=1}^{n} \tilde{g}_k \right| = \left| \prod_{k=1}^{n} h_k - \prod_{k=1}^{n} \widehat{G}_k \right| \leq \exp\{-1.5(n-1)p^2 \sin^2(t/2)\} \sum_{k=1}^{n} |h_k - \widehat{G}_k|$$

$$\leq Cnp^4 |z|^3 \exp\{-1.5np^2 \sin^2(t/2)\}$$

$$= Cnp^4 \exp\{-1.3np^2 \sin^2(t/2)\}|\sin(t/2)|^3 \exp\{-0.2np^2 \sin^2(t/2)\}$$

$$\leq C \min\left(np^4, \frac{p}{\sqrt{n}}\right) \exp\{-1.3np^2 \sin^2(t/2)\}.$$

Similarly,

$$\left| \left(\prod_{k=1}^{n} \tilde{h}_k - \prod_{k=1}^{n} \tilde{g}_k \right)' \right| \leq \sum_{k=1}^{n} |\tilde{h}_k'| \left| \prod_{m \neq k} \tilde{h}_m - \prod_{m \neq k} \tilde{g}_m \right| + \sum_{k=1}^{n} |\tilde{h}_k' - \tilde{g}_k'| \prod_{m \neq k} |\tilde{g}_m|$$

$$\leq Cnp^2 |\sin(t/2)| \min\left(np^4, \frac{p}{\sqrt{n}}\right) \exp\{-1.3np^2 \sin^2(t/2)\}$$

$$+ C \exp\{-1.5np^2 \sin^2(t/2)\} np^4 \sin^2(t/2)$$

$$\leq C \min(np^4, p^2) \exp\{-np^2 \sin^2(t/2)\}.$$

□

Proof of Theorem 13.2 The proof follows from Lemma 5.1 applied with $a = np^2$ and $b = \max(1, \sqrt{n}p)$ and (1.31). □

13.4 Problems

Let $\xi_j, j = 0, 1, 2, \ldots, n$ be independent identically distributed indicator variables, $P(\xi_1 = 1) = p$, $P(\xi_1 = 0) = 1 - p$. Let $\eta_j = \xi_j(1 - \xi_{j-1})$, $S = \eta_1 + \eta_2 + \cdots + \eta_n$. Let $p(1 - p) \leq 1/150$.

13.1 Prove that

$$\widehat{\mathbb{E}}\,(e^{it\eta_1} - 1, \ldots, e^{it\eta_k} - 1) = (-1)^{k+1}(p(1-p))^k.$$

13.2 Prove that, for $n \geq 3$,

$$\| \mathcal{L}(S) - \exp\{np(1-p)(I_1 - I)\} \|_\infty \leq C \min\left(np^2(1-p)^2, \frac{\sqrt{p(1-p)}}{\sqrt{n}} \right).$$

Bibliographical Notes

In a slightly different form the mixed centered moment $\widehat{\mathbb{E}}$ was introduced in [150]. Applications to large deviations can be found in [132], Section 4.1. Heinrich introduced his method in [69, 70] and extended it to Markov chains [71] and random vectors [72], see also an overview in [73]. Heinrich's method for Poisson-type approximations was adapted in [103]. Further results for compound Poisson, binomial and negative binomial approximations of m-dependent integer-valued random variables can be found in [45]. In this section we used special cases of Theorem 1 from [104] and Theorem 2 from [103].

Chapter 14
Other Methods

In this chapter, we give a brief overview of some methods and approaches that might be of interest for further studies. For the reader's convenience, bibliographical notes are presented at the end of each section.

14.1 Method of Compositions

In principle, the method of compositions is a version of the method of convolutions. However, it is not related to compound structure of measures, but strongly depends on the smoothing properties of the normal distribution. We recall that Φ_σ denotes the normal distribution with zero mean and variance σ^2, $\Phi \equiv \Phi_1$ denotes the standard normal distribution and $\Phi(x)$ denotes its distribution function. Note that $\Phi_\sigma(x) = \Phi(x/\sigma)$.

The method of compositions is based on mathematical induction and the following version of the inversion Lemma 9.1.

Lemma 14.1 *Let $G \in \mathcal{F}$. Then for any $\varepsilon > 0$*

$$|G - \Phi_\sigma|_K \leqslant 2|(G - \Phi_\sigma)\Phi_\varepsilon|_K + \frac{C_1\varepsilon}{\sigma}.$$

As an example we consider approximation by Φ in terms of the so-called pseudomoments. Let $\xi_1, \xi_2, \ldots$ be a sequence of iid rvs with $\mathbb{E}\,\xi_1 = 0$, $\mathrm{Var}\,\xi_1 = 1$ and having the same distribution F. The third pseudomoment is defined by

$$\tilde{\beta}_3 = \int_{\mathbb{R}} |x|^3 |\,\mathrm{d}(F(x) - \Phi(x))\,|.$$

© Springer International Publishing Switzerland 2016
V. Čekanavičius, *Approximation Methods in Probability Theory*, Universitext,
DOI 10.1007/978-3-319-34072-2_14

Typical for the CLT is to approximate by the normal law the distribution of the normed sum:

$$S_n = \frac{\xi_1 + \xi_2 + \cdots + \xi_n}{\sqrt{n}} = \frac{\xi_1}{\sqrt{n}} + \frac{\xi_2}{\sqrt{n}} + \cdots + \frac{\xi_n}{\sqrt{n}}.$$

Let V be the distribution of $\xi_1/\sqrt{n}$, i.e. $V(x) = F(\sqrt{n}x)$, $\mathcal{L}(S_n) \equiv V^n$.

Theorem 14.1 *Let $\tilde{\beta}_3 \leqslant 1$. Then*

$$|\mathcal{L}(S_n) - \Phi|_K \leqslant \frac{C_2 \tilde{\beta}_3^{1/4}}{\sqrt{n}}. \tag{14.1}$$

Note that, in general, $C_2 = C_2(F)$. However, assuming that $\xi_1, \xi_2, \ldots$ form a sequence of random variables, we indirectly assumed that F does not depend on n in any way. The main advantage of (14.1) over the standard estimate of Berry-Esseen type lies in the fact that the right-hand side of (14.1) equals zero for $F \equiv \Phi$.

Proof   *Step 1* First we prove (14.1) for $n = 1$. In this case $\mathcal{L}(S_1) \equiv V \equiv F$. Observe that

$$\left| \frac{\partial^3 \Phi_\sigma(x)}{\partial x^3} \right| = \frac{1}{\sqrt{2\pi}\sigma} \left| \frac{\partial^2 \exp\{-x^2/(2\sigma^2)\}}{\partial x^2} \right| \leqslant \frac{C_3}{\sigma^3}. \tag{14.2}$$

Let us recall that the means and variances of F and Φ are equal. Thus,

$$\int_{\mathbb{R}} x \, d(F(x) - \Phi(x)) = \int_{\mathbb{R}} x^2 d(F(x) - \Phi(x)) = 0.$$

Therefore applying Taylor series to $\Phi_\varepsilon(x - y)$ and taking into account (14.2) we obtain

$$|(F - \Phi)\Phi_\varepsilon|_K = \sup_x \left| \int_{\mathbb{R}} \Phi_\varepsilon(x - y) d(F(x) - \Phi(y)) \right|$$

$$= \sup_x \int_{\mathbb{R}} \left| \frac{\partial^3 \Phi_\varepsilon(x - (1 - \theta)y)}{6\partial x^3} \right| \left| d(F(x) - \Phi(y)) \right|$$

$$\leqslant C_4 \sup_x \left| \frac{\partial^3 \Phi_\varepsilon(x)}{\partial x^3} \right| \tilde{\beta}_3 \leqslant \frac{C_5 \tilde{\beta}_3}{\varepsilon^3}.$$

From Lemma 14.1 it follows that

$$|F - \Phi|_K \leqslant 2 \frac{C_6 \tilde{\beta}_3}{\varepsilon^3} + C_1 \varepsilon.$$

Taking $\varepsilon = \tilde{\beta}_3^{1/4}$ we prove that

$$|F - \Phi|_K \leqslant C_7 \tilde{\beta}^{1/4}.$$

Step 2 Next, we write $V^n - \Phi$ in a form convenient for mathematical induction. For the sake of brevity, let $\tilde{\Phi} = \Phi_{n-1/2}$. Observe that

$$(V^n - \Phi)\Phi_\varepsilon = (V^n - \tilde{\Phi}^n)\Phi_\varepsilon = (V - \tilde{\Phi}) \sum_{m=0}^{n-1} V^m \tilde{\Phi}^{n-m-1} \Phi_\varepsilon$$

$$= (V - \tilde{\Phi}) \left(\sum_{m=1}^{n-1} (V^m - \tilde{\Phi}^m) \tilde{\Phi}^{n-m-1} \Phi_\varepsilon + n\tilde{\Phi}^{n-1} \Phi_\varepsilon \right)$$

$$= (V - \tilde{\Phi}) \left(\sum_{m=1}^{n-1} M_m + n\tilde{\Phi}^{n-1} \Phi_\varepsilon \right).$$

Here $M_m = (V^m - \tilde{\Phi}^m)\tilde{\Phi}^{n-m-1}\Phi_\varepsilon$. We can write

$$|(V^n - \Phi)\Phi_\varepsilon|_K \leqslant \sum_{m=1}^{n-1} |M_m(V - \tilde{\Phi})|_K + n|\tilde{\Phi}^{n-1}\Phi_\varepsilon(V - \tilde{\Phi})|_K. \qquad (14.3)$$

Step 3 Let us assume that (14.1) holds for $m = 1, 2, \ldots, n - 1$. More precisely, we assume that

$$|V^m - \tilde{\Phi}^m|_K \leqslant \frac{C_2 \tilde{\beta}_3^{1/4}}{\sqrt{m}}. \qquad (14.4)$$

Step 4 In this step, we will obtain a preliminary estimate for $|M_m(V - \tilde{\Phi})|_K$. Note that V and $\tilde{\Phi}$ have two matching moments. Therefore applying Taylor's expansion to $M_m(x - y)$ we prove that

$$|M_m(V - \tilde{\Phi})|_K = \sup_x \left| \int_{\mathbb{R}} M_m(x - y) d(V - \tilde{\Phi})(y) \right|$$

$$\leqslant \frac{1}{6} \sup_x \left| \frac{\partial^3 M_m(x)}{\partial x^3} \right| \left| \int_{\mathbb{R}} |y|^3 |d(V - \tilde{\Phi})(y)| \right.$$

Taking into account that $V(x) = F(\sqrt{n}x)$ and $\tilde{\Phi}(x) = \Phi(\sqrt{n}x)$, we have

$$\int_{\mathbb{R}} |y|^3| \, d(V - \tilde{\Phi})(y)| = \frac{\tilde{\beta}_3}{n^{3/2}}.$$

Combining the last two estimates, for $m > 0$, we can write

$$| M_m(V - \tilde{\Phi}) |_K \leqslant \frac{\tilde{\beta}_3}{6n\sqrt{n}} \sup_x \left| \frac{\partial^3 M_m(x)}{\partial x^3} \right|. \qquad (14.5)$$

Step 5 Observe that

$$M_m(x) = \int_{\mathbb{R}} (V^m - \tilde{\Phi}^m)(x - y) \, d\tilde{\Phi}^{n-m-1} \Phi_\varepsilon(y) = \int_{\mathbb{R}} (V^m - \tilde{\Phi}^m)(x - y)\varphi_a(y) \, dy$$

$$= -\int_{\mathbb{R}} (V^m - \tilde{\Phi}^m)(y)\varphi_a(x - y) \, dy.$$

Here

$$a^2 = \frac{n - m - 1}{n} + \varepsilon^2, \qquad \Phi^{n-m-1}\Phi_\varepsilon = \Phi_a, \qquad \varphi_a(x) = \frac{1}{\sqrt{2\pi}a} \exp\left\{ -\frac{x^2}{2a^2} \right\}.$$

Observe that by (1.48)

$$\left| \frac{\partial^3 \varphi_a(x - y)}{\partial x^3} \right| \leqslant \frac{1}{\sqrt{2\pi}a} \left(\frac{|x - y|^3}{a^6} + \frac{3|x - y|}{a^4} \right) \exp\left\{ -\frac{(x - y)^2}{2a^2} \right\}$$

$$\leqslant \frac{C_8}{a^4} \exp\left\{ -\frac{(x - y)^2}{4a^2} \right\}.$$

Therefore taking into account the last estimate and (14.4), we obtain

$$\left| \frac{\partial^3 M_m(x)}{\partial x^3} \right| \leqslant \int_{\mathbb{R}} | V^m(y) - \tilde{\Phi}^m(y) | \left| \frac{\partial^3 \varphi_a(x - y)}{\partial x^3} \right| dy$$

$$\leqslant \frac{C_8}{a^4} | V^m - \tilde{\Phi}^m |_K \int_{\mathbb{R}} e^{-(x-y)^2/(4a^2)} \, dy$$

$$\leqslant \frac{C_9}{a^3} | V^m - \tilde{\Phi}^m |_K \leqslant \frac{C_9 C_2 \tilde{\beta}_3^{1/4}}{\sqrt{m}a^3}.$$

Substituting the last estimate into (14.5) we get

$$|M_m(V - \tilde{\Phi})|_K \leq \frac{\tilde{\beta}_3^{5/4}}{6n\sqrt{n}} \frac{C_9 C_2}{\sqrt{ma^3}} = \frac{C_9 C_2 \tilde{\beta}_3^{5/4}}{6\sqrt{m}(n + n\varepsilon^2 - 1 - m)^{3/2}}. \tag{14.6}$$

Step 6 Let $b = n + n\varepsilon^2 - 1$. Next, we observe that, for any $2 \leq m \leq n - 2$,

$$\frac{1}{\sqrt{m}(b - m)^{3/2}} \leq \frac{1}{\sqrt{x - 1}(b - x)^{3/2}}, \qquad m \leq x \leq m + 1.$$

Therefore

$$\sum_{m=1}^{n-2} \frac{1}{\sqrt{m}(b - m)^{3/2}} = \sum_{m=1}^{n-2} \int_m^{m+1} \frac{1}{\sqrt{m}(b - m)^{3/2}} dx$$

$$\leq \sum_{m=1}^{n-2} \int_m^{m+1} \frac{1}{\sqrt{x - 1}(b - x)^{3/2}} dx = \int_1^{n-1} \frac{1}{\sqrt{x - 1}(b - x)^{3/2}} dx$$

$$= \frac{2\sqrt{x - 1}}{(b - 1)\sqrt{b - x}} \Big|_1^{n-1} = \frac{2\sqrt{1 - 2/n}}{n\varepsilon(1 - 2/n + \varepsilon^2)}$$

$$\leq \frac{2}{n\varepsilon} \sup_{n \geq 3} \frac{1}{\sqrt{1 - 2/n}} \leq \frac{2\sqrt{3}}{n\varepsilon}.$$

Taking into account (14.6) we observe that

$$\sum_{m=1}^{n-1} |M_m(V - \tilde{\Phi})|_K \leq \sum_{m=1}^{n-2} \frac{C_9 C_2 \tilde{\beta}_3^{5/4}}{6\sqrt{m}(n + n\varepsilon^2 - 1 - m)^{3/2}} + \frac{C_9 C_2 \tilde{\beta}_3^{5/4}}{6\sqrt{n - 1}n\sqrt{n\varepsilon^3}}$$

$$\leq \frac{C_9 C_2 \tilde{\beta}_3^{5/4}}{6} \left(\frac{\sqrt{2}}{n^2 \varepsilon^3} + \frac{\sqrt{3}}{n\varepsilon} \right). \tag{14.7}$$

Step 7 Similarly to the above argument and taking into account (14.2) we prove that

$$n| \tilde{\Phi}^{n-1} \Phi_\varepsilon(V - \tilde{\Phi})|_K \leq \frac{n}{6} \sup_x \left| \frac{\partial^3 \tilde{\Phi}^{n-1} \Phi_\varepsilon(x)}{\partial x^3} \right| \frac{\tilde{\beta}_3}{n^{3/2}} \leq \frac{8 C_3 \tilde{\beta}_3}{6\sqrt{n}}.$$

Step 8 Combining the last estimate, (14.7) and (14.3) and applying Lemma 14.1 with $\varepsilon = \tilde{\beta}_3^{1/4} h / \sqrt{n}$ (the quantity $h \geqslant 1$ will be chosen later) we obtain

$$
\begin{aligned}
|V^n - \Phi|_K &\leqslant \frac{C_9 C_2 \tilde{\beta}_3^{5/4}}{3} \left(\frac{\sqrt{2}}{n^2 \varepsilon^3} + \frac{\sqrt{3}}{n\varepsilon} \right) + \frac{8 C_3 \tilde{\beta}_3}{3\sqrt{n}} + C_1 \varepsilon \\
&= \frac{\tilde{\beta}_3^{1/4}}{\sqrt{n}} \left(\frac{C_9 C_2 \tilde{\beta}_3^{1/4} \sqrt{2}}{3h^3} + \frac{C_9 C_2 \tilde{\beta}_2^{3/4} \sqrt{3}}{3h} + \frac{8 C_3 \tilde{\beta}_3^{3/4}}{3} + C_1 h \right) \\
&\leqslant \frac{\tilde{\beta}_3^{1/4}}{\sqrt{n}} \left(\frac{4 C_9 C_2}{h} + 3 C_3 + C_1 h \right).
\end{aligned}
$$

For the last estimate we used the fact that $\tilde{\beta}_3 \leqslant 1$. The theorem will be proved if we choose $h \geqslant 1$ and C_2 so that

$$
\frac{4 C_9 C_2}{h} + 3 C_3 + C_1 h \leqslant C_2.
$$

To show that such a choice is possible let us rewrite the last inequality in the following way

$$
C_2 \left(1 - \frac{4 C_9}{h} \right) \geqslant 3 C_3 + C_1 h.
$$

Let $h = 1 + 8 C_9$. Then

$$
C_2 \left(1 - \frac{4 C_9}{h} \right) \geqslant \frac{C_2}{2}
$$

and it suffices to take any C_2 satisfying $C_2 \geqslant 6 C_3 + 2 C_1 (1 + 8 C_9)$. $\square$

Bibliographical Notes

The method of compositions is one of the classical methods. Though rarely used for one-dimensional distributions, it is quite popular for the multivariate CLT, see [24, 26, 133–135, 154]. The method is discussed, in detail, in [136], Chapters 4–5. Lemma 14.1 is taken from [133]. Theorem 14.1 is taken from [102].

14.2 Coupling of Variables

Coupling means that for a set of random variables $\xi_1, \xi_2, \ldots, \xi_n$ a new set of random variables $\eta_1, \eta_2, \ldots, \eta_n$ is constructed in such a way that η_i has the same distribution as ξ_i and useful interdependence of new variables can be employed in the process of estimating. Finding such a useful interdependence is the essence of the coupling technique.

We illustrate the coupling approach by considering the so-called maximal coupling in the case of two independent random variables X and Y concentrated on $\mathbb{Z}$. Then the maximal coupling result can be formulated as follows: there exist random variables $\tilde{X}$ and $\tilde{Y}$ such that $\tilde{X}$ has the same distribution as X, $\tilde{Y}$ has the same distribution as Y and

$$P(\tilde{X} = \tilde{Y}) = \sum_{k=-\infty}^{\infty} \min(P(X = k), P(Y = k)). \qquad (14.8)$$

Observe that (14.8) is the same as (1.9), written in terms of probabilities. For the proof of (14.8) we use the following notation: we denote by A the event that $\tilde{X} = \tilde{Y}$, denote by a the right-hand side of (14.8) and introduce the auxiliary random variables:

$$P(\mathbb{I} = 1) = a = 1 - P(\mathbb{I} = 0), \quad P(V = k) = \frac{1}{a} \min(P(X = k), P(Y = k)),$$

$$P(Z_X = k) = \frac{1}{1 - a}(P(X = k) - aP(V = k)),$$

$$P(Z_Y = k) = \frac{1}{1 - a}(P(Y = k) - aP(V = k)), \quad k \in \mathbb{Z}.$$

Next, we introduce the coupling variables:

$$\tilde{X} = \begin{cases} V & \text{if } \mathbb{I} = 1, \\ Z_X & \text{if } \mathbb{I} = 0, \end{cases} \qquad \tilde{Y} = \begin{cases} V & \text{if } \mathbb{I} = 1, \\ Z_Y & \text{if } \mathbb{I} = 0. \end{cases}$$

Direct calculation shows that for all $k \in \mathbb{Z}$

$$P(\tilde{X} = k) = P(V = k)P(\mathbb{I} = 1) + P(Z_X = k)P(\mathbb{I} = 0) = P(X = k)$$

and $P(\tilde{Y} = k) = P(Y = k)$. Moreover,

$$P(\tilde{X} = \tilde{Y}) = P(\mathbb{I} = 1) = a$$

and, therefore, (14.8) holds.

Example 14.1 Let $p \in (0, 1)$, $P(X = 1) = p = 1 - P(X = 0)$ and let Y have a Poisson distribution with parameter p. Then $1 - p \leq e^{-p}$, $p \geq pe^{-p}$ and $P(\tilde{X} = \tilde{Y}) = 1 - p + pe^{-p} \geq 1 - p^2$. Therefore

$$\| \mathcal{L}(X) - \mathcal{L}(Y) \| = 2P(X \neq Y) \leq 2p^2.$$

It is interesting to note that to construct a suitable coupling for *sums* of Bernoulli variables is a difficult task, and the typical approach is a rough simplification of the initial problem leading to Le Cam's type estimate $2np^2$, see [151], p. 12.

Apart from the maximal coupling there exist other couplings (the Ornstein coupling, the Mineka coupling, to name a few) related to specific stochastic structures. Coupling is usually combined with some other techniques, such as the Stein method.

Bibliographical Notes

Various couplings are widely used to investigate the closeness of stochastic processes, see the comprehensive studies of various applications in [96] and [151]. The coupling idea also plays a central role in the Stein method, see the short overview in [117] and the exhaustive discussion in [49]. Note that the Barbour-Xia inequality from Sect. 5.4 was obtained via the Mineka coupling of random walks.

14.3   The Bentkus Approach

In [17], Bentkus proposed to unite approximated and approximating random variables with an additional parameter. We illustrate the idea by considering independent variables. Let $X_1, X_2, \ldots, X_n$ and $Y_1, Y_2, \ldots, Y_n$ be independent random variables, $\mathbb{E} X_i = \mathbb{E} Y_i = 0$ and $\mathbb{E} X_i^2 = \mathbb{E} Y_i^2$ and $\mathbb{E} |X_i|^3 < \infty$, $(i = 1, \ldots, n)$. Let $f : \mathbb{R} \to \mathbb{R}$ be a thrice differentiable function, such that $\| f''' \| = \sup_x |f'''(x)| < \infty$.

Proposition 14.1 *Let* $S_X = X_1 + X_2 + \cdots + X_n$, $S_Y = Y_1 + Y_2 + \cdots + Y_n$ *and suppose that all the above assumptions. Then*

$$|\mathbb{E}f(S_X) - \mathbb{E}f(S_Y)| \leq \pi \|f'''\| \sum_{i=1}^{n} (\mathbb{E}|X_i|^3 + \mathbb{E}|Y_i|^3). \tag{14.9}$$

Proof The main tool in the proof is the introduction of additional random variables:

$$Z_i(\alpha) = X_i \sin \alpha + Y_i \cos \alpha, \quad T(\alpha) = \sum_{i=1}^{n} Z_i(\alpha), \quad T_i(\alpha) = T(\alpha) - Z_i(\alpha).$$

It can be easily checked that

$$Z_i'(\alpha) = X_i \cos \alpha - Y_i \sin \alpha, \quad \mathbb{E} Z_i(\alpha) = \mathbb{E} Z_i'(\alpha) = 0 \qquad (14.10)$$

and

$$\mathbb{E} Z_i(\alpha)Z_i'(\alpha) = \mathbb{E}\left[(X_i^2 - Y_i^2)\sin\alpha\cos\alpha + X_iY_i(\cos^2\alpha - \sin^2\alpha)\right] = 0, \qquad (14.11)$$

since $\mathbb{E} X_i^2 = \mathbb{E} Y_i^2$, $\mathbb{E} X_iY_i = \mathbb{E} X_i \mathbb{E} Y_i = 0$.

Note that $f(S_X) = f(T(\pi/2))$ and $f(S_Y) = f(T(0))$. Next, we apply the standard formula

$$f(\pi/2) - f(0) = \int_0^{\pi/2} f'(\alpha) d\alpha$$

to $\mathbb{E} f(T(\alpha))$, obtaining

$$\mathbb{E} f(S_X) - \mathbb{E} f(S_Y) = \int_0^{\pi/2} \mathbb{E} f'(T(\alpha))T'(\alpha) d\alpha. \qquad (14.12)$$

Observing that $T'(\alpha) = \sum_{i=1}^n Z_i'(\alpha)$ and $T(\alpha) = T_i(\alpha) + Z_i(\alpha)$, we can rewrite (14.12) in the following way

$$\mathbb{E} f(S_X) - \mathbb{E} f(S_Y) = \sum_{i=1}^n \int_0^{\pi/2} \mathbb{E} f'(T_i(\alpha) + Z_i(\alpha))Z_i'(\alpha) d\alpha.$$

Next, we apply Taylor's expansion and (14.10) and (14.11)

$$\mathbb{E} f(S_X) - \mathbb{E} f(S_Y) = \sum_{i=1}^n \int_0^{\pi/2} \mathbb{E} f'(T_i(\alpha)) \mathbb{E} Z_i'(\alpha) d\alpha$$

$$+ \sum_{i=1}^n \int_0^{\pi/2} \mathbb{E} f''(T_i(\alpha)) \mathbb{E} Z_i(\alpha)Z_i'(\alpha) d\alpha$$

$$+ \sum_{i=1}^n \int_0^{\pi/2} \int_0^1 (1-\tau)\mathbb{E} f'''(T_i(\alpha) + tZ_i(\alpha))Z_i^2(\alpha)Z_i'(\alpha) dt d\alpha$$

$$= \sum_{i=1}^n \int_0^{\pi/2} \int_0^1 (1-\tau)\mathbb{E} f'''(T_i(\alpha) + tZ_i(\alpha))Z_i^2(\alpha)Z_i'(\alpha) dt d\alpha.$$

Therefore

$$|\mathbb{E} f(S_X) - \mathbb{E} f(S_Y)| \leq \frac{1}{2}\|f'''\| \int_0^{\pi/2} \mathbb{E} Z_i^2(\alpha)|Z_i'(\alpha)| d\alpha.$$

The propositions proof is easily completed by observing that

$$Z_i^2(\alpha)|Z_i'(\alpha)| \leqslant (|X_i| + |Y_i|)^3 \leqslant 4(|X_i|^3 + |Y_i|^3).$$

□

Remark 14.1 Note a certain similarity between (14.9) and (11.4), and between their proofs. However, further development of the Stein method was intertwined with research on suitable couplings. Meanwhile, for the Bentkus approach no serious research on suitable couplings was ever attempted.

Example 14.2 Let $\xi_1, \xi_2, \ldots, \xi_n$ be non-degenerate iid rvs with finite third absolute moment and zero mean. Let $S = (\xi_1 + \cdots + \xi_n)/\sqrt{n}$ and $Y \sim \mathcal{N}(0,1)$. Then from (14.9) we obtain the following version of the CLT

$$|\mathbb{E}f(S) - \mathbb{E}f(Y)| \leqslant \frac{A}{\sqrt{n}}\|f'''\|.$$

Here A is a constant depending only on the distribution of ξ_1. Indeed, it suffices to take in (14.9) $X_i = \xi_i/\sqrt{n}$ and $Y_i = \eta_i/\sqrt{n}$, where $\eta_i \sim \mathcal{N}(0,1)$ and all rv are independent.

Bibliographical Notes

In [17], it is shown that the same or similar approach works for asymptotic expansions, vectors and even general operators, see also [20, 149]. In [23], the Bentkus approach was extended to random fields.

14.4   The Lindeberg Method

In this section, we give a brief outline of the discrete version of the Lindeberg method as it is presented in [27]. The primary field of application of the method is estimation of the expectation of *unbounded* functions.

Let $\xi_1, \xi_2, \ldots, \xi_n$ be independent Bernoulli random variables with the success probabilities $p_j = P(\xi_j = 1), j = 1, \ldots, n$. Let $\eta_1, \eta_2, \ldots, \eta_n$ be independent Poisson random variables with parameters $p_1, \ldots, p_n$, respectively. Put $S = \xi_1 + \xi_2 + \cdots + \xi_n$, $Z = \eta_1 + \eta_2 + \cdots + \eta_n$ and let f be an arbitrary function on the nonnegative integers. Let $\tilde{p}_k = \max(p_1, p_2, \ldots, p_k), \tilde{p}_0 = 0, \tilde{p} = \tilde{p}_n$. Let $\Delta f(k) = f(k+1) - f(k)$, $\Delta^r f(k) = \Delta(\Delta^{r-1} f(k))$.

Theorem 14.2 *Let* $\mathbb{E}\,|f(Z)\,| < \infty$. *Then*

$$\mathbb{E}f(Z) - \mathbb{E}f(S) = \sum_{k=1}^{n}\sum_{r=2}^{\infty}\frac{p_k^r}{r!}\mathbb{E}\,\Delta^r f(T_k),\qquad (14.13)$$

where $T_k = \xi_1 + \cdots + \xi_{k-1} + \eta_{k+1} + \cdots + \eta_k$, *and, for each k, the corresponding series in (14.13) absolutely converges. Moreover, if* $\mathbb{E}\,Z_{s+1}|f(Z)\,| < \infty$ *then*

$$\left|\sum_{r=s+1}^{\infty}\frac{p_k^r}{r!}\mathbb{E}\,\Delta^r f(T_k)\right| \leqslant \frac{e^{p_k}}{(1 - \tilde{p}_{k-1})^2}\frac{p_k^{s+1}}{(s+1)!}\mathbb{E}\,|\,\Delta^{s+1}f(Z)\,|,\qquad s \geqslant 1.$$

Corollary 14.1 *If* $\mathbb{E}\,Z^2|f(Z)\,| < \infty$, *then*

$$|\,\mathbb{E}f(S) - \mathbb{E}f(Z)\,| \leqslant \frac{1}{2}\frac{e^{\tilde{p}}}{(1 - \tilde{p})^2}\sum_{j=1}^{n}p_j^2\mathbb{E}\,|\,\Delta^2 f(Z)\,|.$$

The main benefit of Corollary 14.1 is that all estimates are through a Poisson variable and it is applicable to unbounded functions. To get the better idea of the limitations and advantages of the method let us consider the case $f(k) = e^{hk}$, $h > 0$. Then

$$\mathbb{E}\,|\,\Delta^2 f(Z)\,| \leqslant \sum_{k=0}^{\infty}|\,e^{(k+2)h} - 2e^{(k+1)h} + e^{kh}\,|\frac{\lambda^k}{k!}e^{-\lambda}$$

$$\leqslant 2e^{2h}\sum_{k=0}^{\infty}\frac{(\lambda e^h)^k}{k!}e^{-\lambda} = 2e^{2h}\exp\{\lambda(e^h - 1)\}.$$

Here $\lambda = p_1 + p_2 + \cdots + p_n$. Therefore Corollary 14.1 can be written in the following form

$$\sum_{k=0}^{\infty}e^{hk}|\,P(S = k) - P(Z = k)\,| \leqslant \frac{e^{\tilde{p}}}{(1 - \tilde{p})^2}e^{2h}\exp\{\lambda(e^h - 1)\}\sum_{j=1}^{n}p_j^2.\qquad (14.14)$$

It is obvious that (14.14) can be used to estimate the rate of convergence to the Poisson limit (when $\max p_j \to 0$ and $\lambda \to \gamma$) in a norm stronger than total variation.

Bibliographical Notes

Note that typically by the Lindeberg method one understands a version of the CLT with Lindeberg's condition. However, Borisov and Ruzankin [27] used the term in the context of this section. Note also that in [27] more general results than Theorem 14.2 are proved. The Lindeberg method was further extended to the case of weakly dependent summands in [130].

14.5   The Tikhomirov Method

The Stein method for continuous approximations involves the solution of some differential equation for densities. Tikhomirov proposed to use exactly the same approach for characteristic functions, combining the resulting estimate with Esseen type smoothing inequalities. Roughly the Tikhomirov method (or, as it is sometimes called, the Stein-Tikhomirov method) means that one can write the characteristic function in a form similar to the normal one, by solving an appropriate differential equation.

As an example let us consider the characteristic function of a centered and normed Poisson random variable with parameter $\lambda > 0$, that is, let

$$\widehat{F}(t) = \exp\left\{\lambda\left(e^{it/\sqrt{\lambda}} - 1 - \frac{it}{\sqrt{\lambda}}\right)\right\}.$$

Then

$$\widehat{F}'(t) = i\sqrt{\lambda}\left(e^{it/\sqrt{\lambda}} - 1\right)\widehat{F} = i\sqrt{\lambda}\left(e^{it/\sqrt{\lambda}} - 1 - \frac{it}{\sqrt{\lambda}} + \frac{it}{\sqrt{\lambda}}\right)\widehat{F}$$

$$= \left(-t + i\sqrt{\lambda}\left(e^{it/\sqrt{\lambda}} - 1 - \frac{it}{\sqrt{\lambda}}\right)\right)\widehat{F}(t).$$

Since $|e^{ix} - 1 - ix| \leqslant x^2/2$, $x \in \mathbb{R}$, we can then rewrite the last equation in the following form

$$\widehat{F}'(t) = \left(-t + \theta_1(t)\frac{t^2}{2\sqrt{\lambda}}\right)\widehat{F}(t). \tag{14.15}$$

Here $|\theta_1(t)| \leqslant 1$.

The form of equation (14.15) is essential for normal approximation. We recall some facts about differential equations. If we have a differential equation

$$A(y)dy + B(t)dt = 0, \quad y(t_0) = y_0,$$

then its solution can be found from

$$\int_{y_0}^{y} A(s)ds + \int_{t_0}^{t} B(u)du = 0.$$

Note that $\widehat{F}(0) = 1$, since this is a general property of all characteristic functions. Observe also that for $t > 0$

$$\left| \int_0^t \theta_1(u) u^2 du \right| \leqslant \int_0^t u^2 du = \frac{t^3}{3}$$

and, consequently,

$$\left| \int_0^t \theta_1(u) u^2 du \right| = \theta_2(t) \frac{t^3}{3}, \quad |\theta_2(t)| \leqslant 1.$$

A similar estimate holds for $t < 0$. Therefore solving (14.15) we get

$$\widehat{F}(t) = \exp\left\{ -\frac{t^2}{2} + \theta_2(t) \frac{|t|^3}{6\sqrt{\lambda}} \right\}, \quad |\theta_2(t)| \leqslant 1. \tag{14.16}$$

The estimate (14.16) can then be used to prove

$$\left| \widehat{F}(t) - e^{-t^2/2} \right| \leqslant \frac{|t|^3}{6\sqrt{\lambda}} e^{-t^2/4}, \quad |t| < 3\sqrt{\lambda} 2,$$

which can be substituted into (9.4) or other inversion estimates from Sect. 9.1.

The important fact is that, in principle, the exact form of $\theta_1(t)$ in (14.15) is unnecessary. The next lemma gives a more precise formulation of this idea.

Lemma 14.2 *Let for $|t| \leqslant T_1$ the characteristic function $\widehat{G}(t)$ satisfy the differential equation*

$$\widehat{G}'(t) = [-t + \theta_3(t) a(t)] \widehat{G}(t) + \theta_4(t) b(t), \quad \widehat{G}(0) = 1.$$

Here $|\theta_3(t)|, |\theta_4(t)| \leqslant 1$, $a(t) = a_0 + a_1|t| + a_2 t^2 + a_3|t|^{s-1}$, $2 < s \leqslant 3$, $b(t) = b_0 + b_2 t^2$ and the coefficients $a_i \geqslant 0$, $(i = 0, 1, 2, 3)$ and $b_j \geqslant 0$, $(j = 0, 2)$ do not depend on n. Then for $|t| \leqslant \min(T_1, T_2)$ and $a_1 \leqslant 1/6$

$$|\widehat{G}(t) - e^{-t^2/2}| \leqslant C(a_0|t| + a_1 t^2 + a_2|t|^3 + a_3|t|^s) e^{-t^2/4}$$
$$+ C(b_0 \min(|t|^{-1}, |t|) + b_2|t|).$$

Here

$$T_2 = \min\left\{ \frac{1}{a_0}, \frac{1}{6a_2}, \left(\frac{1}{6a_3} \right)^{1/(s-2)} \right\}.$$

Bibliographical Notes

The method was introduced in [152] and was applied to the normal approximation of dependent random variables in [146, 147]. Lemma 14.2 was proved in [146]. More recent applications and modifications of the method can be found in [61, 62, 148]. The above example is taken from [148].

14.6 Integrals Over the Concentration Function

Convolution of measures can be combined with the properties of the concentration function to prove estimates in the Kolmogorov norm. We illustrate this approach by proving (2.31).

Theorem 14.3 *Let $G \in \mathcal{F}_s$, $n \in \mathbb{N}$. Then*

$$|G^{n+1} - G^n|_K \leq \frac{C}{\sqrt{n}}.$$

Proof First we investigate absolutely continuous distributions. Let $G = F\Phi_\varepsilon$. Here Φ_ε, $\varepsilon > 0$ denotes the normal distribution with zero mean and variance equal to ε. Let $G(x) = G\{(-\infty, x)\}$. Then by (1.21)

$$|G^{n+1}(x) - G^n(x)| = \left| \int_{\mathbb{R}} [G^n(x-y) - G^n(x)]G\{dy\} \right|$$

$$\leq \int_{\mathbb{R}} Q(G^n, |y|)G\{dy\} \leq C \int_{\mathbb{R}} \frac{G\{dy\}}{\sqrt{n(1 - Q(G, |y|))}}.$$

If $y \leq 0$, then $Q(G, |y|) \leq 1 - G(y)$. If $y > 0$, then $Q(G, |y|) \leq G(y)$. Consequently,

$$\int_{\mathbb{R}} \frac{G\{dy\}}{\sqrt{n(1 - Q(G, |y|))}} \leq \frac{C}{\sqrt{n}} \int_{y \leq 0} \frac{dG(y)}{\sqrt{G(y)}} + \frac{C}{\sqrt{n}} \int_{y > 0} \frac{dG(y)}{\sqrt{1 - G(y)}} \leq \frac{C}{\sqrt{n}}.$$

The general case can be proved by taking $\varepsilon_m \to 0$ and applying

$$|G^{n+1} - G^n|_K \leq \overline{\lim}_{m \to \infty} |(G\Phi_{\varepsilon_m})^{n+1} - (G\Phi_{\varepsilon_m})^n|_K.$$

$\square$

Bibliographical Notes

The technique of integration of concentration functions was mostly developed by A. Zaĭtsev. Theorem 14.3 is a special case of a more general Theorem 4.1 from [5]. Other results proved in a similar way can be found in Section 4 of Chapter 5 in [5] and [162].

14.7   Asymptotically Sharp Constants

In this section, we discuss how asymptotically sharp constants can be calculated for compound Poisson and compound binomial approximations. Let us assume that for some $M \in \mathcal{M}$

$$\| M \| \leqslant C_1 \varepsilon(n), \qquad \varepsilon(n) \to 0 \text{ as } n \to \infty. \tag{14.17}$$

If

$$\lim_{n \to \infty} \frac{\| M \|}{\varepsilon(n)} = \tilde{C}_1,$$

then $\tilde{C}_1$ is an asymptotically sharp constant for the estimate (14.17). An asymptotically sharp constant gives an idea about the magnitude of C_1 for large n.

Set

$$\varphi_0(x) = \frac{1}{\sqrt{2\pi}} \, e^{-x^2/2}, \quad \varphi_k(x) = \frac{d^k}{dx^k} \varphi_0(x) \quad (k \in \mathbb{N}, \ x \in \mathbb{R}),$$

$$\| \varphi_k \|_1 = \int_{\mathbb{R}} | \varphi_k(x) | \, dx, \quad \| \varphi_k \|_\infty = \sup_{x \in \mathbb{R}} | \varphi_k(x) | \quad (k = 0, 1, 2, \dots).$$

Then

$$\| \varphi_1 \|_1 = \sqrt{\frac{2}{\pi}}, \quad \| \varphi_1 \|_\infty = \frac{1}{\sqrt{2\pi e}}, \quad \| \varphi_2 \|_1 = \frac{4}{\sqrt{2\pi e}}, \quad \| \varphi_2 \|_\infty = \frac{1}{\sqrt{2\pi}},$$

$$\| \varphi_3 \|_1 = \sqrt{\frac{2}{\pi}} (1 + 4e^{-3/2}), \quad \| \varphi_3 \|_\infty = \sqrt{\frac{3}{\pi}} \exp\left\{ \sqrt{\frac{3}{2}} - \frac{3}{2} \right\} \sqrt{3 - \sqrt{6}},$$

$$\| \varphi_4 \|_1 = 4e^{-3/2} \sqrt{\frac{3}{\pi}} \left[\exp\left\{ \sqrt{\frac{3}{2}} \right\} \sqrt{3 - \sqrt{6}} + \exp\left\{ -\sqrt{\frac{3}{2}} \right\} \sqrt{3 + \sqrt{6}} \right],$$

$$\| \varphi_4 \|_\infty = \frac{3}{\sqrt{2\pi}}, \quad \| \varphi_5 \|_1 = \frac{2(3e^{5/2} - 32 \sinh(\sqrt{5/2}) + 16\sqrt{10} \cosh(\sqrt{5/2}))}{\sqrt{2\pi} \, e^{5/2}}.$$

The following lemmas summarize the facts about sharp constants for compound Poisson and compound binomials distributions. Note that, due to (5.16) and (3.23), they can also be used for estimates in the Kolmogorov and Wasserstein norms.

Lemma 14.3 *For $j \in \mathbb{Z}_+$ and $t \in (0, \infty)$, we have*

$$\left| \| (I_1 - I)^j \exp\{t(I_1 - I)\} \| - \frac{\| \varphi_j \|_1}{t^{j/2}} \right| \le \frac{C(j)}{t^{(j+1)/2}},$$

$$\left| \| (I_1 - I)^j \exp\{t(I_1 - I)\} \|_\infty - \frac{\| \varphi_j \|_\infty}{t^{(j+1)/2}} \right| \le \frac{C(j)}{t^{j/2+1}}.$$

Lemma 14.4 *For $j \in \mathbb{Z}_+$, $p = 1 - q \in (0, 1)$ and $n \in \mathbb{N}$, we have*

$$\left| \| (I_1 - I)^2 (qI + pI_1)^n \| - \frac{\| \varphi_2 \|_1}{npq} \right| \le \frac{C}{(npq)^2},$$

$$\left| \| (I_1 - I)^2 (qI + pI_1)^n \|_\infty - \frac{\| \varphi_2 \|_\infty}{(npq)^{3/2}} \right| \le \frac{C}{(npq)^{5/2}},$$

$$\left| \| (I_1 - I)^j (qI + pI_1)^n \| - \frac{\| \varphi_j \|_1}{(npq)^{j/2}} \right| \le \frac{C(j)}{(npq)^{(j+1)/2}} \quad (j \ne 0),$$

$$\left| \| (I_1 - I)^j (qI + pI_1)^n \|_\infty - \frac{\| \varphi_j \|_\infty}{(npq)^{(j+1)/2}} \right| \le \frac{C(j)}{(npq)^{j/2+1}}.$$

Finally we present some results on the simplest case of symmetric integer-valued random variables.

Lemma 14.5 *Let $j \in \mathbb{Z}_+$, $t \in (0, \infty)$, and $F = 2^{-1}(I_{-1} + I_1)$. Then*

$$\left| \| (F - I)^j \exp\{t(F - I)\} \| - \frac{\| \varphi_{2j} \|_1}{(2t)^j} \right| \le \frac{C(j)}{t^{j+1/2}}, \qquad (j \ne 0),$$

$$\left| \| (F - I)^j \exp\{t(F - I)\} \|_\infty - \frac{\| \varphi_{2j} \|_\infty}{2^j t^{j+1/2}} \right| \le \frac{C(j)}{t^{j+1}}.$$

Lemma 14.6 *Let $j \in \mathbb{Z}_+$, $p = 1 - q \in (0, 1)$, $n \in \mathbb{N}$, and $F = 2^{-1}(I_{-1} + I_1)$. Then*

$$\left| \| (F - I)^j (qI + pF)^n \| - \frac{\| \varphi_{2j} \|_1}{(2np)^j} \right| \le \frac{C(j)}{q(npq)^{j+1/2}}, \qquad (j \ne 0),$$

$$\left| \| (F - I)^j (qI + pF)^n \|_\infty - \frac{\| \varphi_{2j} \|_\infty}{2^j (np)^{j+1/2}} \right| \le \frac{C(j)}{q(npq)^{j+1}}.$$

Asymptotically sharp constants are usually calculated by combination of the second-order approximations, Lemmas 14.3, 14.4, 14.5 and 14.6 and the following simple version of the triangle inequality.

Lemma 14.7 *Let $M_1, M_2 \in \mathcal{M}$, $a > 0$. Then*

$$| \| M_1 \| - a | \leq \| M_1 - M_2 \| + | \| M_2 \| - a |. \tag{14.18}$$

The total variation in (14.18) can be replaced by the local, Kolmogorov or Wasserstein norms.

Proof It suffices simply to check all four cases. If $\| M_1 \|, \| M_2 \| > a$ or $\| M_1 \|, \| M_2 \| \leq a$, then (14.18) becomes a simple triangle inequality for M_1 or M_2. If $\| M_1 \| > a$ and $\| M_2 \| \leq a$, then (14.18) is equivalent to $\| M_1 \| + \| M_2 \| \leq \| M_1 - M_2 \| + 2a$, which is obviously correct, since $\| M_1 \| + \| M_2 \| \leq \| M_1 - M_2 \| + 2\| M_2 \| \leq \| M_1 - M_2 \| + 2a$. If $\| M_1 \| \leq a$ and $\| M_2 \| > a$, then (14.18) is equivalent to $2a \leq \| M_1 \| + \| M_1 - M_2 \| + \| M_2 \|$, which is correct, since $2a \leq 2\| M_2 \| \leq \| M_2 - M_1 \| + \| M_1 \| + \| M_2 \|$. $\square$

As an example we consider a signed CP approximation to the binomial distribution. Let

$$Bi = ((1 - p)I + pI_1)^n, \quad G = \exp\left\{ np(I_1 - I) - \frac{np^2(I_1 - I)^2}{2} \right\}.$$

Theorem 14.4 *Let $p \leq 1/5$, $n \in \mathbb{N}$, $np > 1$. Then*

$$\left| \| Bi - G \| - \frac{\| \varphi_3 \|_1 p \sqrt{p}}{3 \sqrt{n}} \right| \leq C\left(\frac{p}{n} + \frac{p^{5/2}}{\sqrt{n}} \right).$$

Corollary 14.2 *Let the assumptions of Theorem 14.4 hold and let $p \to 0$ and $np \to \infty$ as $n \to \infty$. Then*

$$\lim_{n \to \infty} \frac{\| Bi - G \|}{p^{3/2}/\sqrt{n}} = \frac{\| \varphi_3 \|_1}{3} = 0.503 \dots.$$

Proof of Theorem 14.4 Let

$$U = I_1 - I, \quad M = \frac{np^3 U^3}{3}, \quad A = \frac{\| \varphi_3 \|_1 p \sqrt{p}}{3 \sqrt{n}}.$$

Then by Lemma 14.18, problem 5 of Chapter 2 and Lemma 14.3

$$| \| Bi - G \| - A | \leq \| Bi - G - GM \| + | \| GM \| - A |$$

$$\leq \frac{Cp^2}{n} + \| M(G - \exp\{npU\}) \| + \frac{np^3}{3} \left| \| U^3 \exp\{npU\} \| - \frac{\| \varphi_3 \|_1}{(np)^{3/2}} \right|$$

$$\leq \frac{Cp}{n} + \frac{np^3}{3} \| U^3 (G - \exp\{npU\}) \|. \tag{14.19}$$

By (1.35), (2.18) and (2.12)

$$\| U^3(G - e^{npU}) \| = \| U^3 e^{npU}(e^{-np^2 U^2/2} - I) \| \leqslant \frac{np^2}{2} \int_0^1 \| U^5 e^{npU - np^2 \tau U^2/2} \| d\tau$$

$$\leqslant \frac{np^2}{2} \| U^5 e^{0.5npU} \| \int_0^1 \| \exp\{0.5npU + (np/7)U^2 \Theta\} \| d\tau$$

$$\leqslant Cnp^2 \| U^5 e^{0.5npU} \| \leqslant \frac{C}{n\sqrt{np}}.$$

Substituting the last estimate into (14.19) we complete the proof of theorem. $\square$

Bibliographical Notes

Lemma 14.3 is a special case of Proposition 4 from [119]. The first two estimates of Lemma 14.4 were proved in [120], Lemma 8. The second two estimates were obtained in [42], Proposition 3.5. Lemma 14.5 was proved in [43] (Lemma 4.7). Lemma 14.6 was proved in [42] (Proposition 3.4). Asymptotically sharp constants for Poisson approximation to the Markov binomial distribution were calculated in [44]. Asymptotically sharp constants for m-dependent variables were calculated in [45, 103]. Note that asymptotically sharp constants are still calculated for the normal approximation, see [138].

Solutions to Selected Problems

Chapter 1

1.1 By (1.45)

$$2|\sin(kt/2)| = |e^{itk} - 1| \leqslant k|e^{it} - 1| = 2k|\sin(t/2)|.$$

1.2 We have

$$|F_1 F_2 - G_1 G_2|_K = |(F_1 - G_1)F_2 + G_1(F_2 - G_2)|_K \leqslant |F_1 - G_1|_K \|F_2\|$$
$$+ \|G_1\| \|F_2 - G_2\|_K = |F_1 - G_1|_K + |F_2 - G_2|_K \leqslant a_1 + a_2.$$

1.3 The second inequality is obvious. The first follows from

$$\|MV\|_\infty = \sup_{j \in \mathbb{Z}} \left| \sum_{k \in \mathbb{Z}} M\{k\} V\{j-k\} \right| \leqslant \sup_{j \in \mathbb{Z}} |V\{j\}| \sum_{k \in \mathbb{Z}} |M\{k\}| = \|V\|_\infty \|M\|.$$

1.4 Let $a = Re\widehat{F}(t)$, $b = Im\widehat{F}(t)$. Then $a^2 + b^2 = |\widehat{F}(t)|^2 \leqslant 1$. Therefore

$$|1 - \widehat{F}(t)|^2 = (1-a)^2 + b^2 \leqslant (1-a)^2 + 1 - a^2 = 2(1-a).$$

1.5 By (1.37)

$$e^{itj} = 1 + \sum_{m=1}^{s-1} \binom{j}{m} (e^{it} - 1)^m + \sum_{m=s}^{j} \binom{m-1}{s-1} e^{it(j-m)} (e^{it} - 1)^s.$$

© Springer International Publishing Switzerland 2016
V. Čekanavičius, *Approximation Methods in Probability Theory*, Universitext,
DOI 10.1007/978-3-319-34072-2

Therefore, taking into account (1.39), we obtain

$$\widehat{F}(t) = \sum_{j=0}^{\infty} F\{j\} e^{itj} = 1 + \sum_{m=1}^{s-1} (e^{it} - 1)^m \sum_{j=0}^{\infty} \binom{j}{m} F\{j\}$$

$$+ |e^{it} - 1|^s \sum_{j=0}^{\infty} F\{j\} \binom{j}{s} \theta$$

$$= 1 + \sum_{m=1}^{s-1} (e^{it} - 1)^m \nu_m\{F\} + \theta |e^{it} - 1|^s \frac{\nu_s(F)}{s!}.$$

1.6 Let

$$a = \int_{\mathbb{R}} \cos(tx) M\{dx\}, \quad b = \int_{\mathbb{R}} \sin(tx) M\{dx\}.$$

Then $\widehat{M}(t) = a + ib$, $\widehat{M}(-t) = a - ib$ and $\widehat{M}(t)\widehat{M}(-t) = a^2 + b^2 = |\widehat{M}(t)|^2$.

1.7 We have

$$\| e^M - I \| = \Big\| \sum_{j=1}^{\infty} \frac{M^j}{j!} \Big\| \leq \sum_{j=1}^{\infty} \frac{\| M \|^j}{j!} = e^{\| M \|} - 1$$

$$= \| M \| \sum_{j=1}^{\infty} \frac{\| M \|^{j-1}}{j!} = \| M \| \sum_{j=0}^{\infty} \frac{\| M \|^j}{(j+1)!}$$

$$\leq \| M \| \sum_{j=0}^{\infty} \frac{\| M \|^j}{j!} = \| M \| e^{\| M \|}.$$

1.8 By the Jordan-Hahn decomposition $M = M_1 - M_2$, where M_1 and M_2 are nonnegative finite measures. Moreover, M_1 is concentrated on the set A_1, M_2 is concentrated on the set A_2 and $A_1 \cap A_2 = \emptyset$. Then

$$0 = M\{\mathbb{R}\} = M_1\{A_1\} - M_2\{A_2\} \Rightarrow M_1\{A_1\} = M_2\{A_2\},$$

$$\| M \| = M_1\{A_1\} + M_2\{A_2\} = 2M_1\{A_1\}$$

and

$$\sup_{A \in \mathcal{B}} |M\{A\}| = \sup_{A \in \mathcal{B}} |M_1\{A\} - M_2\{A\}| = \max(M_1\{A_1\}, M_2\{A_2\}) = M_1\{A_1\}.$$

We used the definition of total variation based on the Jordan-Hahn decomposition. We show that this definition is equivalent to the supremum over all functions bounded by unity. Let $|f(x)| \leqslant 1$, then

$$\left| \int_{\mathbb{R}} f(x) M\{dx\} \right| \leqslant \int_{\mathbb{R}} 1 |M\{dx\}| = M_1\{A_1\} + M_2\{A_2\} = \|M\|$$

and equality is attained for

$$f(x) = \begin{cases} 1, & x \in A_1, \\ -1, & x \in A_2. \end{cases}$$

1.9 By (1.42) and (1.30)

$$e^{itj} = 1 + \sum_{m=1}^{3} \binom{j}{m} (e^{it} - 1)^j + \theta \binom{j}{4} |e^{it} - 1|^4,$$

$$e^{-itj} = 1 + \sum_{m=1}^{3} \binom{j}{m} (e^{-it} - 1)^m + \theta \binom{j}{4} |e^{it} - 1|^4 = 1 + \sum_{m=1}^{2} \binom{j}{m} (e^{-it} - 1)^m$$

$$- \binom{j}{3} (e^{it} - 1)^3 + \binom{j}{3} [(e^{-it} - 1)^3 + (e^{it} - 1)^3] + \theta \binom{j}{4} |e^{it} - 1|^4$$

$$= 1 + \sum_{m=1}^{2} \binom{j}{m} (e^{-it} - 1)^m - \binom{j}{3} (e^{it} - 1)^3 + \theta |e^{it} - 1|^4 \left(3 \binom{j}{3} + \binom{j}{4} \right).$$

It remains to substitute these expansions into

$$\widehat{F}(t) = 1 + \sum_{j=1}^{\infty} e^{itj} F\{j\} + \sum_{j=1}^{\infty} e^{-itj} F\{-j\}.$$

Chapter 2

2.1 We recall that the norm of any distribution (including CP distribution) equals 1. Therefore by (2.17)

$$\| \exp\{a(F - I)\} - \exp\{b(F - I)\} \|$$

$$\leqslant |a - b| \| F - I \| \sup_{0 \leqslant \tau \leqslant 1} \| \exp\{(\tau a + (1 - \tau)b)(F - I)\} \|$$

$$= |a - b| \| F - I \| \leqslant |a - b| (\| F \| + \| I \|) = 2|a - b|.$$

2.2 The norm of any distribution equals 1. Therefore

$$\| H^n - F^n \| \leqslant \sum_{j=1}^{n} \| H - F \| \| H^{n-j} F^{j-1} \| = n \| H - F \|. \tag{14.1}$$

Applying (2.2) to H with $s = 1$ we obtain

$$H = I + (F - I) \sum_{j=0}^{\infty} \frac{j}{2^{j+1}} + \frac{(F - I)^2}{2} \sum_{j=0}^{\infty} \frac{j(j-1)}{2^{j+1}} \Theta.$$

It is not difficult to check that, for $0 < x < 1$,

$$\sum_{j=1}^{\infty} j x^{j-1} = \sum_{j=1}^{\infty} (x^j)' = \left(\sum_{j=0}^{\infty} x^j \right)' = \left(\frac{1}{1-x} \right)' = \frac{1}{(1-x)^2}$$

and

$$\sum_{j=2}^{\infty} j(j-1) x^{j-2} = \left(\sum_{j=0}^{\infty} x^j \right)'' = \left(\frac{1}{1-x} \right)'' = \frac{2}{(1-x)^3}.$$

Therefore

$$\sum_{j=0}^{\infty} \frac{j}{2^{j+1}} = \frac{1}{2^2} \sum_{j=1}^{\infty} \frac{j}{2^{j-1}} = 1, \quad \sum_{j=0}^{\infty} \frac{j(j-1)}{2^{j+1}} = \frac{1}{2^3} \sum_{j=2}^{\infty} \frac{j(j-1)}{2^{j+1}} = 2$$

and $H = F + (F - I)^2 \Theta$. Now it suffices to substitute this expression into (14.1).

2.3 By (1.35)

$$e^M = I + \| M \| e^{\| M \|} \Theta, \quad e^M = I + M + \frac{M^2}{2} + \frac{\| M \|^3 e^{\| M \|}}{6} \Theta.$$

Therefore

$$e^M \left(I - \frac{M^2}{2} \right) = I + M + \frac{M^2}{2} + \frac{\| M \|^3 e^{\| M \|}}{6} \Theta - \frac{M^2}{2} (I + \| M \| e^{\| M \|} \Theta)$$

$$= I + M + \Theta e^{\| M \|} \left(\frac{\| M \|^3}{6} + \frac{\| M \| \| M^2 \|}{2} \right)$$

$$= I + M + \frac{2}{3} \Theta \| M \|^3 e^{\| M \|}.$$

2.4 For the sake of brevity set $U = F - I$ and

$$M_1 = npU - \frac{np^2U^2}{2} + \frac{np^3U^3}{3}, \quad M_2 = M_1 + \frac{np^4U^4}{4(1-2p)}\Theta.$$

Then from Example 2.4 it follows that $((1-p)I + pF)^n = \exp\{nM_2\}$ and by (2.17)

$$\| \exp\{M_2\} - \exp\{M_1\} \| \leq \frac{1}{4(1-2p)} \sup_{0 \leq \tau \leq 1} \left\| np^4U^4 \exp\left\{ nM_1 + \tau\frac{np^4U^4}{4(1-2p)}\Theta \right\} \right\|.$$

Observe that $\| U \| \leq \| F \| + \| I \| = 2$ and, therefore,

$$-\frac{np^2U^2}{2} + \frac{np^3U^3}{3} + \tau\frac{np^4U^4}{4(1-2p)}\Theta = npU^2\left(\frac{p}{2} + \frac{2p^2}{3} + \frac{2^2p^2}{4(1-2p)} \right)\Theta$$

$$= npU^2\frac{0.7}{5}\Theta = \frac{npU^2}{7}\Theta.$$

Consequently, applying Lemma 2.5 with $a = 0.5np$ we obtain

$$\| \exp\{M_2\} - \exp\{M_1\} \| \leq Cnp^4\| U^4 \exp\{0.5npU\} \|.$$

For the estimate Cnp^4, it suffices to observe that $\exp\{0.5npU\} \in \mathcal{F}$ and, therefore, $\| \exp\{0.5npU\} \| = 1$. The second estimate follows from (2.12).

2.5 We use the same notation M_1, U as in previous solution. Let

$$M_3 = \frac{np^3U^3}{3}, \quad M_4 = npU - \frac{np^2U^2}{2}.$$

Then by (1.35)

$$\| \exp\{M_1\} - \exp\{M_4\}(I + M_3) \| \leq \int_0^1 (1-\tau)\| \exp\{M_4 + \tau M_3\}M_3^2 \| d\tau.$$

Arguing similarly to the previous solution and applying (2.18) and (2.12) we obtain

$$\| \exp\{M_4 + \tau M_3\}M_3^2 \| \leq C\| \exp\{0.5npU\}M_3^2 \| \leq C\frac{p^3}{n}.$$

It remains to apply the estimate from the previous problem and the triangle inequality.

2.6 As in previous solutions, we use the abbreviation $U = F - I$. Without loss of generality we can assume that $n > 9$. Indeed, if $n \leqslant 9$, then we simply use the fact that the norm of any distribution equals 1 and observe that

$$| F^n - \exp\{nU\}(I - nU^2/2) |_K \leqslant \| F^n \| + \| \exp\{nU\} \| (1 + n\| U^2 \|/2)$$

$$\leqslant 1 + 1 + 9 \cdot 4/2 = 20 \leqslant 20/(9n^2).$$

By the triangle inequality

$$| F^n - e^{nU}(I - nU^2/2) |_K = | F^n - e^{nU} + e^{nU} nU^2/2 |_K$$

$$\leqslant | F^n - e^{nU} - ne^{(n-1)U}(F - e^U) |_K$$

$$+ n| (e^{(n-1)U} - e^{nU})(F - e^U) |_K$$

$$+ n| e^{nU}(I + U + U^2/2 - e^U) |_K. \qquad (14.2)$$

Applying consequently to the first estimate of (14.2) Bergström's identity (1.37), (1.35), (1.39), (2.29) and (2.30) we prove that it is less than or equal to

$$\sum_{m=2}^{n} \binom{m-1}{2} | (I + U - e^U)^2 F^{n-m} e^{(m-2)U} |_K$$

$$\leqslant C \sum_{m=2}^{n} \binom{m-1}{2} | U^4 F^{n-m} e^{(m-2)U} |_K$$

$$\leqslant C \sum_{m \leqslant n/3} \binom{m-1}{2} \| e^{(m-2)U} \| | F^{n-m} U^4 |_K$$

$$+ C \sum_{m > n/3} \binom{m-1}{2} \| F^{n-m} \| | e^{(m-2)U} U^4 |_K$$

$$\leqslant Cn^2(| F^{\lfloor 2n/3 \rfloor} U^4 |_K + | e^{nU/9} U^4 |_K) \leqslant Cn^{-2}.$$

Applying (2.37), (1.35) and (2.30) to the second estimate of (14.2) we prove that

$$n| (e^{(n-1)U} - e^{nU})(F - e^U) |_K \leqslant Cn| U^3 e^{nU} |_K \leqslant Cn^{-2}.$$

Similarly,

$$n| e^{nU}(I + U + U^2/2 - e^U) |_K \leqslant Cn| U^3 e^{nU} |_K \leqslant Cn^{-2}.$$

It remains to substitute these estimates into (14.2).

2.7 The norm of the difference of two distributions is less that 2. Therefore, without loss of generality, we further assume that $n > 10$. Observe that

$$\frac{1}{2}I + \frac{1}{2}F^2 = I + (F - I) + \frac{(F - I)^2}{2}.$$

Therefore by (1.35)

$$\frac{1}{2}I + \frac{1}{2}F^2 - \exp\{F - I\} = (F - I)^3 C\Theta$$

and

$$\left| \left(\frac{1}{2}I + \frac{1}{2}F^2 \right)^n - \exp\{n(F - I)\} \right|_K$$

$$\leqslant C \sum_{j=1}^{n} \left| (F - I)^3 \left(\frac{1}{2}I + \frac{1}{2}F^2 \right)^{j-1} \exp\{(n - j)(F - I)\} \right|_K$$

$$\leqslant C \sum_{j>n/2}^{n} \left| (F - I)^3 \left(\frac{1}{2}I + \frac{1}{2}F^2 \right)^{j-1} \right|_K \| \exp\{(n - j)(F - I)\} \|$$

$$+ C \sum_{j\leqslant n/2}^{n} \left\| \left(\frac{1}{2}I + \frac{1}{2}F^2 \right)^{j-1} \right\| |(F - I)^3 \exp\{(n - j)(F - I)\}|_K \quad (14.3)$$

$$\leqslant Cn \left| (F - I)^3 \left(\frac{1}{2}I + \frac{1}{2}F^2 \right)^m \right|_K + Cn| (F - I)^3 \exp\{(n/2)(F - I)\}|_K.$$

From (2.30) it follows that the last expression in the above is less than Cn^{-2}. Applying (2.29) we get

$$\left| (F - I)^3 \left(\frac{1}{2}I + \frac{1}{2}F^2 \right)^m \right|_K \leqslant \frac{1}{2^m} \sum_{j=0}^{m} \binom{m}{j} |(F - I)^3 F^{2j}|_K$$

$$\leqslant \frac{C}{2^m} + \frac{C}{2^m} \sum_{j=1}^{m} \binom{m}{j} \cdot \frac{1}{(2j)^3} \leqslant \frac{C}{2^m} + \frac{C}{2^m} \sum_{j=1}^{m} \binom{m}{j} \cdot \frac{6}{(j + 1)(j + 2)(j + 3)}$$

$$\leqslant \frac{C}{2^m} + \frac{C}{2^m(m + 1)(m + 2)(m + 3)} \sum_{j=1}^{m} \binom{m + 3}{j + 3} \leqslant \frac{C}{2^m(m + 1)^3} \sum_{j=0}^{m+3} \binom{m + 3}{j}$$

$$= \frac{2^{m+3}C}{2^m(m + 1)^3} \leqslant \frac{C}{n^3}.$$

Substituting the last estimate into (14.3) we complete the proof.

2.8 By definition and (2.36)

$$|\varphi(F) - \psi(F)|_K = \Big| \sum_{k,j=0}^{\infty} p_k q_j F^k - \sum_{k,j=0}^{\infty} p_k q_j F^j \Big|_K \leqslant \sum_{k,j=1}^{\infty} p_k q_j |F^k - F^j|_K$$

$$\leqslant C \sum_{k,j=1}^{\infty} p_k q_j \frac{|k-j|}{\max(k,j)} \leqslant C \sum_{k,j=0}^{\infty} p_k q_j \frac{|k-j|}{\max(k,j)+1}.$$

2.9 Let $q = 1 - p$. By (2.13)

$$\| (qI + pF)^n - (qI + pF)^m \| = \Big\| (qI + pF)^n \sum_{j=1}^{m-n} (qI + pF)^{j-1} p(F - I) \Big\|$$

$$\leqslant (m - n) p \| (qI + pF)^n (F - I) \| \leqslant (m - n) p \frac{C}{\sqrt{np}}.$$

2.10 $F \in \mathcal{F}_s, p \leqslant 1/2$. Therefore $\widehat{F}(t) \in \mathbb{R}$ and $q + p\widehat{F}(t) \geqslant q - p = 1 - 2p \geqslant 0$. In other words, $(qI + pF) \in \mathcal{F}_+$ and it remains to apply (2.35).

Chapter 3

3.1 By (3.2), (1.45) and (1.16)

$$\| F^n - G^n \|_\infty \leqslant \frac{1}{2\pi} \int_{-\pi}^{\pi} |\widehat{F}^n(t) - \widehat{G}^n(t)| dt$$

$$\leqslant \frac{n}{2\pi} \int_{-\pi}^{\pi} |\widehat{F}(t) - \widehat{G}(t)| dt \leqslant \frac{n(\nu_s(F) + \nu_s(G))}{2\pi s!} \int_{-\pi}^{\pi} |e^{it} - 1|^s dt$$

$$\leqslant \frac{n(\nu_s(F) + \nu_s(G)) 2^s}{s!}.$$

3.2 By (3.2), (1.14), (1.48) and (3.20)

$$\| (F - I)^k \exp\{n(F - I)\} \|_\infty \leqslant \frac{1}{2\pi} \int_{-\pi}^{\pi} |\widehat{F}(t) - 1|^k| \exp\{n(\widehat{F}(t) - 1)\}| dt$$

$$\leqslant C(k) \int_{-\pi}^{\pi} (1 - \mathrm{Re}\widehat{F}(t))^{k/2} \exp\{n(\mathrm{Re}\widehat{F}(t) - 1)\} dt$$

$$\leqslant \frac{C(k)}{n^{k/2}} \int_{-\pi}^{\pi} \exp\{(n/2)(\mathrm{Re}\widehat{F}(t) - 1)\} dt \leqslant \frac{C(k)}{n^{(k+1)/2}}.$$

3.3 Observe that $|\widehat{F}(t) - 1| = 1 - \widehat{F}(t)$, $\widehat{F}(t) = 1 + (\widehat{F}(t) - 1) \leqslant \exp\{\widehat{F}(t) - 1\}$. Therefore by (1.48) (3.2), (1.48) and (3.20)

$$\| (F - I)^k F^n \|_\infty \leqslant \frac{1}{2\pi} \int_{-\pi}^{\pi} (1 - \widehat{F}(t))^k \exp\{n(\widehat{F}(t) - 1)\}dt$$

$$\leqslant \frac{C(k)}{n^{k/2}} \int_{-\pi}^{\pi} \exp\{(n/2)(\widehat{F}(t) - 1)\}dt \leqslant \frac{C(k)}{n^{k+1}\sqrt{1 - q_0}}.$$

3.4 It is not difficult to check that

$$\sum_{m=1}^{\infty} \frac{m}{2^{m+1}} = 1, \qquad \sum_{m=2}^{\infty} \frac{m(m-1)}{2^{m+1}} = 2.$$

Therefore by the analogue of (2.2) for Fourier transforms, we get

$$\widehat{H}(t) = 1 + (\widehat{F}(t) - 1) + \theta|\widehat{F}(t) - 1|^2.$$

Let $\widehat{F}(t) = 0.7 + 0.3\widehat{V}(t)$, where $V \in \mathcal{F}_Z$ is concentrated on $\mathbb{N}$. Then by (1.14)

$$|\widehat{F}(t) - 1|^2 = 0.3^2|\widehat{V}(t) - 1|^2 \leqslant 2 \cdot 0.3^2 (1 - Re\widehat{V}(t)) = 0.6(1 - Re\widehat{F}(t)).$$

Next, observe that $H\{0\} > 0.2$. Therefore by (3.18)

$$|\widehat{H}(t)| \leqslant \exp\{0.5(Re\widehat{H}(t) - 1)\} \leqslant \exp\{0.5(Re\widehat{F}(t) - 1)(1 - 0.6)\} = \exp\{0.2(Re\widehat{F}(t) - 1)\}$$

and

$$|\widehat{F}(t)| \leqslant \exp\{0.7(Re\widehat{F}(t) - 1)\} \leqslant \exp\{0.2(Re\widehat{F}(t) - 1)\}.$$

Note also that combining the above estimates and applying (1.45) and (1.48) we prove

$$|\widehat{H}^n(t) - \widehat{F}^n(t)| \leqslant 0.6n \exp\{0.2(n-1)(Re\widehat{F}(t) - 1)\}(1 - Re\widehat{F}(t))$$
$$\leqslant C \exp\{0.1n(Re\widehat{F}(t) - 1)\}.$$

It remains to apply (3.20).

3.5 Let $b \leqslant a$. Then by (1.34), (1.14) and (1.48)

$$|\exp\{a(\widehat{F}(t) - 1)\} - \exp\{b(\widehat{F}(t) - 1)\}| \leqslant \exp\{b(Re\widehat{F}(t) - 1)\}|a - b||\widehat{F}(t) - 1|$$
$$\leqslant \sqrt{2}|a - b|\sqrt{1 - Re\widehat{F}(t)} \exp\{b(Re\widehat{F}(t) - 1)\}$$
$$\leqslant \frac{C|a - b|}{\sqrt{b}} \exp\{(b/2)(Re\widehat{F}(t) - 1)\}.$$

The required estimate follows from (3.20). Next, assume that $a < b \leqslant 2a$. Then, similarly,

$$\| \exp\{a(F - I)\} - \exp\{b(F - I)\} \|_\infty \leqslant C\frac{|a - b|}{a\sqrt{1 - F\{0\}}} \leqslant 2C\frac{\lceil a - b\rceil}{b\sqrt{1 - F\{0\}}}.$$

Let $2a \leqslant b$. Then

$$1 < \frac{2|a - b|}{b}$$

and

$$\| \exp\{a(F - I)\} - \exp\{b(F - I)\} \|_\infty \leqslant 2 \leqslant \frac{4|a - b|}{b} \leqslant \frac{4|a - b|}{b\sqrt{1 - F\{0\}}}.$$

3.6 We have

$$\| F^n - F^{n+1} \|_\infty \leqslant \| F^n - \exp\{n(F - I)\} \|_\infty + \| F^{n+1} - \exp\{(n + 1)(F - I)\} \|_\infty$$
$$+ \| \exp\{n(F - I)\} - \exp\{(n + 1)(F - I)\} \|_\infty.$$

It remains to use Theorem 3.4 and the estimate of the previous problem.
3.7 By (3.27), (1.11) and (3.24)

$$\| \exp\{a(F - I)\} - \exp\{b(F - I)\} \|_\infty$$
$$\leqslant \sup_{0 \leqslant \tau \leqslant 1} \| \exp\{(1 - \tau)(b - a)(F - I)\} \|\,\|(b - a)\|\,(F - I) \exp\{a(F - I)\} \|_\infty$$
$$= (b - a)\|\,(F - I) \exp\{a(F - I)\} \|_\infty \leqslant \frac{C(b - a)}{a\sqrt{a}}.$$

Chapter 4

4.1 By (3.8) or (3.18) $|(1 - p) + pe^{it}| \leqslant \exp\{-2p(1 - p) \sin^2(t/2)\}$. It is easy to check that

$$\left| \exp\left\{ p(e^{it} - 1) - \frac{p^2}{2}(e^{it} - 1)^2 \right\} \right| \leqslant \exp\left\{ -2p \sin^2(t/2) + \frac{p^2 |e^{it} - 1|^2}{2} \right\}$$
$$= \exp\{-2p(1 - p) \sin^2(t/2)\}$$

and

$$\exp\left\{p(e^{it} - 1) - \frac{p^2(e^{it} - 1)^2}{2}\right\} = 1 + p(e^{it} - 1) + \theta Cp^3 |e^{it} - 1|^3.$$

Applying (1.45) and noting that $2p(1 - p)\sin^2(t/2) \leqslant 1$, we get

$$\left| ((1 - p) + pe^{it})^n - \exp\left\{np(e^{it} - 1) - \frac{np^2(e^{it} - 1)^2}{2}\right\} \right|$$

$$\leqslant Cnp^3 |\sin(t/2)|^3 \exp\{-2np(1 - p)\sin^2(t/2)\}.$$

It remain to apply (4.1) and (1.31).

4.2   It is sufficient to use (4.1) and

$$|\widehat{F}(t) - 1||\widehat{F}(t)|^n \leqslant 2p|\sin(t/2)|\exp\{-2np(1 - p)\sin^2(t/2)\}.$$

4.3   Observe that

$$|\widehat{F}(t) - 1| \leqslant \sum_{k=-\infty}^{\infty} |e^{itk} - 1||F\{k\} \leqslant \sum_{k=-\infty}^{\infty} |k||e^{it} - 1||F\{k\} \leqslant |e^{it} - 1|\sqrt{b},$$

$$|\exp\{\lambda(\widehat{G}(t)-1)\}| = \exp\left\{-2\lambda a \sin^2(t/2) - 2\lambda \sum_{j=2}^{\infty} \sin^2(tj/2)\right\} \leqslant \exp\{-2\lambda a \sin^2(t/2)\}$$

and apply (4.1).

4.4   In the solution to problem 3.4 it was shown that

$$|\widehat{H}(t)| \leqslant \exp\{0.2(Re\widehat{F}(t) - 1)\} = \exp\{-0.2\sin^2(t/2)\}$$

and

$$\widehat{H}(t) = \widehat{F}(t) + \theta|\widehat{F}(t) - 1|^2 = \widehat{F}(t) + 4\theta\sin^4(t/2).$$

Moreover

$$|\exp\{\widehat{F}(t) - 1\}| = \exp\{-2\sin^2(t/2)\}, \quad \exp\{\widehat{F}(t) - 1\} = \widehat{F}(t) + \theta C\sin^4(t/2).$$

Next, apply (1.45), (4.1) and (1.31).

4.5   Observe that by (1.16)

$$Re\widehat{F}(t) - 1 = -2v_1(F)\sin^2(t/2) + 2\theta v_2(F)\sin^2(t/2) \geqslant -2b\sin^2(t/2)$$

and a similar estimate holds for $ReG(t)$. Moreover, $|\widehat{F}(t) - \widehat{G}(t)| \leqslant Ca\sin^4(t/2)$. Therefore by (1.45) and (3.3)

$$| \exp\{n(\widehat{F}(t) - 1)\} - \exp\{n(\widehat{G}(t) - 1)\}| \leqslant Ca\sin^4(t/2)\exp\{-2nb\sin^2(t/2)\}.$$

The required estimate then follows from (4.1) and (1.31).

4.6 By Example 1.8

$$\widehat{F}(t) = \exp\left\{\frac{16}{4}(e^{it} - 1) + \frac{16}{2 \cdot 16}(e^{it} - 1)^2 + \theta C|e^{it} - 1|^3\right\}$$

$$= \exp\left\{3(e^{it} - 1) + \frac{1}{2}(e^{2it} - 1) + \theta C|e^{it} - 1|^3\right\}.$$

On the other hand, by (1.5)

$$|\widehat{F}(t)| = \exp\left\{-32\sum_{j=1}^{\infty}\frac{0.2^j}{j}\sin^2(tj/2)\right\} \leqslant \exp\{-6.4\sin^2(t/2)\}.$$

A similar estimate holds for $|\widehat{G}(t)|$. Therefore by (3.3)

$$|\widehat{F}^n(t) - \widehat{G}^n(t)| \leqslant Cn|\sin(t/2)|^3\exp\{-6n\sin^2(t/2)\}.$$

It remains to use (4.1) and (1.31).

4.7 By Example (1.8) the characteristic function of $CNB(\gamma, 0.2, F)$ is equal to

$$\exp\left\{\gamma\sum_{j=1}^{\infty}\frac{(\widehat{F}(t) - 1)^j}{j4^j}\right\} = \exp\left\{\gamma\sum_{j=1}^{\infty}\frac{0.2^j(\widehat{F}^j(t) - 1)}{j}\right\}$$

and a similar expression holds for $CNB(\gamma, 0.2, G)$. From (1.45) it follows that

$$|(\widehat{F}(t) - 1)^j - (\widehat{G}(t) - 1)^j| \leqslant j2^{j-1}|\widehat{F}(t) - \widehat{G}(t)|$$

and, therefore,

$$\sum_{j=1}^{\infty}\frac{1}{j4^j}|(\widehat{F}(t) - 1)^j - (\widehat{G}(t) - 1)^j| \leqslant C|\widehat{F}(t) - \widehat{G}(t)| \leqslant C(s)a|\sin(t/2)|^s.$$

Just as in the solution to problem 4.5 we show that

$$Re\widehat{F}(t) - 1,\ Re\widehat{G}(t) - 1 \geqslant -2b\sin^2(t/2).$$

Thus, the characteristic functions of $CNB(\gamma, 0.2, F)$ and $CNB(\gamma, 0.2, G)$ by absolute value are less than $\exp\{0.2\gamma(Re\widehat{F}(t) - 1)\} \leqslant \exp\{-0.4b\gamma \sin^2(t/2)\}$. From (3.3) it follows that the difference of both compound negative binomial characteristic functions is less than

$$C(s)\gamma a|\sin(t/2)|^s \exp\{-0.4b\gamma \sin^2(t/2)\}$$

and it remains to apply (4.1).

Chapter 5

5.1 The left-hand side of inequality is majorized by 2. Therefore it suffices to consider $\lambda \geqslant 1$. We also consider the case $k = 1$ only, since by the properties of the norm

$$\| (I_1 - I)^k \exp\{\lambda(I_1 - I)\} \| \leqslant \| (I_1 - I) \exp\{(\lambda/k)(I_1 - I)\} \|^k.$$

Let $\widehat{M}(t) = (e^{it} - 1) \exp\{\lambda(e^{it} - 1)\}$. Then by (1.48)

$$|\widehat{M}(t)| = 2|\sin(t/2)|\exp\{-2\lambda \sin^2(t/2)\} \leqslant \frac{C}{\sqrt{\lambda}} \exp\{-\lambda \sin^2(t/2)\},$$

$$\left(\widehat{M}(t)e^{-it\lambda}\right)' = ie^{-it\lambda}e^{\lambda(e^{it}-1)}(e^{it}+\lambda(e^{it}-1)^2), \quad |(\widehat{M}(t)e^{-it\lambda})'| \leqslant C\exp\{-\lambda \sin^2(t/2)\}.$$

It remains to apply Lemma 5.1 with $b = \sqrt{\lambda}$, $a = \lambda$.

5.2 We have $\widehat{F}(t) = q + pe^{it}$, $q = 1 - p$. Let $\widehat{M}(t) = \widehat{F}^n(t)(e^{it} - 1)$. By (3.7) $|\widehat{F}(t)| \leqslant \exp\{-2pq \sin^2(t/2)\}$. Therefore, taking into account (1.48), we prove

$$|\widehat{M}(t)| \leqslant 2p|\sin(t/2)|\exp\{-2pq \sin^2(t/2)\} \leqslant C\min\left(p, \sqrt{\frac{p}{n}}\right)\exp\{-npq \sin^2(t/2)\}.$$

Observe that

$$\left(e^{-itnp}\widehat{M}(t)\right)' = e^{-itnp}i\left(\widehat{F}^n(t)pe^{it} + np^2q(e^{it} - 1)^2\widehat{F}^{n-1}(t)\right).$$

Therefore

$$\left|\left(e^{-itnp}\widehat{M}(t)\right)'\right| \leqslant pe^{-2npq \sin^2(t/2)}(1+4npq \sin^2(t/2)e^{2pq}) \leqslant Cp\exp\{-npq \sin^2(t/2)\}.$$

It remains to apply Lemma 5.1 with $b = \max(1, \sqrt{np})$, $a = np$.

5.3 For the sake of brevity set $q = 1 - p$, $A = |\sin(t/2)|$, $\widehat{F} = q + pe^{it}$, $\widehat{G} = \exp\{p(e^{it} - 1) - (p^2/2)(e^{it} - 1)^2\}$, $f = \exp\{-itp\}\widehat{F}$, $g = \exp\{-itp\}\widehat{G}$. From the solution to problem 4.1 we have

$$|\widehat{F}|, |\widehat{G}| \leqslant \exp\{-2pqA^2\}, \quad |\widehat{F} - \widehat{G}| \leqslant Cnp^3A^3,$$

$$|\widehat{M}(t)| = |\widehat{F}^n - \widehat{G}^n| \leqslant Cnp^3A^3 \exp\{-2npqA^2\} \leqslant C\left(np^3, p\sqrt{\frac{p}{n}}\right)\exp\{-npqA^2\}.$$

Observe that

$$f' = ipq(e^{it} - 1)e^{-itp},$$
$$g' = ip(e^{it} - 1)(1 - pe^{it})e^{-itp}\widehat{G} = ip(e^{it} - 1)e^{-itp}(q + \theta p^2 A).$$

Therefore $|f' - g'| \leqslant Cp^3A^2$ and

$$|(e^{-itp}\widehat{M}(t))'| = |(f^n - g^n)'| \leqslant n|f^n|\,|f' - g'| + n|g'|\,|f^{n-1} - g^{n-1}|$$
$$\leqslant C\exp\{-2npqA^2\}(np^3A^2 + n^2p^4A^4) \leqslant C\min(np^3, p^2)\exp\{-npqA^2\}.$$

To complete the proof Lemma 5.1 should be applied with $a = np$, $b = \max(1, \sqrt{np})$.

5.4 The left-hand side of the inequality is majorized by $2 + 2n$. Therefore, without loss of generality, we can assume that $n > 6$. By (1.37) and (2.3)

$$\|F^n - G^n - nG^{n-1}(F - G)\| \leqslant \sum_{m=2}^{n}(m - 1)\|F^{n-m}(F - G)^2G^{m-2}\|$$

$$\leqslant C\sum_{m=2}^{n}(m - 1)\|F^{n-m}(V - I)^4G^{m-2}\|$$

$$\leqslant Cn^2\|F^{\lfloor n/2\rfloor}(V - I)^4\| + Cn^2\|G^{\lfloor n/2\rfloor}(V - I)^4\|.$$

It remains to apply Lemma 5.3.

5.5 Let $F = (1 - 2p)I + pI_1 + pI_{-1}$. Observe that

$$\|F^n(I_1 - I)^2(I_{-1} - I)\| = \|F^n(I_1 - I)^3I_{-1}\| = \|F^n(I_1 - I)^3\|$$

and it remains to use (5.14).

5.6 Let $M(k) = M\{(-\infty, k]\}$. By definition $MI_1(k) = M(k - 1)$ and

$$\|M(I_1 - I)\|_W = \sum_{k=-\infty}^{\infty}|MI_1(k) - M(k)| = \sum_{k=-\infty}^{\infty}|M\{k\}| = \|M\|.$$

Chapter 6

6.1 Taking into account the proof of Theorem 6.1 we obtain

$$|\widehat{M}(t)| \leqslant Cnr^2(t)\exp\{-2nr(t)\} \leqslant \frac{C}{n}\exp\{-nr(t)\},$$

$$|\widehat{M}''(t)| \leqslant C\sigma^2(n^3r^3(t) + n^2r^2(t) + nr(t))\exp\{-2nr(t)\} \leqslant C\sigma^2\exp\{-nr(t)\}.$$

The required estimate follows from (3.2), (6.2) and (6.8).

6.2 Applying the estimates from Theorem's 6.1 proof and (1.48) we obtain

$$|(\widehat{F}^n(\widehat{F} - 1))''|$$
$$= |\widehat{F}''(n\widehat{F}^{n-1}(\widehat{F} - 1) + \widehat{F}^n) + (\widehat{F}')^2(n(n-1)\widehat{F}^{n-2}(\widehat{F} - 1) + 2n\widehat{F}^{n-1})|$$
$$\leqslant C\sigma^2\exp\{-2nr(t)\}(nr(t) + 1 + n^2r^2(t)) \leqslant C\sigma^2\exp\{-nr(t)\}.$$

The proof is completed by application of (6.2) and (6.8).

6.3 It suffices to apply (1.46) to (6.9).

6.4 Let $\widehat{M} = \widehat{F}^n(t) - \widehat{G}^n(t)$, $u(t) = \widehat{M}/(e^{-it} - 1)$. From the solution to problem 5.3 it follows that

$$|\widehat{M}| \leqslant Cnp^3|\sin(t/2)|^3\exp\{-2npq\sin^2(t/2)\},$$

$$|(\widehat{M}e^{-itnp})'| \leqslant C(np^3\sin^2(t/2) + n^2p^4\sin^4(t/2))\exp\{-2npq\sin^2(t/2)\}$$
$$\leqslant Cnp^3\sin^2(t/2)\exp\{-1.5npq\sin^2(t/2)\}.$$

These estimates allow us to prove that

$$|(u(t)e^{-itnp})'| \leqslant Cnp^3\exp\{-npq\sin^2(t/2)\}.$$

The proof now follows from (6.5), (1.31) and the statement of problem 4.1.

Chapter 7

7.1 Let $u = (0, 1, a)$. Then $supp\, F \subset K_1(u)$ and $supp\, F^n \subset K_n(u)$. Taking in Lemma 7.2 $a = 1$, $V = n(F - I)$, $b_2 = 2n$ and $y = 7n$ we get $\delta(\exp\{n(F - I)\}, 7n, u) \leqslant \exp\{-n\}$. It is easy to check that $|\exp\{\widehat{F}(t) - 1\}| = \exp\{Re\widehat{F}(t) - 1\}$. Arguing exactly as in the proof of (3.18) we prove that

$|\widehat{F}(t)| \leqslant \exp\{q_0(Re\widehat{F}(t) - 1)\}$. Next, observe that

$$Re\widehat{F}(t) - 1 = -2(q_1 + q_2)\sin^2(ta/2) - 2q_3\sin^2(t/2),$$

$$|\widehat{F}(t) - 1| = |(q_1 - q_2)(e^{ita} - 1) + q_2(e^{-ita} - 1 + e^{ita} - 1) + q_3(e^{it} - 1)|$$

$$\leqslant C(|q_1 - q_2||\sin(t/2)| + q_2\sin^2(t/2) + q_3|\sin(t/2)|).$$

From (2.3) it follows that $|\widehat{F}(t) - \exp\{\widehat{F}(t) - 1\}| \leqslant |\widehat{F}(t) - 1|^2/2$. Let $\widehat{U}(t) = \exp\{0.5nq_0(Re\widehat{F}(t) - 1)\}$. Then by (1.45) and (1.48)

$$|\widehat{F}^n(t) - \exp\{n(\widehat{F}(t) - 1)\}| \leqslant Cn|\widehat{F}(t) - 1|^2\widehat{U}^2(t) \leqslant C\widehat{U}(t)\epsilon.$$

Let $h = 2a/\pi$ and let $|t| \leqslant 1/h$. Then, taking into account (1.50), we obtain

$$Re\widehat{F}(t) - 1 \leqslant -\frac{2t^2}{\pi^2}(a(q - 1 + q_2) + q_3),$$

$$|\widehat{F}(t) - 1|^2 \leqslant Ct^2(a^2(q_1 - q_2)^2 + q_2^2 + q_3^2).$$

Substituting all above estimates into Lemma 7.1 we get

$$|F^n - \exp\{n(F - I)\}|_K \leqslant C\epsilon(1 + Q(U, h)\ln n) + e^{-n}.$$

Thus, it remains to estimate $Q(U, h)$. Applying (1.20) and (1.22) we obtain

$$Q(U, h) \leqslant (4a/\pi + 1)Q(U, 0.5) \leqslant \frac{Ca}{\sqrt{q_0(1 - q_0)n}}$$

completing the proof.

7.2 Let in Lemma 7.1, $M = (F - I)^k\exp\{n(F - I)\}$, $U = \exp\{(n/2)(F - I)\}$, $\mathbf{u} = (0, x_1, x_2, \ldots, x_N)$, $m = k + 7n$, $h = \min_{1 \leqslant i \leqslant N}|x_i|/2$.
     By taking in Lemma 7.2 $V = n(F - I)$, $b_2 = 2n$, $W = (F - I)^k$, $b_1 = 2^k$, $a = 1$, $s = k$ and $y = 7n$ we prove that

$$\delta(M, k + 7n, \mathbf{u}) \leqslant 2^k e^{-n} \leqslant \frac{C(k)}{n^k}.$$

Next, observe that by (1.48)

$$|\widehat{M}(t)| = (1 - \widehat{F}(t))^k\exp\{(n/2)(\widehat{F}(t) - 1)\}\widehat{U}(t) \leqslant \frac{C(k)}{n^k}\widehat{U}(t).$$

By (1.12) $|\widehat{F}(t) - 1| \leqslant \sigma^2 t^2/2$, where σ^2 is the variance of F. Therefore

$$|\widehat{M}(t)\,||\,t| \leqslant C\sigma\,|\widehat{F}(t) - 1|^{k-1/2}\widehat{U}^2(t) \leqslant \frac{C\sigma}{n^{k-1/2}}\widehat{U}(t)$$

and by (1.24)

$$\int_{|t|<1/h} \frac{|\widehat{M}(t)|}{|t|}\,dt \leqslant \frac{C\sigma}{hn^{k-1/2}}.$$

From (1.22) it follows that

$$Q(U, h) \leqslant \frac{C}{\sqrt{n(1 - F\{0\})}} \leqslant \frac{C(F)}{\sqrt{n}}.$$

It remain to substitute all the estimates into Lemma (7.1) and to note that $\ln m \leqslant C(k) \ln n$, $\ln n / \sqrt{n} \leqslant C$.

Chapter 8

8.1 Let $M = FF^{(-)}$. Then $\widehat{M}(t) = \widehat{F}(t)\widehat{F}(-t) = |\widehat{F}(t)|^2$ and M has density

$$p(y) = \int_{-\infty}^{\infty} f(y + x)f(x)dx.$$

It remains to apply (8.2) for $y = 0$.

8.2 The mean of F is zero and the variance equals to $1/2$. Therefore, for approximation we use $\Phi_{n/2}$. It is not difficult to check that, for $|t| \leqslant \pi$,

$$|\widehat{F}(t)| = \widehat{F}(t) = 1 - \sin^2(t/2) \leqslant \exp\{-\sin^2(t/2)\} \leqslant \exp\{-t^2/\pi^2\},$$

$$\widehat{F}(t) = 1 - \tfrac{t^2}{4} + \theta Ct^4, \quad e^{t^2/4} = 1 - \tfrac{t^2}{4} + \theta Ct^4, |\widehat{F}(t) - e^{-t^2/4}| \leqslant Ct^4.$$

Now we can formulate the analogue of Theorem 8.2.

$$\sup_{m \in \mathbb{N}} \left| F^n\{m\} - \frac{1}{\sqrt{\pi n}}\,e^{-m^2/n} \right| \leqslant \frac{1}{2\pi}\int_{|t|<\pi} |\widehat{F}^n(t) - e^{-nt^2/4}|\,dt + \frac{1}{2\pi}\int_{|t|\geqslant\pi} e^{-nt^2/4}dt$$

$$\leqslant C\int_{-\pi}^{\pi} ne^{-(n-1)t^2/\pi^2}|\widehat{F}(t) - e^{-t^2/4}|\,dt + Ce^{-n}$$

$$\leqslant C\int_{-\pi}^{\pi} nt^4 e^{-nt^2/\pi^2}\,dt + Ce^{-n} \leqslant \frac{C}{n\sqrt{n}}.$$

8.3 We apply (8.5). Then, for $|t| \leqslant \varepsilon$,

$$\widehat{F}(t) \leqslant \exp\{-C_1(F)t^2\}, \quad \widehat{F}(t) = 1 - \frac{\sigma^2 t^2}{2} + \theta C t^4,$$

$$|\widehat{F}^n(t) - e^{-n\sigma^2/2}| \leqslant Cne^{-(n-1)C(F)t^2}|\widehat{F}(t) - e^{-\sigma^2 t^2/2}| \leqslant Cnt^4 e^{-C(F)nt^2}.$$

By (8.6), (8.3) and (8.5)

$$\sup_x \left| f_n(x) - \frac{1}{\sqrt{2\pi n\sigma}} e^{-x^2/2n\sigma^2} \right| \leqslant \frac{1}{2\pi} \int_{|t| \leqslant \varepsilon} |\widehat{F}^n(t) - e^{-n\sigma^2 t^2/2}|\, dt$$

$$+ \frac{1}{2\pi} \int_{|t| > \varepsilon} \left(|\widehat{F}(t)|^2 e^{-(n-2)C_2(F)} + e^{-n\sigma^2 t^2} \right) dt$$

$$\leqslant C \int_{\mathbb{R}} nt^4 e^{-nC(F)t^2}\, dt + C(F)e^{-nC(F)} \leqslant C(F)n^{-3/2}.$$

8.4 In the solution of the previous problem (8.6) must be replaced by (8.10).
8.5 Let $M = F^n(F - I)$. Without loss of generality we assume that $n > 4$, since $\|M\| \leqslant 2$. By (8.5) there exists an $\varepsilon = \varepsilon(F)$ such that

$$|\widehat{F}(t)^n| \leqslant \begin{cases} \exp\{-nC(F)t^2\}, & \text{if } |t| \leqslant \varepsilon, \\ \exp\{-n\tilde{C}(F)\}, & \text{if } |t| > \varepsilon. \end{cases}$$

Moreover, by (1.12) and (1.13) and (8.13)

$$|\widehat{F}(t) - 1| \leqslant \frac{\sigma^2 t^2}{2}, \quad |\widehat{F}'(t)| \leqslant \sigma^2 |t|, \quad |\widehat{F}'(t)| \leqslant C(F).$$

Then

$$M' = n\widehat{F}^{n-1}(t)\widehat{F}'(t)(\widehat{F}(t) - 1) + \widehat{F}^n(t)\widehat{F}'(t)$$

and, for $|t| \leqslant \varepsilon$,

$$|\widehat{M}(t)| \leqslant C(F)t^2 \exp\{-nC(F)t^2\} \leqslant \frac{C(F)}{n} \exp\{-nC(F)t^2\},$$

$$|\widehat{M}'(t)| \leqslant C(F)(n|t|^3 + |t|) \exp\{-(n-1)C(F)t^2\} \leqslant \frac{C(F)}{\sqrt{n}} \exp\{-nC(F)t^2\}.$$

We recall that in the above estimates the same symbol $C(F)$ is used for different constants depending on F. Now let $|t| > \varepsilon$. Then

$$|\widehat{M}(t)| \leqslant \widehat{2F}(t)^2 \exp\{-(n-2)\tilde{C}(F)\} \leqslant \frac{C(F)}{n}\widehat{F}(t)^2,$$

$$|\widehat{M}'(t)| \leqslant C(F)\widehat{F}(t)^2(n\exp\{-(n-3)\tilde{C}(F)\} + \exp\{-n\tilde{C}(F)\}) \leqslant \frac{C(F)}{n}\widehat{F}(t)^2.$$

Applying (8.11) with $b = \sqrt{n}$ we obtain

$$2\|M\|^2 \leqslant \frac{C(F)}{n^2}\int_{|t|\leqslant\varepsilon}\sqrt{n}\exp\{-nC(F)t^2\}dt + \frac{C(F)}{n^2}\int_{|t|>\varepsilon}\widehat{F}^2(t)dt \leqslant \frac{C(F)}{n^2}.$$

8.6 Without loss of generality we assume that $n > 1$. Observe that $0 \leqslant \widehat{F}_1(t) \leqslant \widehat{F}_2(t) \leqslant \cdots \leqslant \widehat{F}_n(t)$. Let $|t| \leqslant \lambda_1/2$.

$$\widehat{F}_i(t) = \exp\{-\ln(1 + t^2/\lambda_i^2)\} = \exp\left\{-\frac{t^2}{\lambda_i^2} + \frac{t^2}{\lambda_i^2}\sum_{j=2}^{\infty}\frac{(-1)^j}{j}\left(\frac{t^2}{\lambda_i^2}\right)^j\right\}$$

$$\leqslant \exp\left\{-\frac{t^2}{\lambda_i^2} + \frac{t^2}{\lambda_i^2}\sum_{j=2}^{\infty}\frac{(-1)^j}{j}\left(\frac{1}{4}\right)^j\right\} \leqslant \exp\left\{-\frac{5t^2}{6\lambda_i^2}\right\} \leqslant \exp\left\{-\frac{5t^2}{6\lambda_n^2}\right\}$$

and

$$\widehat{F}_i(t) = 1 - \frac{t^2}{\lambda_i^2} + \theta Ct^4, \quad |\widehat{F}_i(t) - \exp\{-t^2/\lambda_i^2\}| \leqslant Ct^2, \quad \exp\left\{-\frac{t^2}{\lambda_i^2}\right\} \leqslant \exp\left\{-\frac{5t^2}{6\lambda_n^2}\right\}.$$

Let $M = \prod_{i=1}^{n}F_i - \Phi_\sigma$. Then by (1.40)

$$|\widehat{M}(t)| \leqslant \sum_{i=1}^{n}|\widehat{F}_i(t) - \exp\{-t^2/\lambda_i^2\}|\prod_{j=1}^{i-1}\widehat{F}_i(t)\prod_{j=i+1}^{n}\exp\{-t^2/\lambda_i^2\} \leqslant Cnt^4\exp\left\{-\frac{5nt^2}{6\lambda_n^2}\right\}.$$

Thus,

$$\int_{|t|\leqslant\lambda_1/2}\frac{|\widehat{M}(t)|}{|t|}dt \leqslant Cn\int_{\mathbb{R}}|t|^3\exp\left\{-\frac{5nt^2}{6\lambda_n^2}\right\}dt \leqslant \frac{C}{n}.$$

Now let $|t| > \lambda_1/2$. Then

$$\prod_{i=1}^{n}\widehat{F}_i(t) \leqslant \widehat{F}_n^n(t) \leqslant \widehat{F}_n(t)\left(\frac{\lambda_n^2}{\lambda_n^2 + \lambda_1^2/4}\right)^{n-1} \leqslant \widehat{F}_n(t)\frac{C}{n}$$

and

$$\int_{|t|>\lambda_1/2} \frac{\widehat{M}(t)}{|t|} dt \leqslant \frac{C}{n} \int_{\mathbb{R}} \frac{\widehat{F}_n(t)}{\lambda_1/2} dt \leqslant \frac{C}{n}.$$

It remains to apply (8.10).

Chapter 9

9.2 We have $F^{2n}\{(-\infty,0)\} = F^{2n}\{(0,\infty)\}$ and $F^{2n}\{(-\infty,0)\} + F^{2n}\{(0,\infty)\} + F^{2n}\{0\} = 1$. Therefore, taking into account (1.43), we obtain

$$|F^{2n} - \Phi_{2n}|_K \geqslant |F^{2n}\{(-\infty,0)\} - \Phi_{2n}(0)| = |F^{2n}\{(-\infty,0)\} - 1/2|$$

$$= \frac{1}{2} F^{2n}\{0\} = \frac{1}{2} \binom{2n}{n} \frac{1}{2^{2n}} \approx \frac{1}{2\sqrt{n\pi}}.$$

9.3 Let $f(x), g(x)$ denote the densities of F and G. By (8.10)

$$|F - G|_K \leqslant \frac{1}{2\pi} \int_{-T}^{T} \frac{|\widehat{F}(t) - \widehat{G}(t)|}{|t|} dt + \frac{1}{2\pi} \int_{|t|>T} \frac{\widehat{F}(t) + \widehat{G}(t)}{|t|} dt.$$

By the inversion formula

$$\frac{1}{2\pi} \int_{-\infty}^{\infty} \widehat{F}(t) dt = f(0) \leqslant A, \quad \frac{1}{2\pi} \int_{-\infty}^{\infty} \widehat{G}(t) dt = g(0) \leqslant A$$

and, therefore,

$$\frac{1}{2\pi} \int_{|t|>T} \frac{\widehat{F}(t) + \widehat{G}(t)}{|t|} dt \leqslant \frac{1}{2\pi T} \int_{|t|>T} (\widehat{F}(t) + \widehat{G}(t)) dt \leqslant \frac{2A}{T}.$$

9.4 For the sake of brevity we omit the dependency of constants from F and the dependency of characteristic functions from t. By (9.10), (9.16) and (9.17) there exists an ε such that, for $|t| \leqslant \varepsilon$, $\max(|\widehat{F}|, |\exp\{\widehat{F} - 1\}|) \leqslant \exp\{-Ct^2\}$. If $|t| > \varepsilon$, then $\max(|\widehat{F}|, |\exp\{\widehat{F} - 1\}|) \leqslant \exp\{-C_1\}$. By (9.13) $|\widehat{F} - 1| \leqslant Ct^2$. Therefore, for $|t| \leqslant \varepsilon$,

$$|\widehat{F}^n - \exp\{n(\widehat{F} - 1)\}| \leqslant Cn \exp\{-Cnt^2\}|\widehat{F} - 1|^2 \leqslant Cnt^4 \exp\{-Cnt^2\}.$$

Similarly to the proof of (9.12) we obtain

$$\int_{-T}^{T} \frac{|\widehat{F^n} - e^{n(\widehat{F}-1)}|}{|t|} dt \leq 2 \int_{0}^{\varepsilon} \frac{|\widehat{F^n} - e^{n(\widehat{F}-1)}|}{|t|} dt + CTe^{-Cn}$$

$$\leq Cn \int_{0}^{\varepsilon} t^3 e^{-Cnt^2} dt + CTe^{-Cn} \leq \frac{C}{n} + CTe^{-Cn}.$$

By (9.11)

$$\frac{1}{T} \int_{0}^{T} |\widehat{F}|^n dt \leq C\left(\frac{1}{T} + e^{-Cn}\right)$$

and by exactly the same argument a similar estimate can be proved for $\exp\{n(\widehat{F} - 1)\}$. It remains to apply (9.5) with $T = n$.

9.5 We use decomposition (9.22) with $p = a^{-1/2}$. Then

$$| (FI_u - I) \exp\{a(FI_u - I)\} |_K$$
$$\leq | (A - I) \exp\{a(FI_u - I)\} |_K + | p(V - I) \exp\{a(FI_u - I)\} |_K.$$

From Lemma 2.3 it follows that

$$|p(V - I) \exp\{a(FI_u - I)\} |_K \leq \| p(V - I) \exp\{ap(V - I)\} \| \leq \frac{Cp}{\sqrt{ap}} = Ca^{-3/4}.$$

We apply (9.3) with $W = (A - I) \exp\{aq(A - I)/2\}$ and $H = \exp\{a(FI_u - I)/2\}$. Taking into account (9.24), (9.25) and (9.26) we prove

$$| (A - I) \exp\{a(FI_u - I)\} |_K \leq | WH |_K \leq | C |_K \int_{|t| \leq 1/h} \sigma^2 |t| \exp\{-a\sigma^2 t^2/3\} dt$$

$$+ C\| W \| (ap)^{-1/2} \leq \frac{C}{a} + \frac{C}{\sqrt{a}} \cdot \frac{1}{a^{1/4}} \leq \frac{C}{a^{3/4}}.$$

9.6 We use decomposition (9.22) with $p = n^{-1/3}$. Then

$$| (FI_u)^{n+1} - (FI_u)^n |_K \leq | (FI_u - I)[(FI_u)^n - \exp\{n(FI_u - I)\}] |_K$$
$$+ | (FI_u - I) \exp\{n(FI_u - I)\} |_K.$$

Observe that by the proof of Theorem 9.4

$$| (FI_u - I)[(FI_u)^n - e^{n(FI_u - I)}] |_K \leq \| FI_u - I \| | (FI_u)^n - e^{n(FI_u - I)} |_K \leq Cn^{-1/3}.$$

Repeating the proof of the previous problem with $a = n$, $p = n^{-1/3}$, we also show that $| (FI_u - I) \exp\{n(FI_u - I)\} |_K \leq Cn^{-1/3}$.

Chapter 10

10.1 We have

$$|\widehat{M}(t)| = \left| \int_{\mathbb{R}} e^{itx} M\{dx\} \right| \leqslant \int 1|M\{dx\}| = \|M\|.$$

10.2 By (1.49) $(1 + 1/n)^n \leqslant e$. Therefore

$$|\widehat{F}^n(\lambda/\sqrt{n})(\widehat{F}(\lambda/\sqrt{n}) - 1)| = \left(1 + \frac{1}{n}\right)^{-n} \frac{1}{n(1 + 1/n)} \geqslant \frac{1}{2en}.$$

10.3 Let $b = \sqrt{hn}/\lambda^2$. Here $h \geqslant \lambda^4$ and will be chosen later. Observe that $F \in \mathcal{F}_s$ and its variance is $2n/\lambda^2$. Therefore from (1.12) it follows that $\widehat{F}^n(t/b) = 1 + \theta C t^2/h$. Similarly,

$$\widehat{F}(t) - 1 = \frac{t^2}{hn} + \frac{t^4}{h^2 n^2} \cdot \frac{1}{1 + t^2/hn} = \frac{t^2}{hn} + \frac{t^4}{h^2 n^2} \theta C$$

and

$$\widehat{M}(t/b) = \left(1 + \theta C \frac{t^2}{h}\right)\left(\frac{t^4}{h^2 n^2} + \theta \frac{C(t^6 + t^8)}{h^3 n^3}\right) = \frac{t^4}{h^2 n^2} + \theta C \frac{t^6 + t^8 + t^{10}}{h^3 n^2}.$$

It remains to apply Lemma 10.2 and to choose a sufficiently large constant h.

10.4 Let μ denote the mean of F. Then by (1.12)

$$\widehat{F}(t) - 1 = it\mu + \theta C t^2, \quad \exp\{n(\widehat{F}(t) - 1 - it\mu)\} = 1 + \theta C n t^2.$$

Therefore

$$(\widehat{F}(t) - 1) \exp\{n(\widehat{F}(t) - 1 - it\mu)\} = it\mu + \theta C t^2 + \theta C n |t|^3.$$

It remains to apply Lemma 10.3 with $j = 2$, $a = \mu$, $b = h\sqrt{n}$ and to choose a sufficiently large absolute constant h.

10.5 Let $G = \exp\{F - I\}$ and let σ^2 denote the variance of F. For the sake of brevity we omit the dependence on t. In the proof of 10.2 it was shown that

$$|\widehat{F} - 1| \leqslant C t^2, \quad |\widehat{F} - 1 + \sigma^2 t^2/2| \leqslant C t^4.$$

Moreover,

$$|\widehat{G}^{n-1} - 1| \leqslant n C t^2, \quad |\widehat{F} - \widehat{G}| \leqslant |\widehat{F} - 1|^2 \leqslant C t^4$$

and by (1.37)

$$
\begin{aligned}
\widehat{F}^n - \widehat{G}^n &= n\widehat{G}^{n-1}(\widehat{F} - \widehat{G}) + \theta Cn^2|\widehat{F} - 1|^4 \\
&= n(\widehat{F} - \widehat{G}) + n\theta|\widehat{G}^{n-1} - 1||\widehat{F} - 1|^2 + \theta Cn^2|\widehat{F} - 1|^3 \\
&= -n\frac{(\widehat{F} - 1)^2}{2} + \theta Cn|\widehat{F} - 1|^3 + \theta Cn^2 t^6 = -n\frac{(\widehat{F} - 1)^2}{2} + C\theta n^2 t^6 \\
&= -\frac{n\sigma^4 t^4}{8} + \theta Cn^2(t^6 + t^8).
\end{aligned}
$$

The proof is completed by putting $b = h\sqrt{n}\sigma$ in Lemma 10.10, $j = 1$ and choosing a sufficiently large absolute constant h.

Chapter 11

11.1 Let $p_j := P(\eta = j)$. Since $\Gamma(x + 1) = x\Gamma(x)$ we get the following recursion $(j + 1)p_{j+1} = (\gamma + j)p_j, j \in \mathbb{Z}_+$.

11.2 Let $G_n(z) = (q + pz)^N = \sum_{k=0}^N p_k z^k$. Then, formally,

$$
G'_n(z) = Np(q+pz)^{N-1} = Np(q+pz)^N/(q+pz) = \frac{Np}{q+pz}\sum_{k=0}^N p_k z^k = \sum_{k=0}^N kp_k z^{k-1}p.
$$

Consequently,

$$
Np\sum_{k=0}^N p_k z^k = q\sum_{k=0}^N kp_k z^{k-1} + p\sum_{k=0}^N kp_k z^{k-1}
$$

and $p(N-k)p_k = (k+1)qp_{k+1}, (k = 0, 1, \ldots, N)$. Since $g(k) = 0$, for $k > N$, it is easy to check that (11.5) holds.

11.3 One can directly check that $P(\eta = j + 1) = qP(\eta = j), j \in \mathbb{N}$. This fact leads to the first Stein operator. On the other hand, the geometrical distribution is a special case of the negative binomial distribution and (11.15) can be applied with $\gamma = 1$.

11.4 Combining (11.11) with the proof of Theorem 11.2 we obtain

$$
\mathbb{E}(\mathcal{A}g)(S) = \sum_{i=1}^n p_i\Big(\mathbb{E}g(S + 1) - \mathbb{E}g(S^i + 1) - p_i\mathbb{E}\,\Delta g(S + 1) + p_i^2\mathbb{E}\,\Delta^2 g(S + 1)\Big)
$$

$$
= \sum_{i=1}^n p_i^3(-\mathbb{E}\,\Delta^2 g(S^i + 1) + \mathbb{E}\,\Delta^2 g(S + 1)) = \sum_{i=1}^n p_i^4\mathbb{E}\,\Delta^3 g(S^i + 1).
$$

We can assume that $\lambda > 8$, since $\| \mathbb{E} \, \Delta^3 g(S^i + 1) \| \leqslant C$. Then similarly to Example 11.1 it can be shown that

$$| \mathbb{E} \, \Delta^3 g(S^i + 1) | \leqslant \| \Delta g \|_\infty \| \mathcal{L}(S^i)(I_1 - I)^2 \| \leqslant C\lambda^{-2}.$$

11.5 Observe that

$$(q + pz)^n = \exp\left\{ \sum_{j=1}^{\infty} \frac{N(-1)^{j+1}}{j} \left(\frac{p}{q} \right)^j (z^j - 1) \right\}.$$

Then

$$\upsilon = \frac{p}{(1 - 2p)^2} < \frac{1}{2},$$

and Lemma 11.5 can be applied.

11.6 Observing that $nq = \sum_1^n p_i p$ we obtain

$$\mathbb{E} \, (\mathcal{A}g)(S) = \mathbb{E} \, \{ nqg(S + 1) + qSg(S + 1) - Sg(S) \}$$

$$= \sum_{i=1}^n p_i \{ p\mathbb{E} \, g(S + 1) + q\mathbb{E} \, g(S^i + 2) - \mathbb{E} \, g(S^i + 1) \}$$

$$= \sum_{i=1}^n p_i \{ p(\mathbb{E} \, g(S + 1) - \mathbb{E} \, g(S^i + 1)) + q\mathbb{E} \, \Delta g(S^i + 1) \}$$

$$= \sum_{i=1}^n p_i \{ p((1 - p_i)\mathbb{E} \, g(S^i + 1) + p_i\mathbb{E} \, g(S^i + 2) - \mathbb{E} \, g(S^i + 1))$$

$$+q\mathbb{E} \, \Delta g(S^i + 1) \} = \sum_{i=1}^n p_i(pp_i + q)\mathbb{E} \, \Delta g(S^i + 1).$$

Now Lemma 11.2 can be applied.

11.7 We will apply (11.40) with $b = 20/\lambda$. Then, on one hand,

$$\lambda(2e^{-2/3} + be^{-1}) \leqslant 2\lambda + 20e^{-1} \leqslant 10.$$

On the other hand,

$$3\lambda + 7 \leqslant 10, \quad 1 - (3\lambda + 7)/(b\lambda) \geqslant 0.5.$$

Therefore

$$10\| \mathcal{L}(S) - \mathcal{L}(\eta) \| \geqslant 0.5 \sum_{j=1}^n p_j^2.$$

Chapter 13

Set $z = e^{it} - 1$, $a = p(1-p)$. We also write h_k instead of $h_k(t)$.

13.1 We use induction. Equality obviously holds for $k = 1$. Let us assume that equality holds for $k = 2, \ldots, m$. Then

$$\widehat{\mathbb{E}}\,(e^{it\eta_1} - 1, \ldots, e^{it\eta_{m+1}} - 1) = 0 - \sum_{j=1}^{m} (-1)^{j+1} a^j \mathbb{E}\,(e^{it\eta_{j+1}} - 1) \cdots (e^{it\eta_{m+1}} - 1)$$

$$= -(-1)^{m+1} a^m \mathbb{E}\,(e^{it\eta_{m+1}} - 1) = (-1)^{m+2} a^{m+1}.$$

13.2 The proof is divided into 5 steps.
Step 1. We will prove by induction that

$$|h_k - 1 - az| \leqslant 2a^2|z|^2, \quad |h_k - 1| \leqslant \frac{1}{50}, \quad \frac{1}{|h_k|} \leqslant \frac{50}{49}. \tag{14.4}$$

Observe that the last two estimates follow from the first one. The estimate is trivial for $k = 1$. Let us assume that (14.4) holds for $k = 2, \ldots, m$. Then

$$|h_k - 1 - az| \leqslant \sum_{j=1}^{k-1} \frac{|z|^{k+1-j} a^{k+1-j}}{|h_j \cdots h_{k-1}|} \leqslant a^2 |z|^2 \frac{50}{49} \sum_{j=1}^{k-1} \left(2a \cdot \frac{50}{49}\right)^{k-j-1},$$

$$\leqslant a^2 |z|^2 \frac{59}{49} \cdot \frac{1}{0.986} \leqslant 2a^2 |z|^2.$$

Step 2. Applying Taylor's expansion and (14.4) we get

$$|h_k - e^{az}| \leqslant |h_k - 1 - az| + |1 + az - e^{az}| \leqslant Ca^2 |z|^2. \tag{14.5}$$

Step 3. Obviously, $|\exp\{az\}| = \exp\{-2a\sin^2(t/2)\}$. Similarly to the proof of Lemma 13.8 we then prove

$$|h_k| \leqslant |1 + az| + |h_k - 1 - az| \leqslant 1 - 2a(1-5a)\sin^2\frac{t}{2} \leqslant \exp\left\{-1.9a\sin^2\frac{t}{2}\right\}.$$

Step 4. Applying the results from previous steps we get

$$\left|\prod_{k=1}^{n} h_k - \exp\{naz\}\right| \leqslant \exp\{-1.9(n-1)a\sin^2(t/2)\} \sum_{k=1}^{n} |h_k - e^{az}|$$

$$\leqslant Cna^2 \sin^2(t/2) \exp\{-1.9na\sin^2(t/2)\}.$$

Step 5. It remains to apply (3.2) and (1.31).

Bibliography

1. Aleškevičienė, A.K. and Statulevičius, V.A. (1998). Inversion formulas in the case of a discontinuous limit law. *Theory Probab. Appl.* **42**(1), 1–16.
2. Arak, T.V. (1980). Approximation of n-fold convolutions of distributions having a non-negative characteristic functions with accompanying laws. *Theory Probab. Appl.* **25**(2), 221–243.
3. Arak, T.V. (1981). On the convergence rate in Kolmogorov's uniform limit theorem. I. *Theory Probab. Appl.* **26**(2), 219–239.
4. Arak, T.V. (1981). On the convergence rate in Kolmogorov's uniform limit theorem. II. *Theory Probab. Appl.* **26**(3), 437–451.
5. Arak, T.V. and Zaïtsev, A.Yu. (1988). *Uniform limit theorems for sums of independent random variables.* Proc. Steklov Inst. Math. **174**, 1–222.
6. Barbour, A.D. and Čekanavičius, V. (2002). Total variation asymptotics for sums of independent integer random variables. *Ann. Probab.* **30**, 509–545, 2002.
7. Barbour, A.D. and Chen, L.H.Y. (eds) (2005). *An introduction to Stein's method.* IMS Lecture Note Series **4**, World Scientific Press, Singapore.
8. Barbour, A.D. and Chen, L.H.Y. (2014). Stein's (magic) method. *arXiv:1411.1179v1 [math.PR].*
9. Barbour, A.D. and Chryssaphinou, O. (2001). Compound Poisson approximation: a user's guide. *Ann. Appl. Prob.* **11**, 964–1002.
10. Barbour, A.D. and Hall, P. (1984). On the rate of Poisson convergence. *Math. Proc. Camb. Phil. Soc.* **95**, 473–480.
11. Barbour, A.D. and Jensen, J.L. (1989). Local and tail approximations near the Poisson limit. *Scand. J. Statist.* **16**, 75–87.
12. Barbour, A.D. and Nietlispach, B. (2011). Approximation by the Dickman distribution and quasi-logarithmic combinatorial structures. *Electr. J. Probab.* **16**, 880–902.
13. Barbour, A.D. and Utev, S. (1998). Solving the Stein equation in compound Poisson approximation. *Adv. Appl. Prob.* **30**, 449–475.
14. Barbour, A.D. and Xia, A. (1999). Poisson perturbations, *ESAIM: Probab. Statist.* **3**, 131–150.
15. Barbour, A.D., Čekanavičius, V. and Xia, A. (2007). On Stein's method and perturbations. *ALEA,* **3**, 31–53.
16. Barbour, A.D., Holst, L. and Janson, S. (1992). *Poisson approximations.* Oxford. Clarendon Press.
17. Bentkus, V. (2003). A new method for approximations in probability and operator theories. *Lithuanian Math. J.* **43**, 367–388.
18. Bentkus, V. (2004). On Hoeffding's inequalities. *Ann. Probab.* **32**(2), 1650–1673.
19. Bentkus, V. and Götze, F. (1997). On the lattice point problem for ellipsoids. *Acta Arith.* **80**(2), 101–125.
20. Bentkus, V. and Paulauskas, V. (2004). Optimal error estimates in operator-norm approximations of semigroups. *Letters in Mathematical Physics.* **68**, 131–138.
21. Bobkov, S.G., Chistyakov, G.P. and Götze, F. (2011). Non-uniform bounds in local limit theorems in case of fractional moments I. *Mathematical Methods of Statistics.* **20**(3), 171–191.
22. Bobkov, S.G., Chistyakov, G.P. and Götze, F. (2011). Non-uniform bounds in local limit theorems in case of fractional moments II. *Mathematical Methods of Statistics.* **20**(4), 269–287.
23. Bentkus, V. and Sunklodas, J. (2007). On normal approximations to strongly mixing random fields. *Publ. Math. Debrecen.* **70**(3-4), 253–270.
24. Bergström, H. (1944). On the Central Limit Theorem. *Skand. Aktuarietidskr.* **27**, 139–153.
25. Bergström, H. (1951). On asymptotic expansion of probability functions. *Skand. Actuarietidskr.* **1**, 1–34.

26. Bergström, H. (1970). A comparison method for distribution functions of sums of independent and dependent random variables. *Theory Probab. Appl.* **15**(3), 430–457.

27. Borisov, I.S. and Ruzankin, P.S. (2002). Poisson approximation for expectations of unbounded functions of independent random variables. *Ann. Probab.* **30**(4), 1657–1680.

28. Borisov, I.S. and Vorozheikin, I.S. (2008). Accuracy of approximation in the Poisson theorem in terms of the χ^2-distance. *Siberian Math. J.* **49**(1), 5–17.

29. Brown, T.C. and Xia, A. (2001). Stein's method and birth-death processes. *Ann. Probab.* **29**, 1373–1403.

30. Čekanavičius, V. (1989). Approximation with accompanying distributions and asymptotic expansions I. *Lithuanian Math. J.* **29**, 75–80.

31. Čekanavičius, V. (1991). An approximation of mixtures of distributions. *Lithuanian Math. J.* **31**, 243–257.

32. Čekanavičius, V. (1993). Non-uniform theorems for discrete measures. *Lithuanian Math. J.* **33**, 114–126.

33. Čekanavičius, V. (1995). On smoothing properties of compound Poisson distributions. *Lithuanian Math. J.* **35**, 121–135.

34. Čekanavičius, V. (1997). Asymptotic expansions for compound Poisson measures. *Lithuanian Math. J.* **37**, 426–447.

35. Čekanavičius, V. (1998). Estimates in total variation for convolutions of compound distributions. *J. London Math. Soc. (2)* **58**, 748–760.

36. Čekanavičius, V. (1999). On compound Poisson approximations under moment restrictions. *Theor. Probab. Appl.* **44**(1), 74–86.

37. Čekanavičius, V. (2003). Infinitely divisible approximations for discrete nonlattice variables. *Adv. Appl. Probab.* **35**(4), 982–1006.

38. Čekanavičius, V. (2003). Estimates of pseudomoments via the Stein-Chen method. *Acta Applicandae Mathematicae.* **78**, 61–71.

39. Čekanavičius, V. and Elijio, A. (2005). Lower-bound estimates for Poisson type approximations. *Lithuanian Math. J.* **45**(4), 405–423.

40. Čekanavičius,V. and Elijio, A. (2014). Smoothing effect of compound Poisson approximations to the distributions of weighted sums, *Lithuanian Math. J.* **54**(1), 35–47.

41. Čekanavičius, V. and Roos, B. (2004). Two-parametric compound binomial approximations. *Lithuanian Math. J.* **44**(4), 354–373.

42. Čekanavičius,V. and Roos, B. (2006). An expansion in the exponent for compound binomial approximations. *Lithuanian Math. J.* **46**(1), 54–91.

43. Čekanavičius, V. and Roos, B. (2006). Compound binomial approximations. *Annals of the Institute of Stat. Math.* **58**(1), 187–210.

44. Čekanavičius, V. and Roos, B. (2009). Poisson type approximations for the Markov binomial distribution. *Stochastic Proc. Appl.* **119**, 190–207.

45. Čekanavičius, V. and Vellaisamy, P. (2015). A compound Poisson convergence theorem for sums of m-dependent variables, *ALEA, Lat. Am. J. Probab. Math. Stat.*, **12**, 765–792.

46. Čekanavičius, V. and Wang, Y.H. (2003). Compound Poisson approximations for sums of discrete non-lattice variables. *Adv. Appl. Prob.* **35**(1), 228–250.

47. Chen, L.H.Y. (1975). Poisson approximation for dependent trials. *Ann. Probab.* **3**, 534–545.

48. Chen, L.H.Y. (1998). Stein's method: some perspectives with applications. *Probability towards 2000* (Eds. L. Accardi and C. Heyde), 97–122. Lecture Notes in Statistics No. 128.

49. Chen, L.H.Y. and Röllin A. (2010). Stein couplings for normal approximation. *arXiv:1003.6039v2 [math.PR]*.

50. Chen, L.H.Y. and Röllin A. (2013). Approximating dependent rare events. *Bernoulli.* **19**(4), 1243–1267.

51. Chen, L.H.Y., Goldstein, L. and Shao, Q.-M. (2011). *Normal Approximation by Stein's Method*. Springer, Heidelberg.

52. Daly, F., Lefevre, C. and Utev, S. (2012). Stein's method and stochastic orderings. *Adv. Appl. Prob.* **44**, 343–372.

53. Daly, F. (2013). Compound Poisson approximation with association or negative association via Stein's method. *Electronic Communications in Probability* **18**(30), 1–12.
54. Deheuvels, P. and Pfeifer, D. (1988). On a relationship between Uspensky's theorem and Poisson approximations. *Annals of the Institute of Stat. Math.* **40**, 671–681.
55. Elijio, A. and Čekanavičfus, V. (2015). Compound Poisson approximation to weighted sums of symmetric discrete variables, *Ann. Inst. Stat. Math.* **67**, 195–210.
56. Eliseeva, Yu. S. and Zaitsev, A. Yu. (2012). Estimates for the concentration function of weighted sums of independent random variables. *Theory Probab. Appl.* **57**, 768–777.
57. Eliseeva, Yu. S. and Zaitsev, A. Yu. (2014). On the Littlewood-Offord problem. *Probability and statistics. Part 21, Zap. Nauchn. Sem. POMI.* **431**, 72–81. (In Russian).
58. Eliseeva, Yu. S., Götze, F. and Zaitsev, A. Yu. (2015). Estimates for the concentration functions in the Littlewood-Offord problem. *Journal of Mathematical Sciences.* **206**(2), 146–158.
59. Esseen, C. G. (1945). Fourier analysis of distribution functions. A mathematical study of the Laplace-Gaussian law. *Acta. Math.* **77**, 1–125.
60. Fan, X., Grama, I. and Liu, Q. (2015). Sharp large deviation results for sums of independent random variables. *Sci. China Math.* **58**(9), 1939–1958.
61. Formanov, Sh.K. (2002). The Stein-Tikhomirov method and a nonclassical central limit theorem. *Mathematical.Notes.* **71**(4), 550–555.
62. Formanov, Sh.K. (2007). On the Stein-Tikhomirov method and its applications in nonclassical limit theorems. *Discrete Mathematics and Applications.* **17**(1), 23–36.
63. Franken, P. (1964). Approximation der Verteilungen von Summen unabhängiger nichtnegativen ganzzahliger Zufallsgrösen durch Poissonsche Verteilungen. *Math. Nachr.* **27**, 303–340.
64. Gamkrelidze, N.G. (1982). On the smoothing of probabilities for the sums of integer valued random variables. *Theor. Probab. Appl.* **26**(4), 823–828.
65. Gaponova, M.O. and Shevtsova, I. G. (2009). Asymptotic estimates of the absolute constant in the Berry-Esseen inequality for distributions with unbounded third moment. *Inform. Primen.* **3**(4), 41–56 (In Russian).
66. Goldstein, L. and Reinert, G. (1997). Stein's method and the zero bias transformation with applications to simple random sampling. *Annals Appl. Probab.* **7**, 935–952.
67. Götze, F. and Zaïtsev, A.Yu. (2006). Approximation of convolutions by accompanying laws without centering. *J. Math. Sciences.* **137**(1), 4510–4515.
68. Grigelionis, B. (1999). Asymptotic expansions in the compound Poisson limit theorem. *Acta Applicandae Mathematicae.* **58**, 125–134.
69. Heinrich, L. (1982). Factorization of the characteristic function of a sum of dependent random variables. *Lithuanian Math. J.* **22**(1), 92–100.
70. Heinrich, L. (1982). A method for the derivation of limit theorems for sums of m-dependent random variables. *Z. Wahrscheinlichkeitstheorie verw. Gebiete.* **60**, 501–515.
71. Heinrich, L. (1982). Infinitely divisible distributions as limit laws for sums of random variables connected in a Markov chain. *Math. Nachr.* **107**, 103–121.
72. Heinrich, L. (1985). Some estimates of the cumulant-generating function of a sum of m-dependent random vectors and their application to large deviations. *Math. Nachr.* **120**, 91–101.
73. Heinrich, L. (1987). A method for the derivation of limit theorems for sums of weakly dependent random variables: a survey. *Optimization.* **18**, 715–735.
74. Hipp, C. (1985). Approximation of aggregate claims distributions by compound Poisson distributions. *Insurance Math. Econom.* **4**, 227–232. Correction note: **6**, 165 (1987).
75. Hipp, C. (1986). Improved approximations for the aggregate claims distribution in the individual model. *ASTIN Bull.* **16**, 89–100.
76. Hsiu, K.T. and Chuen, T.S. (2009). Variance inequalities using first derivatives. *Statist. Probab. Lett.* **779**, 1277–1281.
77. Hwang, H.-K. (1999). Asymptotics of Poisson approximation to random discrete distributions: an analytic approach. *Adv. Appl. Prob.* **31**, 448–491.
78. Hwang, H.-K. and Janson, S. (2008). Local limit theorems for finite and infinite urn models. *Annals of Probability.* **36**(3), 992–1022.

79. Ibragimov, I.A. and Linnik Yu.V. (1971). *Independent and stationary sequences of random variables*. Groningen: Wolters-Noordhoff.

80. Ibragimov, I.A. and Presman, E.L. (1974). On the rate of approach of the distributions of sums of independent random variables to accompanying distributions. *Theory Probab. Appl.* **18**(4), 713–727.

81. Ibragimov, R. and Sharakhmetov, Sh. (2002). The exact constant in the Rosenthal inequality for random variables with mean zero. *Theory Probab. Appl.* **46**(1), 127–132.

82. Ito, K. et al. (2000). *Encyclopedic Dictionary of Mathematics: The Mathematical Society of Japan*. **1**. 2nd ed. The MIT Press.

83. Karymov, D.N. (2004). On the accuracy of approximation in the Poisson limit theorem. *Discrete Mathematics and Applications*. **14**(3), 317–327.

84. Kerstan, J. (1964). Verallgemeinerung eines Satzes von Prochorow und Le Cam. *Z. Wahrscheinlichkeitstheorie und Verw. Gebiete* **2**, 173–179.

85. Kolmogorov, A.N. (1956). Two uniform limit theorems for sums of independent random variables. *Theory Probab. Appl.* **1**, 384–394.

86. Korolev, V. and Shevtsova, I. (2012). An upper estimate for the absolute constant in the Berry-Esseen inequality. *Theory Probab. Appl.* **54**(4), 638–658.

87. Korolev, V. and Shevtsova, I. (2012). An improvement of the Berry-Esseen inequality with applications to Poisson and mixed Poisson random sums. *Scandinavian Actuarial J.* **2**, 81–105.

88. Kruopis, J. (1986a). Precision of approximations of the generalized binomial distribution by convolutions of Poisson measures. *Lithuanian Math. J.* **26**, 37–49.

89. Kruopis, J. (1986b). Approximations for distributions of sums of lattice random variables I. *Lithuanian Math. J.* **26**, 234–244.

90. Kruopis, J. and Čekanavičius, V. (2014). Compound Poisson approximations for symmetric vectors. *J. Multivar. Analysis* **123**, 30–42.

91. Le Cam, L. (1960). An approximation theorem for the Poisson binomial distribution. *Pacifc J. Math.* **10**, 1181–1197.

92. Le Cam, L. (1965). On the distribution of sums of independent random variables. In: *Bernoulli, Bayes, Laplace (Anniversary volume)*. Springer-Verlag, Berlin, Heidelberg, New York. 179–202.

93. Le Cam, L. (1986). *Asymptotic Methods in Statistical Decision Theory*. Springer-Verlag, New York.

94. Ley, C. and Swan, Y. (2013). Stein's density approach and information inequalities. *Electronic Communications in Probability*. **18**, 1–14.

95. Ley, C. and Swan, Y. (2013). Local Pinsker inequalities via Stein's discrete density approach. *IEEE Transactions on Information Theory*. **59**(9), 5584–5591.

96. Lindvall, T. (1992). *Lectures on the coupling method*. Wiley, New York.

97. Matskyavichyus, V.K. (1984). A lower bound for the convergence rates in the central limit theorem. *Theory Probab. Appl.* **28**(3), 596–601.

98. Mattner, L. and Roos, B. (2007). A shorter proof of Kanter's Bessel function concentration bound. *Probab. Theory Related Fields*, **139**, 191–205.

99. Michel, R. (1987). An improved error bound for the compound Poisson approximation of a nearly homogeneous portfolio. *ASTIN Bull.* **17**, 165–169.

100. Nagaev, S.V. and Chebotarev, V.I. (2011). On estimation of closeness of binomial and normal distributions. *Theor. Probab. Appl.* **56**(2), 213–239.

101. Nourdin, I. and Peccati, G. (2012). *Normal Approximations with Malliavin Calculus. From Stein's Method to Universality*. Cambridge Tracts in Mathematics No. 192.

102. Paulauskas, V. (1969). On the reinforcement of the Lyapunov theorem. *Litovsk. Matem. Sb.* **9**(2), 323–328 (in Russian).

103. Petrauskienė, J. and Čekanavičius, V. (2010). Compound Poisson approximations for sums of one-dependent random variables I. *Lithuanian Math. J.*, **50**(3), 323–336.

104. Petrauskienė, J. and Čekanavičius, V. (2011). Compound Poisson approximations for sums of one-dependent random variables II. *Lithuanian Math. J.*, **51**(1), 51–65.

105. Petrov, V.V. (1975). *Sums of independent random variables.* Springer-Verlag, Berlin, Heidelberg, New York.

106. Petrov, V.V. (1995). *Limit Theorems of Probability Theory.* Clarendon Press, Oxford.

107. Pinelis, I. (2007). Exact inequalities for sums of asymmetric random variables, with applications. *Probab. Theory Relat. Fields.* **139**, 605–635.

108. Posfai, A. (2009). An extension of Mineka's coupling inequality. *Elect. Comm. in Probab.* **14**, 464–473.

109. Prawitz H. (1972). Limits for a distribution, if the characteristic function is given in a finite domain. *Skand. Aktuarietidskr.*, **55**, 138–154.

110. Presman, É.L. (1982). Two inequalities for symmetric processes and symmetric distributions *Theory Probab. Appl.* **26**(4), 815–819.

111. Presman, É.L. (1983). Approximation of binomial distributions by infinitely divisible ones. *Theory Probab. Appl.* **28**, 393–403.

112. Presman, É.L. (1986). Approximation in variation of the distribution of a sum of independent Bernoulli variables with a Poisson law. *Theory Probab. Appl.* **30**, 417–422.

113. Prohorov, Yu.V. (1953). Asymptotic behaviour of the binomial distribution. (Russian), *Uspekhi Matem. Nauk.* **8**, 3, 135–142. Engl. transl. in *Select. Transl. Math. Statist. and Probability.* **1**, 87–95 (1961).

114. Rachev, S.T. (1991) *Probability metrics and the stability of stochastic models.* Wiley, Chichester-New York-Brisbane-Toronto-Singapore.

115. Rachev, S.T, Stoyanov, S.V. and Fabozzi, F.J. (2011). *A probability metrics approach to financial risk measures.* Wiley-Blackwell.

116. Rachev, S.T, Klebanov, L.B., Stoyanov, S.V. and Fabozzi, F.J. (2013). *The methods of distances in the Theory of Probability and Statistics.* Springer, New York-Heidelberg-Dordrecht-London.

117. Rinott, Y. and Rotar, V. (2000). Normal approximations by Stein's method. *Decisions in Economics and Finance.* **23**, 15–29.

118. Roos, B. (1999). On the rate of multivariate Poisson convergence. *J. Multivar. Analysis* **69**, 120–134.

119. Roos, B. (1999). Asymptotics and sharp bounds in the Poisson approximation to the Poisson binomial distribution. *Bernoulli* **5**(6), 1021–1034.

120. Roos, B. (2000). Binomial approximation to the Poisson binomial distribution: The Krawtchouk expansion, *Theory Probab. Appl.* **45**, 258–272.

121. Roos, B. (2001). Sharp constants in the Poisson approximation, *Statist. Probab. Lett.* **52**, 155–168.

122. Roos, B. (2002). Kerstan's Method in the Multivariate Poisson Approximation: An Expansion in the Exponent. *Theory Probab. Appl.* **47**(2), 358–363.

123. Roos, B. (2003a). Kerstan's method for compound Poisson approximation. *Ann. Probab.* **31**, 1754–1771.

124. Roos, B. (2003b). Improvements in the Poisson approximation of mixed Poisson distributions. *Journal of Statistical Planning and Inference.* **113**, 467–483.

125. Roos, B. (2005). On Hipp\$s compound Poisson approximations via concentration functions. *Bernoulli.* **11**(3), 533–557.

126. Roos, B. (2010). Closeness of convolutions of probability measures. *Bernoulli.* **16**(1), 23–50.

127. Ross, N. (2011). Fundamentals of Stein's method. *Probability Surveys.* **8**, 210–293.

128. Ross, S.M. and Peköz, E.A. (2007). *A second course in Probability.* www.ProbabilityBookstore.com, Boston, MA.

129. Ruzankin, P.S. (2004). On the rate of Poisson process approximation to a Bernoulli process. *J. Appl. Probab.* **41**, 271–276.

130. Ruzankin, P.S. (2010). Approximation for expectations of unbounded functions of dependent integer-valued random variables. *J. Appl. Probab.* **47**, 594–600.

131. Röllin, A. (2005). Approximation of sums of conditionally independent variables by the translated Poisson distribution. *Bernoulli.* **11**, 1115–1128.

132. Saulis, L. and Statulevičius, V.A. (1991). *Limit Theorems for Large Deviations*. Mathematics and its Applications **73**, Kluwer Academic Publishers, Doerdrecht, Boston, London.
133. Sazonov, V.V. (1981). *Normal Approximation – some recent advances*. Lecture Notes in Mathematics **879**, Springer-Verlag Berlin Heidelberg.
134. Sazonov, V.V. and Ulyanov, V.V. (1982). On the accuracy of normal approximation. *Journal of Multivariate Analysis*. **12**(3), 371–384.
135. Sazonov, V.V. and Ulyanov, V.V. (1979). On the Speed of Convergence in the Central Limit Theorem. *Advances Appl. Probab.* **11**(2), 269–270.
136. Senatov, V.V. (1998). *Normal ApproximationŨ new results, methods and problems*. VSP BV, Utrecht.
137. Šiaulys, J. and Čekanavičius, V. (1988). Approximation of distributions of integer-valued additive functions by discrete charges. I. *Lithuanian Math. J.* **28**, 392–401.
138. Shevtsova, I.G. (2010). The lower asymptotically exact constant in the central limit theorem. *Doklady Mathematics*. **81**(1), 83–86.
139. Shevtsova, I.G. (2011). On the asymptotically exact constants in the Berry-Esseen-Katz inequality. *Theory Probab. Appl.* **55**(2), 225–252.
140. Shevtsova, I.G. (2014). On the accuracy of the normal approximation to compound Poisson distributions. *Theory Probab. Appl.* **58**(1), 138–158.
141. Shiryaev, A.N. (1995). *Probability*. Graduate texts in mathematics, v.95, 2nd Edition, Springer.
142. Statulevičius, V.A. (1965). Limit theorems for densities and asymptotic expansions of distributions of sums of independent random variables. *Theory Probab. Appl.* **10**(4), 582–595.
143. Stein, C. (1972). A bound for the error in the normal approximation to the distribution of a sum of dependent random variables. *Proc. of the Sixth Berkeley Symposium on Math. Stat. Probab.* **2**. Univ. California Press, Berkeley, 583–602.
144. Stein, C. (1986). *Approximate computation of expectations*. Institute of Mathematical Statistics, Hayward, CA.
145. Stein, C., Diaconis, P., Holmes, S. and Reinert, G. (2004) Use of exchangeable pairs in the analysis of simulations. In *Stein's method: expository lectures and applications. (P. Diaconis and S. Holmes, eds.)*. IMS Lecture Notes Mon. Ser. **46**, 1–26.
146. Sunklodas, J. (1984). Rate of convergence in the central limit theorem for random variables with strong mixing. *Lithuanian Math. J.* **24**(2), 182–190.
147. Sunklodas, J. (1986). Estimate of the rate of convergence in the central limit theorem for weakly dependent random fields. *Lithuanian Math. J.* **26**(3), 272–287.
148. Sunklodas, J. (2009). On the rate of convergence of L_p norms in the CLT for Poisson and Gamma random variables. *Lithuanian Math. J.* **49**(2), 216–221.
149. Sunklodas, J. (2012). Some estimates of normal approximation for the distribution of a sum of a random number of independent random variables. *Lithuanian Math. J.*, **52**(3), 326–333.
150. Statulevičius, V.A. (1970). On limit theorems for random functions I. *Litovsk. Matem. Sb.*, **10**, 583–592. (In Russian).
151. Thorisson, H. (2000). *Coupling, Stationarity, and Regeneration*. Springer, New York.
152. Tikhomirov, A.N. (1981). On the convergence rate in the Central Limit Theorem for weakly dependent random variables. *Theory Probab. Appl.*, **25**(4), 790–809.
153. Tsaregradskii, I.P. (1958). On uniform approximation of a binomial distribution by infinitely divisible laws. *Theory Probab. Appl.* **3**(4), 434–438.
154. Ulyanov, V.V. (1978) On More Precise Convergence Rate Estimates in the Central Limit Theorem. *Theory Probab. Appl.* **23**(3), 660–663.
155. Upadhye, N.S. and Vellaisamy, P. (2013). Improved bounds for approximations to compound distributions. *Statist. and Probab. Lett.* **83**, 467–473.
156. Upadhye, N.S. and Vellaisamy, P. and Čekanavičius, V. (2014). On Stein operators for discrete approximations. *arXiv:1406.6463v1 [math.PR]* (to appear in Bernoulli).
157. Vellaisamy, P. and Upadhye, N.S. (2009). Compound negative binomial approximations for sums of random variables. *Probab. Math. Statist.* **29**, 205–226.

158. Vellaisamy, P., Upadhye, N.S. and Čekanavičius, V. (2013). On negative binomial approximation. *Theory Probab. Appl.* **57**(1), 97–109.
159. Witte, H.-J. (1990). A unification of some approaches to Poisson approximation. *J. Appl. Probab.* **27**, 611–621.
160. Xia, A. (1997). On using the first difference in the Stein-Chen method, *Ann. Appl. Prob.* **7**, 899–916.
161. Yakshyavichus, Sh. (1998). On a method of expansion of the probabilities of lattice random variables. *Theory Probab. Appl.* **42**(2), 271–282.
162. Zaĭtsev, A.Yu. (1984). Use of the concentration function for estimating the uniform distance. *J. Soviet. Math.* **27**(5), 3059–3070.
163. Zaĭtsev, A.Yu. (1988). Multidimensional generalized method of triangular functions. *J. Soviet Math.* **43**(6), 2797–2810.
164. Zaĭtsev, A.Yu. (1988). Estimates for the closeness of successive convolutions of multidimensional symmetric distributions. *Probab. Theor. Related Fields.* **79**(2), 175–200.
165. Zaĭtsev, A.Yu. (1992). On the approximation of convolutions of multidimensional symmetric distributions by accompanying laws. *J. Soviet. Math.* **61**, 1859–1872.
166. Zaĭtsev, A.Yu. (1999). Approximation of convolutions by accompanying laws under the existence of moments of low orders. *Journal of Mathematical Sciences* **93**(3), 336–340.
167. Zacharovas, V. and Hwang, H.-K. (2010). A Charlier-Parseval approach to Poisson approximation and its applications. *Lithuanian Math. J.*, **50**(1), 88–119.

Index

accompanying law, 41
Arak's lemma, 101
asymptotically sharp constant, 237

centered moments for dependent variables, 207
compound measure, 4, 21
condition
    Cramer's, 11, 127
    Franken's, 57
conjugate complex number, 15
convolution, 2, 4
coupling, 229
cumulant, 11

density, 2, 107
density approach, 157
distribution, 1, 2
    Bernoulli, 2
    binomial, 3
    compound geometric, 4, 35
    compound negative binomial, 5, 35
    compound Poisson, 4
    concentrated at a, 1
    geometric, 10
    negative binomial, 10, 159
    non-degenerate, 11
    normal, 2
    Poisson, 4

expansion in exponent, 25
exponential measure, 3

factorial cumulant, 12, 25
factorial moments expansion, 12

formula
    Abel's summation, 18
    Euler's, 16
    Stirling's, 18
Fourier transform, 9
function
    bounded variation, 135
    characteristic, 10, 15
    concentration, 14
    probability generating, 157
    triangle, 180

Heinrich's lemma, 208

identity
    Bergström's, 17
    Parseval's, 14
imaginary unit, 10
inequality
    Barbour-Xia, 86
    for bounded density, 108
    Hölder's, 5
    Hipp, 28
    Kolmogorov-Rogozin, 14
    Le Cam, 14, 24
    Lyapunov's, 5
    Markov's, 6
    Minkowski's, 5
    Tsaregradskii, 69
    Chebyshev's, 6
    Jensen's, 5
    Rosenthal's, 6
inversion formula
    for densities, 108

© Springer International Publishing Switzerland 2016
V. Čekanavičius, *Approximation Methods in Probability Theory*, Universitext,
DOI 10.1007/978-3-319-34072-2

local, 51
inversion inequalities
  Esseen type, 122
  for continuous distributions, 113
  for densities, 109
  for discontinuous functions, 135
  for total variation, 77, 113
  local, 52

Jordan-Hahn decomposition, 6, 102

Le Cam's trick, 28
lower bound estimates
  for densities, 146
  for Kolmogorov norm, 149
  for probabilities, 147
  for total variation, 144
  via Stein's method, 173

m-dependent variables, 207
method
  Bentkus, 230
  compositions, 223
  convolutions, 21
  Heinrich's, 207
  Kerstan's, 44
  Lindeberg, 232
  Stein's, 153
  Tikhomirov, 234
  triangle function, 179
Mineka coupling, 86
moment
  absolute, 5
  factorial, 11, 21
  of order k, 5
moments expansion, 11

non-uniform estimates
  for continuous approximations, 117
  for distribution functions, 95
  for probabilities, 93
norm
  Kolmogorov, 9
  local, 9
  total variation, 6
  Wasserstein, 9, 89

perturbation, 163
pseudomoment, 167, 223

random variable
  Bernoulli, 2
  binomial, 3
  Poisson, 3

signed measures, 1
smoothing inequalities
  for Kolmogorov norm, 37
  for local norm, 64
  for symmetric distributions, 85
  for total variation, 29
Stein equation, 155
Stein's operator, 154, 155
support of measure, 6

Taylor series, 16
theorem
  Berry-Esseen, 125
  Riemann-Lebesgue, 70
total variation distance, 7
triangular arrays, 19
two-way runs, 214

Printed in the United States
By Bookmasters